The Facts On File

DICTIONARY of BIOTECHNOLOGY and GENETIC ENGINEERING

Third Edition

The Facts On File

DICTIONARY of BIOTECHNOLOGY and GENETIC ENGINEERING

Third Edition

Mark L. Steinberg, Ph.D.
Sharon D. Cosloy, Ph.D.

Facts On File
An imprint of Infobase Publishing

The Facts On File Dictionary of Biotechnology and Genetic Engineering
Third Edition

Copyright © 2006, 2001, 1994 by Mark L. Steinberg, Ph.D., and Sharon Cosloy, Ph.D.

Illustrations © 2006 by Infobase Publishing

Facts On File, Inc.
An imprint of Infobase Publishing
132 West 31st Street
New York NY 10001

Library of Congress Cataloging-in-Publication Data

Steinberg, Mark (Mark L.)
The Facts on File dictionary of biotechnology and genetic engineering/
  Mark L. Steinberg and Sharon D. Cosloy—Third ed.
      p.   cm. — (The Facts On File science library)
    Includes index.
    ISBN 0-8160-6351-6 (alk.paper)
  1. Biotechnology—Dictionaries. 2. Genetic engineering—Dictionaries.
I. Cosloy, Sharon D. II. Title. III. Series.

TP248.16.S84   2000
660.6'03—dc21   00-035463

Facts On File books are available at special discounts when purchased in bulk quantities for businesses, associations, institutions, or sales promotions. Please call our Special Sales Department in New York at (212) 967-8800 or (800) 322-8755.

You can find Facts On File on the World Wide Web at
http://www.factsonfile.com

Text and cover design by Cathy Rincon

Printed in the United States of America

MP FOF   10 9 8 7 6 5 4 3 2

This book is printed on acid-free paper.

This edition is dedicated to the memory of Dr. Sharon Cosloy by her children, Michael and Rebecca, and her husband, Edward. Sharon was a loving mother, a devoted wife, a dedicated mentor, and an accomplished professor and researcher.

And above all, she was a kind and gentle woman with a bright spirit that still lives on today through the people who were fortunate enough to be touched in life by her.

From MLS: To Sharon, in memoriam, a good friend and valued colleague. You are greatly missed.

CONTENTS

PREFACE

The last decades of the 20th century produced a dramatic revolution in the field of biology in which, for the first time, the ability to modify the genetic makeup of higher organisms in the laboratory rather than by the random forces of natural selection was realized. This new era was born out of critical discoveries in the mid-1970s that led to the appearance of new fields of molecular genetics variously known as *gene cloning, genetic engineering,* and *biotechnology.* The central theme of genetic engineering is the introduction of genetic material altered in a laboratory into an organism different from that from which it was originally derived. The introduction of genes from higher organisms into microorganisms made it possible to isolate, amplify, study and ultimately engineer individual genes for a variety of specialized purposes. These techniques have also allowed scientists to look closely at the structure, function, and regulation of genes and their proteins.

Genetic engineering has given rise to technologies that were unthinkable barely two decades ago: recombinant antibodies to fight cancer, the isolation of genes responsible for genetic diseases, the synthesis of unlimited quantities of therapeutic agents, human hormones and critical blood factors in bacterial "factories," the creation of genetically engineered plants and animals, and the decoding of the human genome—only a few examples of technologies that have been realized even at the time of the first printing of this dictionary. Much of the research in biotechnology and genetic engineering has moved from the academic world into the industrial setting. As a consequence, many new and potential applications are in the hands of private enterprises where, fueled by more substantial funding and motivated by the forces of the marketplace, the development of new products has reached an explosive pace. This has also meant that even as the rapidly increasing pace of progress taxes the ability to keep up with new developments, there is an ever-increasing need to understand the legal and ethical issues that inevitably accompany any new technology. However, in contrast to other new technologies, the products of genetic engineering deal directly with fundamental biological processes and are, by their very nature, certain to have an immediate and profound impact on all areas of human health.

The purpose of this dictionary is to provide readers with access to the basic vocabulary of modern biotechnology and genetic engineering so that those with even an elementary knowledge of basic biology and biochemistry will be able to follow the flood of fast-breaking developments in biotechnology and genetic engineering that constantly appears in the media.

At the time of the first printing of *The Facts On File Dictionary of Biotechnology and Genetic Engineering,* molecular cloning of genes had only recently matured. Even then, rapidly accumulating data from large-scale sequence analyses and the development of new techniques for amplification of DNA at the microscale level were already yielding information that allowed for the determination of gene function, including the molecular nature of defects underlying numerous genetic diseases. A revised edition of the dictionary added terminology of the developing biomedical fields of molecular medicine, DNA technology, gene therapy, and genomics. In recent years, new areas of research have elucidated signaling pathways that are now known to regulate essential biochemical pathways, including cell growth, metabolism, and differentiation. Many modern pharmaceuticals are agents that target critical signaling pathways involved in disease processes. Among these are drugs for the treatment of high blood pressure, allergies, sexual dysfunction, anti-inflammatory and anti-viral agents, various cancer chemotherapies, and many others. In a parallel track, the completion of the Human Genome Project in 2003, together with the computer technologies for data mining and relational analyses, created the new area of computational biology known as bioinformatics. The application of bioinformatic methods to burgeoning nucleotide and protein databases has yielded new insights into many genetic diseases and has helped elucidate the relationships between genes and the biochemical pathways that the gene products regulate. Bioinformatics is currently providing new approaches to drug design based on predictive computer models to tailor drugs to act on specific molecular targets. The dictionary was updated to account for these as well as other new developments in this rapidly changing field.

The new "third edition" of the dictionary focuses on the new terminology in the evolving areas of genomics, bioinformatics, cell signaling, and molecular medicine. In addition, there are a number of biochemical terms pertaining to recent advances in medicines for the treatment of viral diseases, mental illness, cholesterol metabolism, plant engineering, and stem cell research.

Since this book addresses an audience from diverse backgrounds and covers a broad field, we attempted to include both basic as well as more technical terminology in a number of areas including plant and animal biology in order to meet the needs of as many readers as possible. There has also been an attempt to make the dictionary self-contained in the sense that, in cases where technical terms appear in definitions, these terms will be defined elsewhere in the book. It is anticipated that the dictionary will be of benefit to a wide-ranging audience, including high school and college students, lawyers, physicians, scientists, or others with a particular need to keep abreast of the rapidly developing areas of biotechnology and genetic engineering.

ACKNOWLEDGMENTS

The authors gratefully acknowledge the support of the RCMI research facilities, where they carry out their research at the City College of New York.

The authors also thank Mr. Frank K. Darmstadt, executive editor, and the production department for their support and insight in the creation of this new edition of the dictionary.

A

ABC transporter The largest class of transmembrane proteins. ABC transporter is an acronym for *A*TP (adenosine triphosphate) *b*inding *c*assette, a region of the protein that is conserved in the transporter in a wide variety of different organisms and is responsible for the binding of ATP. In bacteria these proteins use energy from ATP to transport a wide variety of small molecules including sugars, vitamins, amino acids and ions across the cell membrane. In eukaryotes, ABC transporters generally move molecules either outside the cell or into an organelle such as the endoplasmic reticulum or mitochondrion. Alterations in the ABC transporter genes, particularly duplications, are the basis of resistance to chemotherapeutic drugs that many tumors develop. Inhibitors of the ABC transporters involved in drug resistance is being developed as a strategy to deal with drug resistance in cancer.

ABO blood group A system of antigens expressed at the surface of human red blood cells. Human blood types represented in this group are A, B, AB, or O, depending on which antigen(s), in the form of oligosaccharides, are present at the surface of the erythrocyte membranes. The blood serum of Type A individuals contains anti-B antibodies, those with Type B produce anti-A antibodies, and those with Type AB produce both. Type O individuals produce neither. This system is one of 14 different blood group systems consisting of 100 different antigens. This system is of medical importance because the recipient of a blood transfusion must receive blood that is compatible with his or her own type. Type AB individuals are known as universal acceptors, and Type O individuals as universal donors. In addition, the ABO system can be used in paternity suits to rule out the possibility that a particular male is the father of the child in question.

abscisic acid A plant hormone, lipid in nature, synthesized in wilting leaves. It counteracts the effects of most other plant hormones by inhibiting cell growth and division, seed germination, and budding. It induces dormancy.

absorbance Often referred to as optical density. Absorbance is a unit of measure of the amount of light that is absorbed by a solution or by a suspension of bacterial cells. The absorbance is a logarithimic function of the percent of transmission of a particular wavelength of light through a liquid and is measured by a spectrophotometer or a colorimeter. Absorbance values are used to plot growth of suspensions of bacteria and to determine the concentration and purity of molecules such as nucleic acids or proteins in solutions.

absorption 1. *virology* The entry of a virus or viral genome into a host cell after the virus has absorbed to the cell surface. (See ADSORPTION.)
2. *photometry* When light is neither reflected nor transmitted, it is said to be absorbed. Some biological systems can make use of light energy because they have pigments that absorb light at specific wavelengths. These pigments are able to harness light energy to drive biochemical reactions in vivo. An example can be found in plant pigments, such as chlorophyll, that are used to trap light energy and drive the process of photosynthesis where plants manufacture nutrients.

absorption spectroscopy The use of a spectrophotometer to determine the ability of solutes to absorb light through a range of specified wavelengths. Every compound has a unique absorption spectrum. An absorption spectrum, which is defined as a plot of the light absorbed versus the wavelength, can be derived from a solution (see ABSORBANCE). Absorption spectra are used to identify compounds, determine concentrations, and plot reaction rates.

abzymes Catalytic antibodies that cleave proteins or carbohydrates at specific residues. They are analogous to restriction enzymes that cleave DNA at specific sequences. Catalytic antibodies have the potential to be used as therapeutic agents, attacking specific viral or bacterial surface structures, and as catalysts in reactions in which no enzyme has been found.

acentric fragment A fragment of a chromosome that does not contain a centromere. Because of the absence of a centromere, acentric fragments do not segregate at mitosis and eventually disappear.

Acetobacter A genus of Gram-negative flagella-endowed bacteria that are acid-tolerant aerobic rods. They are also known as the acetic-acid bacteria due to their ability to oxidize ethanol to acetic acid. They are found on fruits and vegetables and can be isolated from alcoholic beverages. They are used commercially in the production of vinegar, but because of their ability to produce acetic acid, they are nuisance organisms in the brewing industry.

Acetobacter aceti An organism used in the commercial production of vinegar. When introduced into wine or cider containing 10 percent–12 percent alcohol, it will convert to acetic acid. See *ACETOBACTER.*

acetone-butanol fermentation The anaerobic fermentation of glucose by *Clostridium acetobutylicum* to form acetone and butanol as end products. At one time, the production of these commercially important chemicals relied on bacterial fermentation, but this has since been replaced by chemical synthesis.

acetylcholine A chemical neurotransmitter that is expelled into the synaptic cleft, or space between two nerve cells. This neurotransmitter permits the transmission of an electrical nerve impulse or action potential from one nerve cell to another by diffusing across the cleft and then binding to a cell-membrane receptor.

acetylcholinesterase An enzyme present in the synaptic cleft, or space between two nerve cells, that hydrolyzes or destroys the unbound neurotransmitter acetylcholine once it has diffused through the cleft. This is required to restore the synaptic cleft to a state that is ready to receive the next nerve impulse. See ACETYLCHOLINE.

acid blobs Certain sequences of amino acids on a protein that bind to a transcriptional regulatory protein and, in so doing, serve to activate transcription.

acid growth hypothesis The hypothesis that elongation of plant cells caused by the plant hormones known as auxins involves a mechanism for creating an acid environment (lowered pH) in the specific region of the cell where growth is to occur. The acidification of a plant cell in a localized region helps account for certain tropic behaviors seen in plants, for example, phototropism.

acidic activation domain In certain types of eukaryotic transcription factors (for example, the GAL4 transcription factor in yeast or the herpes simplex virus VP16 protein), a region containing a number of contiguous acidic amino acid residues that appears to be required for the recruitment of other additional factors needed to regulate the transcription process for different genes.

acidic amino acids The two amino acids that are negatively charged at pH

7.0 are aspartic and glutamic acids. Also referred to as aspartate or glutamate. Both of these amino acids contain in their R or variable groups a second carboxyl group that is ionized under physiological conditions.

acidophile A classification of microorganisms that describes the ability or the necessity of certain species to exist in an acidic environment. These acid-loving organisms can exist at a pH range of 0–5.4, well below the optimum of neutrality for most bacteria. Facultative acidophiles can tolerate a range of pH from low to neutral and include most fungi and yeasts. However, obligate acidophiles including members of the genera *Thiobacillus* and *Sulfolobus* require low pH for growth. A neutral pH is toxic to these species.

acivicin An antibiotic that acts as an inhibitor of the enzyme gamma-glutamyl transpeptidase (GGT), which is necessary for the breakdown and transport of glutathione across the cell membrane. As a glutamine analog, acivicin is also used as an anticancer drug because of its ability to block glutamine metabolism.

acquired immunodeficiency syndrome (AIDS) An infectious disease in humans caused by the human immunodeficiency virus (HIV). The virus attacks the host's immune system leaving him/her susceptible to many other diseases, including certain rare forms of cancer and opportunistic microbial infections that would otherwise be destroyed in an uninfected individual. Most often, AIDS patients die from these secondary infections that run rampant through the body because of the loss of ability to immunologically suppress them. The HIV virus is transmitted through the exchange of body fluids during sexual contact with an infected individual, the sharing of needles among intravenous drug users, transfusion of contaminated blood products (no longer a threat due to the ability to screen donated blood), and from mother to newborn during delivery. It has not been shown to be transmitted through casual contact with infected individuals.

acridine orange One of a group of chemical mutagens known as acridines, including proflavin and acriflavine. The size of the acridines is the same as that of a purine-pyrimidine base pair. For this reason, they can insert or intercalate into the helix between two adjacent base pairs. When DNA that contains an intercalated acridine is replicated, an additional base pair may be added or a base pair may be deleted, disrupting the codon reading frame in the newly synthesized strand. Such a mutation is called a frameshift mutation.

acrosome (process, reaction, vesicle) A vesicle- or membrane-bound compartment covering the sperm head that contains lytic enzymes. The major enzyme found in the mammalian sperm acrosome is hyaluronidase, which promotes the digestion of the tough outer coat of the egg and allows penetration of the sperm.

acrylamide A substance that can polymerize and form a slab gel when poured into a mold in its molten state. It is used as semisolid support medium and is immersed in a conductive buffer through which a current is passed. When solutions containing heterogeneous mixtures of nucleic acid fragments or mixtures of proteins are placed into slots in the gel and subjected to the electrical current, the nucleic acid or protein mixtures may be separated into distinct collections of homogeneous molecules located in different regions of the gel, based on their size or molecular weight. See ELECTROPHORESIS.

ACTH See ADRENOCORTICOTROPIC HORMONE.

actin One of the two major proteins responsible for muscle contraction. Actin and myosin are found in smooth and striated muscle. Actin monomers together with two other proteins, troponin and tropomysin, can polymerize to form long, thin filaments that, together with myosin filaments, can shorten in the presence of ATP (adenosine triphosphate). Actin also plays a role in the shape and structure of cells.

Actinomyces A genus of anaerobic Gram-positive rods that are often found in the mouth and throat. They occasionally display a branched filamentous morphology. Many, such as *A. israelii*, are human pathogens.

actinomycin D An antibiotic produced by *Streptomyces parvullus* that inhibits RNA transcription in both prokaryotes and eukaryotes. It blocks the action of RNA Polymerase I, which synthesizes ribosomal RNA, and forms complexes with DNA by intercatating between G-C pairs, preventing the movement of DNA- and RNA-synthesizing enzymes. Although toxic, it is sometimes used in conjunction with other drugs as a chemotherapeutic agent, due to its antitumor properties

action potential Also called a nerve impulse; sequential wave of depolarization and repolarization across the membrane of a nerve cell (neuron) in response to a stimulus. Depolarization is a reversal in the distribution of charge between the inside and the outside of the neuron membrane.

activated sludge process A secondary sewage-treatment process where biological processing of the sewage by microbial activity is the main method of treatment. In this step, sewage that has been previously treated in settling tanks is aerated in large tanks to encourage growth of microorganisms that oxidize dissolved organics to carbon dioxide and water. Bacteria, yeasts, molds, and protozoans are used. This process proves effective in reducing intestinal pathogens in sewage while encouraging growth of nonpathogens. After activated sludge has been produced, additional processing is required, including anaerobic digestion, filtering, and chlorination.

activation energy The energy required for a chemical reaction to proceed. In biological systems, enzymes lower the activation energy, allowing chemical reactions to occur faster under physiological conditions.

active site The region of an enzyme that contains a special binding site for substrate(s). This site is uniquely shaped for the exclusive binding of the particular substrate molecule(s) and is the site for the catalytic activity of the enzyme. The three-dimensional folding of the enzyme brings distal amino acids in the polypeptide into close proximity, thus forming the active site at the surface of the protein.

active transport The transport, by cells or cellular compartments, of ions and metabolites through cell membranes against a concentration gradient. This type of transport requires cellular energy in the form of ATP (adenosine triphosphate) hydrolysis. One example found in all animal cells is the active transport of Na+ out of cells and the active transport of K+ into cells. This system is known as the sodium-potassium pump. The energy is provided by a specific ATPase located in the plasma membrane. This active transport system is responsible for the generation and maintenance of the electrical potential or voltage gradient across the cell membrane.

acycloguanosine (acyclovir) An antiviral antibiotic used to treat herpes virus infections. Acycloguanosine is a derivative of the normal nucleoside, guanosine, in which the sugar, ribose, has been replaced by an ether chain. Acycloguanosine is an inhibitor of viral DNA synthesis. See HSV.

Acyclovir See ACYCLOGUANOSINE.

acyl carrier protein (ACP) A small protein involved in the synthesis of fatty acids. First isolated from *E. coli* bacteria by Roy Vagelos, it was found to be a 77 amino acid polypeptide chain, capable of binding six other enzymes required for fatty acid synthesis.

adaptation **1.** *sensory* A progressive decrease in the number of impulses that pass over a sensory neuron even when there is continuous or repetitive sensory stimulation to the sense organ involved. Sensory adaptation provides an organism with a way to deal with the constant bombardment of the sense organs with useless information in the environment and the ability to screen for the appropriate stimuli to which to respond. **2.** *evolution* A genetic change in a population of organisms that arises as a result of random chance, involving structures or behaviors that will enable that group and its offspring to be better suited to their environment.

adaptive enzymes Enzymes that are produced by microbes only when their substrates are present. When not needed, they are not produced. This is in contrast to constitutive enzymes, which are always produced.

adaptor molecules A term used to describe transfer RNA due to its role during translation of mRNA. Several properties of the tRNA molecule enable it to act as an adaptor molecule. The highly specific nature of tRNA-amino acid binding, the complementary base pairing of the tRNA anticodon with a specific codon in the message, and its ability to carry its designated amino acid to the mRNA template in the ribosome are all factors that allow the information in the message to be translated into a polypeptide.

adenine One of the four major bases found in nucleic acids. Adenine and guanine are purines; cytosine, thymine, and uracil are pyrimidines. These nitrogenous bases are a component of the basic building blocks of nucleic acids called nucleotides. Within the DNA double helix, adenine forms a double hydrogen bond with thymine.

adenoma A benign tumor of glandular tissue.

adenosine A nucleoside containing the pentose sugar ribose and the purine base adenine. When a phosphate group is attached to the 5' carbon in the ribose, the nucleoside becomes a nucleotide, a basic building block of nucleic acids.

adenosine deaminase deficiency (ADA) A genetic condition in which the lack of the enzyme adenosine deaminase results in the disease severe combined immunodeficiency (SCID). This rare disease leaves individuals with no functioning immune system and results in death at a very early age. This was one of the first diseases to be treated with enzyme replacement therapy and then gene replacement therapy.

adenosine diphosphate (ADP) A product of the hydrolysis or enzymatic cleavage of the terminal phosphate group of adenosine triphosphate (ATP). The other product produced during this reaction is inorganic phosphate. ATP is cleaved to provide energy for cells to do work.

adenosine monophosphate (AMP) A product of the hydrolysis or enzymatic cleavage of the terminal two phosphate group of adenosine triphosphate (ATP). The other product produced during this reaction is inorganic pyrophosphate (PPi). In this reaction, energy is produced both from the release of pyrophosphate from ATP and from the subsequent cleavage of the pyrophosphate to form two molecules of phosphoric acid.

adenosine triphosphate (ATP) The single most important energy source in biological systems; the energy currency of the cell. All cells do work that requires energy. Work can be mechanical, biosynthetic, active transport of molecules into and out of cells and cellular compartments (see ACTIVE TRANSPORT), and all require the hydrolysis of ATP or the breaking of phophate bonds in the energy-rich ATP molecule. ATP is made up of adenine, which is linked to the 5'-carbon sugar ribose. In addition, there are three phosphate groups linked to the ribose in a linear arrangement. The two terminal phosphate groups possess high-energy bonds that, when cleaved, provide energy for the cell.

adenovirus This is a DNA-containing virus whose outer protein coat is in the shape of an icosahedron. There are more than 40 different types of adenoviruses, some of which are among the many viruses that are responsible for the common cold.

adenylate cyclase An enzyme that catalyzes the synthesis of cyclic AMP from ATP. See ADENOSINE MONOPHOSPHATE; ADENOSINE TRIPHOSPHATE; and CYCLIC AMP.

adherens junctions Junctional complexes that occur at (and anchor) the termini of actin (see ACTIN) cytoskeletal (see CYTOSKELETON) elements. Adherens junctions bear a similarity to desmosomes (see DESMOSOME) and hemidesmosomes (see HEMIDESMOSOME) in that adherens junctional complexes on neighboring cells directly oppose one another.

adhering junction A type of cell-cell junction that is a highly specialized region of the cell's surfaces. It is also called a desmosome. Adherent junctions are commonly found in tissues that are subjected to mechanical stress such as the skin. They provide very tight contact between adjacent cells and enable groups of cells to function as a unit in tissues.

adhesion plaque One of specialized regions of the plasma membrane that are involved in the adherence of cells to solid surfaces. Bundles of actin microfilaments called stress fibers attach to the plasma membrane in adhesion plaques. The protein vinculin is localized in adhesion plaques and serves to anchor these microfilaments in place. When cells are transformed into a cancerous state, the adhesion plaques become disordered and cells lose their ability to adhere properly (loss of anchorage dependence), contributing to metastasis.

ADH1 A yeast-transcription termination signal that is incorporated into yeast expression vectors so that cloned yeast genes will form properly terminated mRNA to ensure high amounts of protein expressed. In many cases, mRNA that is not properly terminated is unstable, thus resulting in decreased amounts of protein expressed from the cloned insert.

adjuvent A substance that increases the potency of an immunogen or enhances the ability of a weak antigen to induce an antibody response.

A-DNA One of the several forms that a double helix can assume under different conditions in vivo or in vitro. The molecular characteristics of this helix type differ from the more common B form that is believed to be predominant under physiological conditions. The A form is stable in a less humid milieu and is both the form of a DNA-RNA hybrid helix and the conformation assumed by regions of double-stranded RNA. A-DNA is a right-handed helix; however, it is more compact than the B form. Other forms of DNA include C-DNA and Z-DNA.

adoptive immunity The transfer of immunity to allografts to an animal that was previously tolerant of such allografts. This is done by injection of lymphocytes from an animal that is immune to allografts into the tolerant animal.

ADP See ADENOSINE DIPHOSPHATE.

adrenergic Pertaining to the general class of neurons that utilizes catecholamines (adrenaline, dopamine, and noradrenaline) as a neurotransmitter. See NEUROTRANSMITTER.

adrenergic receptors Membrane receptors for adrenaline (epinephrine). Binding of the adrenergic ligands to their receptors upregulates various cellular processes by activating G-proteins coupled to the receptor(s), which in turn activates the enzyme, adenylyl cyclase, which catalyzes the formation of cAMP. There are

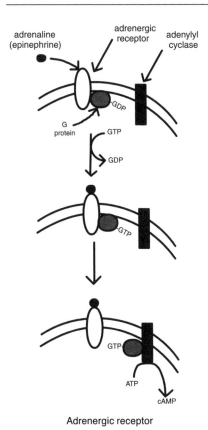

Adrenergic receptor

four types of adrenergic receptors: α_1, α_2, β_1, and β_2. In skeletal muscle and liver, adrenergic receptors stimulate glycogen breakdown (glycogenolysis) to produce glucose for energy. On smooth muscle cells, alpha receptors cause muscle contraction while beta receptors on cardiac muscle cells act to cause more rapid contraction.

adrenocorticotrophic hormone (ACTH)
A hormone, secreted by the anterior lobe of the pituitary gland, that controls the production and secretion of adrenal cortex hormones. It is called a tropic hormone because it regulates the activity of other hormones. It, in turn, is regulated by a factor that is produced by the hypothalamus. Under conditions of stress, the anterior pituitary secretes ACTH that

in turn stimulates the adrenal cortex to secrete glucocorticoids. One way that glucocorticoids aid the body in dealing with the physical consequences of stress is by promoting a metabolic process known as gluconeogenesis, which involves the synthesis of glucose from various noncarbohydrate metabolites in the cell.

adrenoleukodystrophy (X-linked ALD)
A group of disorders caused by the inability to break down very long chain saturated fatty acids (VLCFA) that, as a consequence, accumulate in the adrenal cortex and brain, which leads to the breakdown of the myelin sheath of nerve cells. The X-linked form (X-ALD), in which there is an abnormal gene on the X-chromosome, is the most common form. Symptoms include vision loss, seizures, difficulty swallowing, deafness, and dementia.

adsorption   A step in the replication of bacterial viruses where the virus attaches to a specific receptor located on the outer surface of the cell. The receptor is complementary to the attachment site on the virus. The specificity of a virus for a particular host or a small number of hosts can be explained by the fact that the virus can only adsorb to species of bacterial cells that make the appropriate receptors. Following adsorption, the phage genome penetrates the cell where it is replicated, transcribed, and translated and viral components self-assemble into new viral particles. This is followed by cell lysis, or the bursting open of the cell, and the release of the newly synthesized virions, which can range in number from 50 to 200, depending on the virus.

adult polycystic kidney disease (APKD) A genetic kidney disease that is transmitted in an autosomal dominant pattern of inheritance. The disease is characterized by the formation of large cysts in the kidneys resulting in gradual loss of kidney tissue that can lead to renal failure. The disease is caused mainly by mutations in the genes PKD1 (polycystin 1; gene map locus 16p13.3-p13.12) and PKD2 (polycystin 2; gene map locus 4q21-q23). Polycystin 1 codes for a protein that interacts with polycystin 2, a protein that is involved in cell cycle regulation and intracellular calcium transport. About 85 percent of APKD cases show an abnormality in polycystin 1, while 5–15 percent of the cases involve polycystin 2.

aeration The process whereby small bubbles of air or oxygen are introduced to liquid cultures of bacteria with agitation or stirring to ensure that the cells are receiving a continuous and adequate supply of molecular oxygen. Aeration techniques are applied to growth of microbes in industrial fermentors that have large volume capacities, as well as to ordinary flasks that are grown on a gyrating platform in an incubator or waterbath.

aerobe A microorganism that requires free oxygen for growth. During the process of respiration, oxygen is the final electron acceptor in the electron transport chain in these organisms. There are several different categories of aerobes. Obligate aerobes die in the absence of oxygen. Microaerophiles thrive in the presence of low amounts of oxygen, and facultative anaerobes normally use oxygen but can switch to an anaerobic metabolism when oxygen is depleted. Aerobes are more efficient at producing energy than organisms that do not use oxygen.

aerosol A mist or a cloud of water droplets suspended in air that can carry airborne pathogens and provide a vehicle for transmission. Aerosols may be formed in the environment in numerous ways, such as coughing, sneezing, splashing of falling raindrops, and spray from breaking waves.

aerotolerant Aerotolerant anaerobes are microorganisms that do not use oxygen during metabolism but, unlike obligate anaerobes, can survive in its presence (see ANAEROBE). Members of the genus *Lactobacillus* represent examples of aerotolerant microorganisms.

afferent Refers to the direction in which a nerve impulse is moving toward the central nervous system. Afferent neurons are those nerve cells that carry impulses from sensory organs, such as skin and tongue, toward the central nervous system (i.e., the brain and the spinal cord). This is opposed to efferent neurons, which carry impulses from the central nervous system to effector organs such as muscle.

affinity partitioning A modification of the aqueous two-phase separation technique of using polymers and salts to purify proteins. Affinity partitioning employs polymers with ligands attached to them, thus making them specific for the proteins to be isolated.

affinity tailing Addition of specific residues to the end of a protein so that the protein can be easily identified or isolated. This is accomplished by cloning the gene for a protein into an expression vector having a "tail" sequence at the 3′ end of the insert, thus allowing a fusion of the cloned gene with the tail sequence. A common tail is six histidine residues that bind to a column of a nickel-charged resin. Another molecule that is used to tail a protein is thioredoxin: This allows the protein to be purified on an agarose-based support with phenylarsine oxide (PAO) covalently bound to it or identified with an antithioredoxin antibody.

aflatoxin A highly toxic chemical in a class of compounds called mycotoxins, which is produced by molds. Aflatoxin is produced by *Aspergillus flavus,* which grows on grains and has been found to contaminate many foodstuffs, including beans, cereals, and peanuts. Aflatoxin has been shown to be one of the most potent liver carcinogens in existence.

African sleeping sickness A disease also known as African trypanosomiasis that affects humans and other mammals in central Africa. It is spread by the tsetse fly, which is found in this region of the world. The fly is host to the parasitic protozoans, the trypanosomes (*T. brucei gambiense* and *T. brucei rhodesiense*), which are the causative agents. After being bitten by the fly, the trypanosome enters the victim's bloodstream. Without treatment, the disease is nearly always fatal because the trypanosome enters the central nervous system and coma ensues. The trypanosomes have the ability to evade the host's immune system because they can repeatedly change their coat proteins, against which the host makes antibodies. This pathogen is the subject of intense study due to its devastating effects on humans and livestock and to the unusual characteristics of its molecular biology.

agar A complex polysaccharide made by the red marine alga *Gelidium*. It is used to thicken or solidify bacterial culture media as well as certain foods. It was first applied for the use of culturing microorganisms by the wife of Walter Hesse, a German microbiologist in the late 1800s. The properties of agar make it well suited for use in the culturing of microorganisms. Very few microorganisms can degrade or digest it, so it remains solidified in their presence. It remains solid at high enough temperatures (close to 100°C) to be able to incubate most microbes. When in a molten state, it will solidify when the temperature drops below 42°C, but it can be kept in a liquid state for long periods of time if incubated at 50°C and above. It can be poured into tubes, flasks, petri plates, and any other support and placed in any position to solidify, such as slanted or straight to shape the surface to either maximize or minimize oxygen availablity and surface area. Most solid media are 1.5 percent agar.

agarose A cross-linked polysaccharide that is isolated from red algae and used as a support medium in a number of molecule separation and/or quantification techniques, including gel electrophoresis (see below), electroimmunodiffusion, and immunoelectrophoresis.

agarose gel electrophoresis A procedure that uses agarose gels to separate molecules in solutions of nucleic acids or solutions of proteins according to their size. In this way, molecular weights can be determined or certain specific species of molecules can be isolated and purified. Ranges of sizes of fragments that can be separated are determined in part by the percentage of agarose in the gel. The gels are immersed in a chamber containing a buffer that can conduct a current across the gel. First, the samples are loaded into slots in the top of the gel, then a dye is added, and finally the current is turned on.

agglutination The clumping of cells to one another caused by the binding of molecules (the agglutinin) to the cell surface so that one or more cells are linked to one another by an agglutinin bridge. See HEMAGGLUTINATION.

Agrobacterium A genus of Gram-negative aerobic bacteria that live in soil and cause crown gall disease in broad-leafed plants. This disease is seen as the growth of tumors on the trunks and sometimes the roots of infected plants. The pathogenicity of the organisms is due to the presence of a bacterial plasmid called the Ti plasmid that can be transferred to the plants cells from the bacteria. The plasmid contains genes that direct the plant cells to make nutrients that are useful for the bacteria and gene products that interfere with normal plant cell growth and division. Microbial geneticists and molecular biologists are intensely studying *A. tumefaciens* with the hopes of being able to use this organism to transfer useful genes into crop plants. This can be accomplished by using the Ti plasmid as a vehicle to transfer genes such as those involved in nitrogen fixation into crop plants after the plasmid has been genetically engineered to eliminate its pathogenicity.

Agrobacterium tumefaciens See AGROBACTERIUM.

AIDS See ACQUIRED IMMUNODEFICIENCY SYNDROME.

Akt A component of signaling pathways that involve phosphorylated phosphatidylinositol (PIPi). These pathways are activated by a variety of growth and survival factors that bind to cell membrane receptors, which then produce PIPi, which, in turn, activates Akt. Once activated, Akt can either, 1) inhibit apoptosis or, 2) promote cell survival by activating the transcription factor, NF-kb. Akt, which is a serine/threonine kinase, is the cellular homologue of the retrovirus oncogene, v-Akt.

Alagille syndrome A rare inherited liver disease in which there is a buildup of bile in the liver due to a lack or deficiency of bile ducts. This disease is seen in infants and young children and is characterized by jaundice, stunted growth, and deformities of the face and other internal organs.

alanine One of the 20 amino acids that are incorporated into polypeptides. Alanine has an aliphatic uncharged R group at pH 7.0 that consists of a methyl (CH_3) group as its side chain. Of all the amino acids with aliphatic side groups, it is the least hydrophobic.

alarmones Unusual dinucleotides containing multiple phosphate groups that are produced by bacteria under conditions of stress, such as exposure to oxidative agents (for example, hydrogen peroxide) and which act in a hormonelike fashion to regulate bacterial metabolism under such conditions.

diadenosine 5',5'-P^1,P^4-tetra

An alarmone

albumin The most abundant human blood-plasma protein. It is a heat-coaguable, water-soluble globular protein found in egg white, blood plasma (50 percent of protein content of human plasma), and various other animal and vegetable tissues. Bovine serum albumin is often used in reaction mixtures and storage tubes to stabilize enzymes.

alcohol An organic compound that contains one or more hydroxyl groups (–OH). Also the common name for ethanol. Other alcohols include methanol and propanol.

alcohol dehydrogenase An enzyme that is responsible for the last step in alcoholic fermentation by yeast, which produces the alcohol in alcoholic beverages. This enzyme converts acetaldehyde to ethanol.

alcohol fuel An energy source that is produced by a process known as bioconversion, where organic waste material can be converted into fuel by microorganisms. One example is gasohol (90 percent gasoline and 10 percent ethanol), which is an alternative fuel for automobiles. Another is methane, which is an alternative to fossil fuels and natural gas. Methane is a by-product of the anaerobic treatment of sewage.

aldose A group of monosaccharides that contain an aldehyde group (–CHO).

aldosterone A hormone secreted by the adrenal cortex. Aldosterone is classified as a mineralocorticoid, and it acts mainly on the kidneys to control the water and electrolyte balance in the body. This enzyme ensures the retention of sodium ions and water by causing their reabsorption into the blood before urine excretion. It also causes the excretion of potassium ions in the urine.

algae Algae are photosynthetic eukaryotic organisms. Some are classified as Protista and others as plants according to their morphology, which is varied. Some exist as single-celled organisms, and some are multicellular. Usually, algae are aquatic, occupying both marine and freshwater environments. The dinoflagellates and the diatoms are free floating, and the red and brown algae require a solid substrate to which they attach. They are further classified by their photosynthetic pigments, thus the names brown, red, green, and blue-green algae. Many are of industrial importance providing thickeners for foods and bacterial culture media. See AGAR.

alignment Placement of sequences of unknown genes or proteins side by side to analyze the similarities or differences between them. These alignments are done on computers, utilizing databases in which sequences are stored.

alkali Any basic (high pH) solution or compound. Alkaline conditions denature DNA and have been employed in methods to isolate plasmid DNA from chromosomal DNA. Certain alkali treatments have been used in the isolation of bacterial proteins. Of course, the success of this method depends upon the alkali stability of the protein to be isolated.

alkaline phosphatase An enzyme used in DNA cloning procedures to remove the terminal phosphates from the single-stranded tail of vector molecules that are cleaved with a restriction enzyme, thus preventing recircularization of the vector and enhancing the recovery of vectors with inserts.

alkaloids A class of 3,000 compounds containing nitrogen that are produced by plants but that exert potent physiological effects on animals. They are synthesized from aromatic amino acid precursors such as tyrosine, tryptophan, and phenylalanine. Some examples are morphine, cocaine, nicotine, codeine, and colchicine.

alkalophiles (alkalinophiles) These are microorganisms that flourish in basic environments (base loving). Alkalinophiles exists at a pH range of about 7–12. They include *Vibrio cholerae,* the causitive agent of cholera, whose optimum pH

is 9.0; and the soil bacterium *Agrobacterium,* whose optimum pH is 12.

alkaptonuria The first human genetic disease identified when it was found to follow the laws of Mendelian inheritance. Also known as "dark urine disease," it was studied by Garrod and, in 1902, was recognized to be inherited as a recessive trait. Later, the biochemical nature of the disease was also uncovered. The disease is characterized by a deposits of dark pigment in connective tissue and in the urine after exposure to air. Later stages of the disease result in severe forms of arthritis and possibly death due to blockages in the arteries and valves of the heart. One in a million people is born with this disease; it results from a deficiency in the enzyme homogentisic acid deoxygenase, which results in the accumulation of homogentisic acid in the urine.

alkylating agent A type of mutagen that adds alkyl groups such as the methyl group $(-CH_3)$ and the ethyl group $(-CH_2CH_3)$ to bases in DNA. One such mutagen is ethylmethane sulfonate (EMS), which can alkylate either thymine or guanine residues and cause them to mispair during DNA replication. This causes transition type mutations in DNA.

allele One of several alternative forms of the same gene. A single gene can have as few as one or as many as 100 different alleles. Alleles are differences in the base sequence of a single gene among individuals in a population or on the two homologous chromosomes in one individual. They are the cause of genetic variation or different expressions of a trait in a population of organisms.

allele-specific oligonucleotides A probe designed to detect single base-pair changes in a gene. Under very specific conditions, a nucleotide sequence of about 20 base pairs will hybridize to its complementary sequence, but not to one with a one base-pair change.

allograft immunity The state of the immune system in which grafted tissue originating from a genetically dissimilar animal provokes attack by the immune system of the host animal (i.e., graft rejection).

allolactose A derivative of lactose and the true inducer of the lactose operon in bacteria. Inside the cell, lactose is converted to allolactose, which in turn activates the three structural genes involved in the utilization of lactose as a carbon source. When lactose is present in the medium, the genes required for its breakdown are active; when it is not present, they are shut off. See LAC OPERON.

allopolyploid A hybrid organism, usually a plant, that has been bred from two closely related species and contains one or more extra full sets of chromosomes. For example, if the parents each have two sets of chromosomes, the allopolyploid offspring, instead of having the normal two sets, may have four. The hybrid contains genetic information that is different from either parent.

allopurinol A derivative of the purine base hypoxanthine used to treat gout. As an inhibitor of the enzyme xanthine oxidase, allopurinol acts by preventing the accumulation of uric acid. See URIC ACID.

all-or-nothing phenomenon This phrase refers to the condition in which a nerve cell must receive its threshold level of stimulation to respond and start an action potential. A nerve will either fire an impulse or not fire at all if the stimulus is below threshold. There is no such thing as a weak response to a weak stimulus. See ACTION POTENTIAL.

allosteric enzymes Enzymes that have many subunits and many ACTIVE SITES. They display substrate-induced conformational changes and have different roles or functions in their different conformations. They play an important role in the regulation of metabolic pathways and the regulation of gene expression.

alpha-actinin A protein that binds to the actin fiber in an adhesion plaque, where it is localized.

alpha-amanitin A potent toxin derived from the *Amanita* mushroom, also known as death cap or destroying angel. It has been used to distinquish the three nuclear RNA polymerases of eucaryotes, Polymerase I is insensitive to alpha-amanitin, but RNA polymerase II is highly sensitive, and RNA polymerase III is sensitive but at higher concentrations of the toxin.

alpha-blockers A class of drugs that are used to treat high blood pressure as well as urinary problems related to enlarged prostate (benign prostatic hyperplasia; BPH) by acting as antagonists of alpha adrenergic receptors on smooth muscle. In blood vessels, this inhibits contraction of the muscle, which causes vasodilation, thus lowering blood pressure. In prostate tissue, relaxation of smooth muscle allows increased urinary flow. Cardura, Terazosin, and Doxazosin are examples of alpha-blockers.

alpha-fetoprotein An embryonic protein that is believed to function as the embryonic counterpart of albumin and with which it shows great similarity in amino acid sequence and structure. The presence of alpha-fetoprotein in adult serum is a diagnostic indicator for some types of tumors such as teratomas and liver cancers (hepatomas). Alpha-fetoprotein is also present in the sera of pregnant women. Abnormal levels of alpha-fetoprotein during pregnancy may be indicative of certain fetal disorders.

alpha-helix Refers to one type of three-dimensional conformation that a protein assumes in the cell. An alpha-helix is stabilized by the formation of many hydrogen bonds between nearby amino acids in the protein. Hydrogen bonds form between every three amino acid residues. This provides the regularity to the structure of the helix. Another conformation of proteins is the beta pleated sheet. See BETA-PLEATED SHEET.

Alu elements A family of related DNA sequences that are widely and randomly dispersed through mammalian genomes. They are about 300 base pairs in length and are classified as moderately repetitive DNA sequences. There are about 600,000 copies of these sequences in the human genome. At the ends of these sequences is a cleavage site for the restriction enzyme Alu. Their purpose, if any, in the genome is not known.

Alzheimer's disease (AD) A progressive disease of the brain characterized by memory loss and cognitive dysfunction that was first described in 1907 by Dr. Alois Alzheimer. Alzheimer's disease is caused by the deposition of beta amyloid protein, which forms plaques and causes the death of nerve cells in critical areas of the brain. While the vast majority of AD cases are seen in later life, a small number (<3 percent) of cases show a pattern of inheritance, and these tend to manifest much earlier (early onset Alzheimer's). The inherited form of AD has been traced to dominant mutations in at least three genes: Amyloid Precursor Protein (APP) located on chromosome 21, presenilin 1 located on chromosome 14, and presenilin 2 located on chromosome 1.

amber codon The codon UAG, which is one of the three codons that does not code for an amino acid but represents a stop signal.

amber mutation A type of genetic mutation in a class called nonsense mutations. An amber mutation arises when a three-base-pair sequence in DNA (codon) such as UUG, coding for a specific amino acid, mutates to a UAG codon that does not code for any amino acid. UAG is a termination codon because it causes the termination of protein synthesis. Any mutation that results in a UAG termination codon is called an amber mutation. There are also mutations called opal and ochre that are also nonsense mutations.

amber suppressor Mutations in the anticodon of several different tRNA mol-ecules that allow these mutated tRNAs to recognize the amber mutation UAG (see AMBER MUTATION). Ordinarily, a UAG codon in a message signals the termination of translation, but a tRNA with an amber-suppressor mutation has an anticodon that is complementary (CUA) to the termination codon. It can therefore insert the amino acid that it is carrying at that site in the growing polypeptide chain and avert chain termination.

ambient The physical conditions in an organism's surrounding environment. Microorganisms that live in the human gut, for example, have an ambient temperature of 37°C. Organisms that exist in dust particles in a room have an ambient temperature of about 23°C, or room temperature. Ambient conditions also can include atmospheric pressure, humidity, oxygen levels, and other physical parameters that surround an organism.

Ames test A method for screening potential mutagens and carcinogens. This test was developed by Bruce Ames in the early 1970s and cut down drastically on the time and expense that is involved in animal testing. The Ames test requires the use of bacteria to determine the possible mutagenic potential of a chemical. It relies on the principle that the chemical structure and properties of DNA are universal. In addition, the mechanisms for toxicity in a bacterium mimic those of a mammal if the appropriate liver enzymes are provided to process the chemical in the same way a mammal would. A chemical that causes mutations in bacteria would likely do the same in a mammal, and because 90 percent of all known carcinogens are mutagens, a chemical found to be a mutagen in the Ames test would be a suspected carcinogen.

amide The product of the reaction between an amine compound and carboxylic acid. PEPTIDE BONDS are the bonds that link amino acids together in proteins and are a type of amide bond between two amino acids. The amino group of one amino acid is linked to the carboxyl group of the next amino acid.

amine Compounds that contain an amino group (NH_2). Amino acids are amines. See AMINO ACID and AMINO GROUP.

amino acid The building blocks of proteins. Amino acids contain a free carboxyl group (COOH), a free amino group (NH_2), a hydrogen atom, and a variable side group (R) attached to a single carbon. (One exception to this is proline, whose amino group is involved in a cyclic structure.) The physical and chemical properties vary among the R groups. However, there are several classifications that put certain R groups in the same category because they share similar properties. These are the acidic, basic, aliphatic, aromatic, and hydroxyl-containing or sulfur-containing amino side groups. There are 20 different amino acids that are found in proteins.

aminoacyl-tRNA A tRNA that is carrying its specified amino acid; also called a charged tRNA. The specificity of charging of tRNA molecules is carried out by 20 different enzymes called aminoacyl tRNA synthetases. Each of the 20 amino acids that are incorporated into proteins is the substrate of one of the enzymes. In addition, each enzyme recognizes the appropriate tRNA(s) to charge.

aminoglycoside antibiotics A group of antibiotics that act to kill a broad range of bacteria by interfering with protein synthesis at the bacterial ribosome. They are produced naturally by members of the soil-dwelling genus *Streptomyces* and include streptomycin and kanamycin.

amino group The $-NH_2$ group in a molecule. The presence of an amino group is the defining characteristic of the group of organic compounds known as amines.

aminolevulinic acid (ALA) The first committed intermediate in the synthesis of heme (see HEME) and chlorophylls. Cells that either overproduce ALA or are fed large amounts of ALA overproduce porphyrins, or intermediates in the heme biosynthetic pathway. Porphyrins produce toxic compounds to the cell when they react with oxygen. Thus ALA is being tested as a weed killer and as a photodynamic agent in the treatment of skin lesions.

6-aminopenicillic acid A chemical structure that is found in the different natural and semisynthetic penicillins. This common nucleus of the penicillins contains the beta-lactam ring structure. In addition to the common core, 6-aminopenicillic acid, all penicillins contain a variable side group that distinguishes them from one another.

2-aminopurine A purine derivative that is a potent mutagenic agent because it becomes incorporated into DNA in place of adenine. As a result, it induces mistakes in DNA during DNA replication.

amino side groups See AMINO ACID.

amino sugars These are derivatives of simple sugars that have been modified to form amines because of the addition of an NH_2 group in place of the hydroxyl

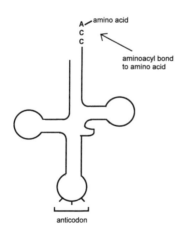

A amino acid
C
C

aminoacyl bond to amino acid

anticodon

Aminoacyl-tRNA

group normally found at carbon 2. Two commonly found amino sugars are D-glucosamine, which is a major component of chitin (the outer hard covering of insects), and D-galactosamine, which is found in cartilage.

amino terminal Also called the N-terminus; one of the ends of a polypeptide chain. This end of the polypeptide consists of an unreacted amino group. The other end is called the carboxyl terminus, or C-terminus.

ammonium sulfate precipitation "Salting out" of proteins in solution. A first step in the purification of proteins from cell extracts; ammonium sulfate, which promotes hydrophobic interactions, is the most common salt used to fractionate proteins according to their solubility in the salt solution.

amniocentesis A procedure for testing the karyotype of a fetus in utero. Cells from the amniotic fluid surrounding the fetus are taken from the mother and cultured in the lab. Karyotype analysis of the cells will determine the sex of the fetus, including any gross deformities of the chromosomes or a chromosome number that signals certain genetic diseases.

AMP See ADENOSINE MONOPHOSPHATE.

amphibolic A metabolic pathway, one that is catalytic and anabolic, that can both degrade metabolites as well as synthesize them. These pathways allow breakdown products of one pathway to be used as substrates in the synthesis of a compound in another pathway.

amphipathic compound A compound that contains both polar and nonpolar groups. Polar groups are soluble in water (hydrophilic), and nonpolar groups are not (hydrophobic). In water or aqueous environments, amphipathic compounds form micelles, or small vesicles with polar regions in contact with water and nonpolar regions regions sequestered in the cen-

ter of the micelle away from water. Fatty acids are amphipathic. Amphipathic molecules are responsible for the properties of biological membranes.

amphiphysin A protein found in nerve terminals, particularly at the synapse where it is believed to be involved in the recruitment of dynamin at sites of endocytosis during the process of synaptic transmission. Amphiphysin autoantibodies are found in patients with Stiffman syndrome (SMS). Amphiphysin autoantibodies are also associated with breast cancer and small cell lung carcinoma.

amphoteric The description of a substance that has both acidic and basic groups and has properties of both acids and bases.

ampicillin A semisynthetic form of the antibiotic penicillin; an antimicrobial agent that kills bacteria by inhibiting the formation of bacterial cell walls. The addition of an amino group to penicillin makes ampicillin effective against gram negative organisms, thus widening the antibiotics spectrum of activity.

amplifiable selection Exploitation of a natural phenomenon in which some transformed cell lines undergo local repeated DNA replication to produce many copies of genes at those locations. A commonly used system is the use of METHOTREXATE to amplify the region around the dihydrofolate reductase (DHFR) gene.

amplification Repeated replication of PLASMIDS or sequences. Plasmid amplification is a process that is used to increase the replication of plasmids over that of chromosomes so that the plasmid isolation from whole cells is facilitated. The process involves growing cells with amplifiable plasmids in the antibiotic chloramphenicol that stops chromosomal DNA replication but does not affect plasmid DNA replication. Specific sequences of DNA can be amplified using the technique of PCR (polymerase chain reaction).

amplification refractory mutation system (ARMS) A PCR (polymerase chain reaction) technique that is used to differentially amplify specific alleles of a gene. Primers (oligonucleotides) for the PCR are constructed so that the 3' base contains the specific base change of the allele. Only DNA targets that hybridize (see HYBRIDIZATION) to the primer will be amplified, and all other alleles with mismatch base pairs and that do not pair with the 3' base of the primer will not be amplified.

amyloid protein The protein forming the core of the characteristic plaques seen in ALZHEIMER'S DISEASE. The protein is composed of 39–43 amino acids and exhibits a tendency to form insoluble precipitates in solution. The formation of plaques is believed to reflect the tendency to aggregate out of the cell fluid, causing interruption of neural transmission.

amylose A starch made up of a long, unbranched chain of glucose. A polymer of monosaccharides is called a polysaccharide. Amylose is the principle storage starch of plants.

anabolic A type of metabolic pathway in which complex molecules are synthesized from smaller precursors, usually in a series of steps; the type of metabolism that builds molecules as opposed to catabolic metabolism, which is degradative. Energy is usually required for anabolic metabolism. An example would be the synthesis of polypeptides from amino acids or the synthesis of nucleic acids from nucleotides. See CATABOLISM.

anaerobe A microorganism that does not or cannot use oxygen during respiration. Obligate anaerobes such as the genus *Clostridium* will die in the presence of oxygen. Others such as *E. coli* are classified as facultative anaerobes because they will use oxygen when present but can switch to anaerobic respiration in its absence.

analog A compound that has important biochemical similarities in structure and/or function to another compound or biomolecule.

anaphase A stage during mitosis or cell division where chromosomes split at the centromere and the resulting chromatids move to opposite ends of the cell. During meiosis, or reduction division, there are two anaphase stages. During anaphase I, homologous pairs of chromosomes separate from each other with their centromeres intact and move to opposite ends of the cell. Anaphase II, during the second meiotic division, resembles the anaphase stage in mitosis, as described above.

anaphylotoxin A substance released by the body as part of an immunological response to the presence of a foreign antigen. Anaphylotoxins stimulate the release of histamines, which cause inflammation in tissues.

androgens A group of male sex hormones that are responsible for the development and the maintainence of masculine features and organs. Testosterone is an androgen.

aneuploid An unbalanced set of chromosomes that results from either the loss or the gain of chromosomes. An individual with the normal complement of two copies of each chromosome (diploid) is disomic. Trisomy is the condition of having one extra chromosome, and monosomy is a loss of one chromosome.

Angelman syndrome (AS) A neurological condition first described by Dr. Harry Angelman in 1965 that is characterized by small head size, severe learning difficulties, fine tremors, jerky limb movements, and epileptic seizures. Cytogenetically, the disease is associated with a deletion of chromosome 15 and is now known to involve a cluster of genes involved in the regulation of ubiquitin at gene map locus 15q11-13.

angiogenesis The process by which new capillaries are formed from endothelial B cells. Angiogenesis is stimulated by a signal in the form of a growth factor(s) and is comprised of at least four components: (1) the production of proteases that allows endothelial cells to invade the surrounding tissue, (2) directed movement of endothelial cells toward the source of the stimulating growth factors, (3) proliferation, and (4) formation of tubules (i.e., capillaries).

angiopoietins A group of secreted factors (Ang-1, Ang-2, Ang-3, Ang-4) that, together with VEGF, regulates endothelial cell survival and capillary formation through the receptor tyrosine kinase, Tie-2, on the endothelial cell surfaces. The angiopoietins are found in the mammalian metanephros, the precursor of the kidney, and they are implicated in deregulated vessel growth in Wilms' kidney tumors and in blood vessel remodeling kidney tissue following toxic injury.

angstrom (Å) A unit of measurement usually used for wavelengths or cellular structures.
$$1 = 10^{-10} \text{ meters, or } 10^{-6} \text{ millimeters,}$$
$$\text{or } 10^{-4} \text{ (0.0001) micrometers,}$$
$$\text{or } 0.1 \text{ nanometers.}$$

anion A negatively charged ion.

anneal Complementary single strands of DNA or DNA and RNA that form hydrogen bonds between complementary base pairs to form double-stranded DNA or DNA-RNA hybrids.

antennapedia complex A genetic locus in the homeotic box that is defined by mutations that cause developmental defects in the thoracic and head segments of the fruit fly, *Drosophila melanogaster*. See HOMEOBOX.

antibiotic A substance usually made by a microorganism that inhibits the growth or kills another microorganism, for example, penicillin. There are many synthetic or manufactured antibiotics that are derivatives of naturally occurring antibiotics and are available for medicinal or research purposes.

antibiotic resistance Microorganisms may have a natural resistance or develop resistance to an antibiotic so that the drug is not effective in inhibiting growth or killing.

antibiotic-resistance genes Genes that confer antibiotic resistance to a microorganism. Examples are genes that encode enzymes that destroy the antibiotic; genes that code for the target of the antibiotic but that become mutated so that the target no longer responds to the drug; or genes that encode proteins that prevent the antibiotic from being taken up by the mircroorganism.

antibodies Proteins that circulate in the bloodstream and bind to foreign invading substances (antigens, e.g., bacteria, toxins, certain viruses) with a great deal of specificity. Antibodies are the mediators of the immune response to soluble antigens. Immunoglobulins.

antibody-producing cell An activated B lymphocyte or plasma cell secretes antibodies. Each plasma cell secretes an antibody with specificity for one antigen.

anticoagulant A chemical substance that prevents the coagulation of blood.

anticodon A three-nucleotide base-pair sequence that is antiparallel and complementary to a codon. The anticodon is found on a tRNA and interacts with a specific codon on the mRNA so that an amino acid will be placed in the correct position according to the mRNA during translation or protein synthesis.

antidiuretic A chemical substance that counteracts a diuretic.

antifungicide A substance or drug that kills fungi.

antigen A substance that will stimulate the production of specific neutralizing antibodies in an immune response. Any chemical substance, usually a protein, that will interact with an antibody.

antigenic determinant A small portion of the antigen that determines the specifically of the antigen-antibody reaction.

antigenic variation A sequential change in the structure of an antigen of microorganisms and viruses so that antigens on these pathogens will not be recognized by antibodies already produced in the host. The disease-relapsing fever, which is characterized by cyclic infections by the same bacterium, is due to the ability of the bacterium to change its antigenic makeup and thus avoid immunity built up by the host. A more subtle type of antigenic variation is seen in the antigenic shift (major antigenic change) and antigenic drift (minor antigenic change) seen in the influenza virus that result in loss of immunity by populations and influenza pandemics and epidemics every number of years.

antigen-processing/-presenting cell Any of a heterogeneous group of cells that bind foreign antigens to their surface and then interact with helper T cells, a process that is required for T-cell activation. Antigen-presenting cells include dendritic cells in lymphoid tissue, Langerhans cells found in skin, and some macrophages.

antihelminthic agent A substance or drug that inhibits the growth or kills helminth parasites.

antihistamine A substance or drug that blocks the effects of histamines in the inflammatory process; a drug that relieves allergy symptoms.

anti-idiotype antibodies An antibody made in response to a unique antigen-combining site of an antibody. The resulting antibody may have a structural similarity to the original antigen and may stimulate antibodies against it, thus serving as a vaccine.

antimetabolite A chemical that inhibits the growth of microorganisms because it blocks the synthesis of some metabolite needed by the microorganism, for example, sulfa drugs that block the synthesis of the vitamin folic acid.

antimorphic allele A mutant allele that has an antagonistic reaction to the normal, wild type of allele. A person who has both an antimorphic allele and wild type of allele for a particular gene will have less of that gene product than an individual who has a deletion for that gene and the wild type of allele.

antimutator A gene that decreases the spontaneous mutation rate of an organism. These genes are usually involved in some DNA repair or metabolism process.

anti-oncogene A tumor suppressor gene, or a gene whose absense is needed for an oncogenic event. Loss or inactivation of a tumor suppressor gene by either mutation or deletion is believed to be an important event in the development of a tumor.

antiparallel Refers to the structure of DNA. The two strands of complementary DNA are antiparallel, that is, the 5' end of one stand is paired with the 3' end of the other and vice versa.

antiparasite A substance or chemical that inhibits or kills parasites.

antiport The transport of two substances across the cell membrane that are coupled but in opposite directions.

antisense RNA A strand of RNA that is complementary to that of a messenger RNA. An antisense RNA would bind to the messenger and prevent synthesis of the protein encoded by the message. Antisense RNA is being explored as a possible therapeutic agent for viral infections and to prevent certain cancer genes from being expressed into proteins.

antisense strand Of the two DNA strands in a double-stranded DNA molecule, the antisense strand is the one that is not used as the template for RNA synthesis.

antiseptic Any chemical that is commonly used to kill microorganisms to prevent infection.

antitermination factor A protein that prevents the termination of transcription. It is involved in certain mechanisms of gene expression control.

apaf-1 A cytoplasmic protein that initiates the process of apoptosis by cleaving, and thereby activating, caspase 9. Activation of caspase 9 causes subsequent activation of other caspases in a reaction chain that ultimately commits the cell to undergo apoptosis. Cleavage of caspase 9 requires the formation of a complex between apaf-1, cytochrome c, and dATP to form an oligomeric structure called an apoptosome.

APC syndrome Familial *a*denomatous *p*olyposis *c*oli; a genetic disease characterized by the development of benign polyps in the colon, a condition that frequently precedes the development of malignant colon cancer. The genetic locus of APC has been shown to reside on human chromosome 5. Research aimed at mapping and then cloning the causative gene(s) via chromosome walking is currently under way. See GENETIC DISEASE.

APH Aminoglycoside 3' phosphotransferase; a bacterial gene that codes for an enzyme that confers resistance to the antibiotic neomycin. The APH gene is commonly used as a selectable marker in transfection experiments in that cells that do not contain the gene can be eliminated from a population by exposure to neomycin. See NEGATIVE SELECTION.

apoenzyme The protein moiety or part of an enzyme without its cofactor; normally inactive.

apoptosis Programmed cell death, or a regulated set of reactions that results in cell death. Apoptosis regulates the balance between cell growth and multiplication and eliminates unnecessary cells. Although apoptosis can be brought about in a variety of ways, the main pathway is initiated by the binding of a ligand to the Fas receptor or by binding of tumor necrosis factor (TNF) to its receptor. Activation of the receptors sets off a cascade by which proteases called caspases are activated, ultimately resulting in degradation of DNA by DNase, proteolysis, and cell death.

aptamer An OLIGONUCLEOTIDE of DNA or RNA or a peptide that binds to and inactivates proteins. Often, libraries of sequences are used to inhibit the protein, and the sequences that are successful are amplified and identified. Aptamers can be used to study the active site of the protein or they can be developed into therapeutics.

apurinic site A site on the DNA in which a purine is missing but the phosphodiester sugar backbone is still intact.

apyrimidinic site A site on the DNA in which a pyrimidine is missing but the phosphodiester sugar backbone is still intact.

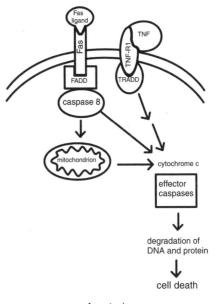

Apoptosis

aqueous Pertaining to water; for example, the aqueous phase after separation with an organic solvent would be the water phase.

aqueous two-phase separation A method to partition proteins during purification using solutions of polyethylene glycol and dextran or polyethylene glycol and certain salts such as a phosphate.

arabinosyladenine (AraA) An antiviral anitibiotic that is used to treat viral encephalitis. AraA is a derivative of the normal purine nucleoside, adenosine, in which the sugar, ribose, has been replaced with one of its optical isomers, arabinose.

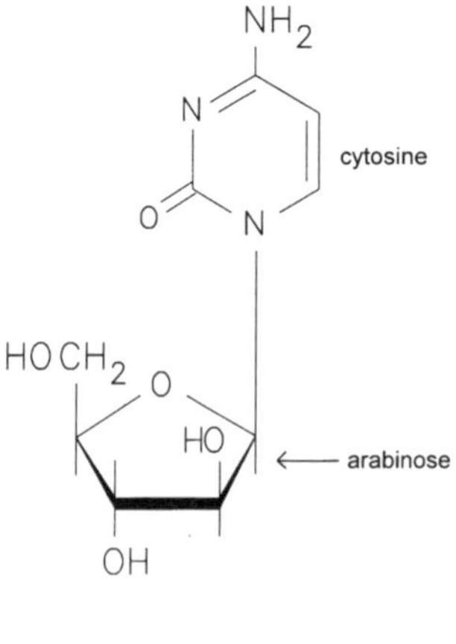

arabinosylcytosine (AraC) An antibiotic that acts by blocking DNA synthesis. AraC is a derivative of the normal pyrimidine nucleoside, cytosine, in which the sugar, ribose, has been replaced with one of its optical isomers, arabinose.

arachidonic acid A 20-carbon fatty acid with four double bonds. It serves as a precursor for the synthesis of prostaglandins.

ARC AIDS-related complex, or a series of symptoms related to an active

human immunodeficiency virus (HIV) infection, including general malaise, night sweats, dementia, wasting, and opportunistic diseases associated with immunodeficiency such as Kaposi's sarcoma (a rare form of skin cancer), pneumonia (generally caused by *Pneumocystis carinii),* and retinitis (caused by cytomegalovirus).

Archaebacteria A group of bacteria including those that produce methane from carbon dioxide and hydrogen (methanogens) and those that live in high-salt environments (halophiles) that appear to be very different from and more primitive than other living bacteria. It is believed that *Archaebacteria* have evolved separately from the true bacteria *(Eubacteria)* and from the eukaryotes and that they constitute a third group of organisms.

arginine An amino acid with the side chain:

$$- (CH_2)_3 -NH-C=NH$$
$$\setminus$$
$$NH_2$$

argininemia A disease caused by a deficiency of the enzyme arginase that catalyzes the last step of the urea cycle in which arginine is hydrolyzed to urea and ornithine. The syndrome is characterized by hyperammonemia, encephalopathy, and respiratory alkalosis. The disease gene is an autosomal recessive, at gene map locus 6q23.

ARMS See AMPLIFICATION REFRACTORY MUTATION SYSTEM.

aromatic An organic compound that contains a benzene-derived ring.

ARS element Autonomously replicating sequences (ARS) found on yeast plasmids that are initiation sites for plasmid replication. Plasmids that lack such sites will not replicate.

Arthus reaction An inflammatory response caused by the production or

Arabinosylcytosine

depositing of antigen-antibody complexes in tissues.

artifact The appearance of a structure in microscopy or an experimental result that is not real but is due to experimental procedures used.

Ascomycetes A class of fungi that is distinguished by a structure, the ascus.

ascorbic acid Vitamin C. Lack of this vitamin in the diet results in the disease scurvy. Ascorbic acid is a reducing agent that keeps the enzyme prolyl hydrolase in active form. Collagen synthesized in the absence of ascorbic acid is insufficiently hydroxylated, can not form fibers properly, and causes the skin lesions that are associated with scurvy.

ascus A saclike structure that produces ascospores, or the sexual spores of the ascomycetes.

aseptic Without germs; sterile.

asexual reproduction Reproduction in the absence of any sexual process, or the reproduction of a unicellular organism by cell division, where a single parent is the sole contributer of genetic information to its offspring.

asparagine An amino acid with the side chain:

$$-(CH_2)-C=O$$
$$\backslash$$
$$NH_2$$

aspartame An artificial sweetener that uses the amino acid phenylalanie as a precursor.

aspartic acid An amino acid with the side chain:

$$-(CH_2)-C=O$$
$$\backslash$$
$$OH$$

Aspergillus A genus of fungi that are important economically because they are used in a number of industrial fermentations.

assay A test. In an enzyme assay, an enzyme is tested for activity under specific conditions.

ataxia telangiectasia (AT) A rare human disease associated with a defect in the DNA repair system. It is a fatal disease that is characterized by a damaged immune system, premature aging and a predisposition to some cancers. Individuals who have only one copy of the gene ATM do not have the disease but are very sensitive to X-rays or chemicals, which cause DNA damage, and these individuals are prone to developing cancer. The gene involved in AT (called ATM) is one of a class of genes called tumor suppressors.

AT content The fraction of the total nucleotides in a DNA molecule that are either adenine or thymine nucleotides; generally given as a percentage.

AT/GC ratio The ratio of adenine plus thymine base pairs to guanine plus cytosine base pairs in a molecule of DNA.

ATM A tumor suppressor gene that is activated by DNA strand breaks. Activated ATM in turn activates the chk2 kinase, which results in cell cycle arrest in G2. ATM is an acronym for *a*taxia *tel*angiectasia *m*utated because both copies of this gene are mutated in patients with this disease. Mutations in the ATM gene result in hypersensitivity to radiation and a tendency to accumulate mutations in other genes, which can lead to cancer, particularly breast cancer, leukemia, and lymphoma. Also, mutations in ATM can cause cells to die, particularly in the cerebellum, which results in the problems with limb movement seen in ataxia telangiectasia. The ATM gene is located on the long (q) arm of chromosome 11 at position 22.3 (gene map locus 11q22.3).

atomic-force microscopy (AFM) A device for visualizing objects with a maximum resolution of about 10 pm, about the size of small molecules. Unlike traditional microscopes, the AFM does not contain a lens but utilizes a probe that measures the attractive or repulsive forces between the probe tip and the molecular structure being visualized as the tip is moved along the surface of the specimen. The movement of the probe tip gives rise to an electrical signal, which is translated into an image by computer. Unlike electron microscopes, AFMs can image samples either in air or in liquids.

ATP See ADENOSINE TRIPHOSPHATE.

ATPase Any of a class of enzymes that acts to remove one or more phosphate groups from ATP to produce ADP or AMP and inorganic phosphate by hydrolysis. The release of phosphate is accompanied by the release of energy that is used to power various cellular functions.

atrial natruiretic factor (ANF) A hormone produced by the right atrium of the heart that stimulates sodium excretion by the kidneys and is involved in the regulation of blood pressure. ANF is believed to play a key role in cardiovascular homeostasis by acting on receptors that stimulate the formaton of cyclic GMP (cGMP). ANF is currently the target of research on new antihypertensive and diuretic drugs.

attenuation 1. A decrease in virulence of a pathogen.
2. A mechanism of gene regulation in bacteria in which availability of certain amino acids will control the expression of genes for their own synthesis by causing premature termination of transcription of the genes involved in the synthesis.

att site A site on the *Escherichia coli* bacterial chromosome that interacts with the bacteriophage lambda genome and at which the bacteriophage genome integrates into the bacterial genome resulting in lysogenization of the bacterium. See LYSOGENIC.

autoclave An apparatus that uses steam under pressure to sterilize materials.

autogenous control Control of gene expression by the gene's product or protein encoded by the gene.

autoimmune The inability to distinquish self from nonself, or a state where the body produces antibodies to its own cells.

autolysin An enzyme that causes cellular self-destruction of the same cells that synthesize it.

autolysis The self-degradation of a cell by release of hydrolytic enzymes of the lysosome. In the case of bacteria, autolysis is brought about by self-destruction of the cell wall by a specific enzyme.

autonomic nervous system The part of the nervous system that regulates involuntary responses.

autonomously replicating sequences (ARS) Special nucleotide sequences in the DNA of chromosomes that serve as sites where DNA replication begins.

autoradiography A technique that involves using a radioactively labeled compound to localize a reaction in a cell or to study a process and using photographic film to visualize the location of the label.

autosomal dominant A mutant allele found on one of the autosomes that will always produce a specific trait or disease. Therefore the chance of passing the gene or the disease to progeny in a pregnancy is 50 percent.

autosome A chromosome that is not a sex chromosome.

autotroph An organism that can make its own nutrients and organic carbon compounds from inorganic carbon in the form of carbon dioxide.

auxin A plant hormone that regulates cell reproduction and cell elongation in certain tissues.

auxotroph A bacterial mutant that can no longer make some required nutrient.

Avastin An anticancer drug that acts by blocking the formation of new blood vessels that feed tumors. Avastin is a recombinant antibody to vascular endothelial growth factor (VEGF), which has been engineered by the insertion of certain human sequences to avoid rejection by the patient's immune system. Avastin acts by blocking VEGF, a protein that plays a key role in tumor angiogenesis. Avastin is used in combination with 5-Fluorouracil chemotherapy to treat patients with primary metastatic cancer of the colon or rectum but is currently being tested for treatment of other types of cancer including renal cell, breast, and non-small cell lung cancers.

axenic culture Pure culture or the growth of one organism.

axin A membrane-bound intermediate in the Wnt signaling pathway that activates the transcription of D-type cyclins. Axin interacts with the adenomatosis polyposis coli (apc) protein, beta-catenin, and glycogen synthase kinase 3b in specific ways that ultimately regulate the entry of beta-catenin into the nucleus. The axin protein contains three domains: a regulation of G-protein signaling (RGS) domain, a disheveled domain, and an axin (DIX) domain. Mutations in axin are associated with liver and ovarian cancers.

axis polarity The orientation of the body in space, depending upon three axes: the anterior/posterior body axis, the dorsal/ventral axis, and the medial/lateral axis. During the development of the embryo, genes that control axis polarity are the axis formation genes that establish embryonic body axis and the axis polarity genes that control anterior/posterior and dorsal/ventral body orientation.

axon Extention of a nerve cell that conducts impulses away from the cell body.

axoneme The structural core of a cilia or eucaryotic flagellum that is made up of nine outer doublets of microtubules and an inner pair of microtubules.

3′-azido-3′-deoxythymidine (azidothymidine [AZT]) An antiviral antibiotic used to treat HIV infection (the AIDS virus). AZT is a derivative of the normal deoxyribonucleoside thymidine in which an azide group is attached to the deoxyribose sugar at the 3′ position. AZT is an inhibitor of the virus reverse transcriptase enzyme that blocks viral replication at the point where viral RNA is copied into DNA.

Azotobacter A genus of free-living microorganisms that are capable of biological nitrogen fixation, or the ability to use nitrogen of the atmosphere for synthesis of nitrogen-containing compounds.

BAC *B*acterial *a*rtificial *c*hromosome; a laboratory-constructed plasmid that is capable of replicating in bacteria, usually *E. coli,* with a very large insert of up to 300 kb of foreign DNA.

Bacillus A genus of free-living rod-shaped bacteria that produce extremely resistant spores, that ensures the organism's survival under harsh environmental conditions. Some species produce antibiotics.

Bacillus Calmette-Guérin (BCG) A nonvirulent form of *Mycobacterium bovis,* an organism that causes tuberculosis in cows. It was isolated by Calmette and Guérin of the Pasteur Institute and has been used since 1928 as a vaccine, primarily in Europe and Japan, against tuberculosis.

bacitracin An antibiotic that is effective against Gram-positive bacteria. It inhibits cell-wall synthesis.

backbone **1.** The spinal column of a vertebrate organism.
2. A structural feature of a molecule that arises from its primary structure. Protein backbones arise from the linking of amino acids through the peptide bond between the carboxyl group of one amino acid and the amino group of the other. Nucleic acid backbones are formed from the joining of nucleotides through sugar-phosphate linkages.

back cross A genetic cross between a heterozygote and one of the its parental homozygotes.

back mutation A mutation that reverts a previous mutation, so the mutant phenotype is changed back to the wild type.

bacteria A group of single celled pro-caryotic organisms that divide by binary fission, are haploid or contain one copy of a chromosome, do not possess organelles such as mitochondria and chloroplasts, and do not have a membrane-bound nucleus.

bacterial transformation A genetic transfer process where cell-free, isolated DNA is taken up by a recipient cell and incorporated into its genome.

bacterial virus A bacteriophage, or a virus that uses a bacterium as its host to reproduce.

bacteriocidal Describing a chemical or drug that can kill bacteria.

bacteriophage A bacterial virus that utilizes the bacterial host replicative systems for its own replication, after which the host cell is usually destroyed—releasing progeny bacteriophage. Many bacteriophage particles consist of an icosohedral-shaped head that carries the bacteriophage DNA genome. The bacteriophage attaches to its bacterial host by means of a cylindrical tail that then serves as a conduit to inject the DNA into the host through a hollow core.

bacteriophage, transducing A phage that acts as a vector in a gene transfer process by injecting donor bacterial DNA into a recipient on viral infection.

bacteriophage lambda A DNA-containing bacterial virus that infects

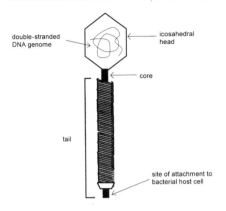

double-stranded DNA genome

icosahedral head

core

tail

site of attachment to bacterial host cell

Bacteriophage

Escherichia coli and has a complex set of regulatory mechanisms governing whether the virus will reproduce itself and lyse its host or lysogenize its host by integration of its genome into its host's genome. Derivatives of lambda are used as cloning vectors to introduce foreign DNA into *E. coli*.

bacteriophage mu A DNA virus that is capable of transposition, or inserting its DNA randomly into the genome of its host. This virus is used in the process of insertional mutagenesis.

bacteriophage φX174 A single-stranded DNA virus that has been used to study the process of DNA replication.

bacteriophage Qβ A single-stranded RNA bacteriophage.

bacteriophage T4 A large DNA virus.

bacteriophage T7 A DNA virus with a very strong promoter that responds to specific T7 RNA polymerase. A number of cloning vectors have been constructed so that foreign DNA is situated next to a T7 promoter, so that expression of the gene can be regulated and amplified by addition of T7 RNA polymerase.

bacteriorhodopsin A transmembrane protein of the "purple membrane" of *Halobacterium halobium* that is capa-

ble of transporting protons across the bacterial membrane, thereby creating a light-dependent electrochemical proton gradient.

bacteriostatic A chemical or drug that inhibits the growth of bacteria but does not kill them.

bacteroid A group of anaerobic, Gram-negative, small-rod bacteria.

baculovirus An insect cell virus that is used as a cloning vector. Proteins made from cloned DNA in baculovirus are glycosylated, a process that does not occur when cloning in bacteria.

baffles Structures on the bottom of some culture flasks that increase aeration when growing a culture of organisms in a shaking water bath or incubator.

baker's yeast *Saccharomyces cerevisiae,* a common yeast, or unicellular budding eukaryotic organism, that ferments sugars and produces carbon dioxide, which is used to leaven bread.

Balbiani rings A very large puff indicating transcriptional activity that is seen at a site on the polytene chromosome of the certain larval insects.

Baltimore, David (b. 1938) A molecular biologist and virologist who won the Nobel Prize in physiology or medicine in 1975 for the discovery that retroviruses, a group of viruses that have an RNA genome produce an enzyme, REVERSE TRANSCRIPTASE. He was founding director of the Whitehead Institute for Biomedical Research at MIT and held that position from 1982 to 1990. He headed the National Institutes of Health AIDS Vaccine Research Committee in 1996.

BamHI A restriction enzyme that recognizes a specific six-base pair sequence (GGATCC) and cuts in a staggered manner, thus creating single-stranded overhangs (sticky ends) at the cut sites.

Bam islands Repeated sequences of fixed length in a nontranscribed spacer region. The designation comes from the fact that these sequences were first isolated by digestion of the spacer region with the restriction enzyme, BamHI.

barophile An organism that grows under conditions of high hydrostatic pressure but cannot grow under normal atmospheric pressure. Such organisms have been isolated from deep seas where the hydrostatic pressure exists at less than 100 atmospheres.

barotolerant An organism that can tolerate high hydrostatic pressure.

Barr body A condensed X chromosome seen in the interphase. The genes on it are not expressed; thus the chromosome is inactive.

basal body Centriole.

basal lamina The thin layer that underlies epithelial cells, which consists of various extracellular matrix proteins including laminin and collagen. The thin membrane surrounding the ovarian follicle is also referred to as a basal lamina.

base 1. A substance that decreases the concentration of H+ ions in solution, or an alkaline substance.
2. A purine or pyrimidine found in nucleic acids.

base analog A purine or pyrimidine base other than the ones normally found in nucleic acids.

base pair (bp) Complementary relationships between purine and pyrimidine molecules that allow adenine to form two hydrogen bonds with thymine or uracil and guanosine to form three hydrogen bonds with cytosine. Base pairing enables nucleic acids to recognize each other and plays an important role in reactions involving nucleic acids such as DNA replication, transcription, and translation.

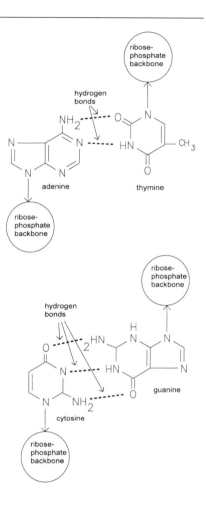

Base pair

base substitution A type of mutation in which one base or base pair is different in the mutant than in the wild type.

basket centrifuges Instruments that operate at very low centrifical forces and act as centrifical filters, collecting large particulate matter. These are useful to collect proteins that have been absorbed to materials such as ion exchange supports in the batch adsorption method.

basophile An organism that lives in alkaline environments.

batch centrifuges Those centrifuges that can accommodate solutions varying from less than 10 ml to liters at a wide range of centrifical forces.

batch culture Growth of microorganisms in a closed system under proscribed conditions of medium, temperature, and aeration.

B cells See B LYMPHOCYTES.

bcl2 An anti-apoptotic factor found in the mitochondrial outer membrane that acts to block the release of cytochrome c from the mitochondrion. The inhibition of cytochrome c blocks the activation of caspase 9 by apaf-1. Bcl2 was originally discovered as an oncogene activated by chromosomal translocations in lymphomas.

bcl-x/bax A bcl2-related gene that can either mimic the function of bcl2 as a repressor of apoptosis or, in an alternatively spliced form, act to promote apoptosis. The alternative splicing products of the gene are characterized in terms of the length of the transcripts with the larger transcript giving rise to the apoptosis repressor form and the smaller transcript coding for the apoptosis promoting form. The same gene characterized in chicken is known as bcl-x, and that described in humans is known as bax.

bcr A region on human chromosome 22 known as the *b*reakpoint *c*luster *r*egion, at which chromosome breakage and translocation occurs in cases of chronic myelogenous leukemia (CML) and acute lymphoblastic leukemia (ALL). In these cancers, there is a reciprocal translocation between chromosomes 22 and 9 that results in the formation of a hybrid chromosome (the Philadelphia Chromosome) in which the abl oncogene is fused to a gene in the bcr region. The bcr-abl fusion product contains an activated form of abl that results in transformation of the cell to a cancerous state.

Beadle, George W. (1903–1991) A geneticist who, in collaboration with

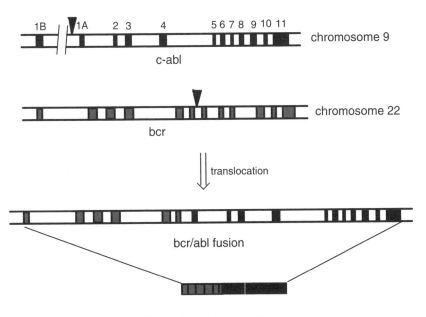

Edward Tatum, showed that genes control enzyme production. Beadle and Tatum shared the 1958 Nobel Prize in physiology or medicine with J. Lederberg.

bectoplasm An archaic term for the outer portion of the cytoplasm of a cell.

Beer-Lambert law The equation that states that the molar concentration of a substance is proportional to how much light of a certain wavelength is absorbed by a solution of the substance:

$$A = ECL$$

Where
A = the absorbance at a given wavelength
E = the molar extinction coefficient
C = the molar concentration of the solution
L = the length of the light path

Bence-Jones protein Part of an antibody molecule (the light chain) that is found in the urine of individuals who have the disease multiple myeloma, a tumor of the bone marrow. These fragments were instrumental in determining the structure of the antibody.

benign Referring to a tumor that does not proliferate and does not invade surrounding tissues.

Benzer, Seymour (b. 1921) A geneticist who studied, and then employed, the process of recombination in bacteriophages to create the first fine structure maps of genes. He is credited with establishing the relationship between genetic units (genes) and proteins as formulated in the "one gene–one protein" hypothesis.

Berg, Paul (b. 1926) A biochemist who gained fame for his work with recombinant DNA. He was a member of the National Academy of Sciences who helped formulate National Institutes of Health policy on recombinant DNA in the mid-1970s. Berg was awarded the Nobel Prize in chemistry in 1980, along with Walter Gilbert and Frederick Sanger, for work on recombinant DNA. He became head of the NIH Scientific Advisory Committee for the Human Genome Project in 1991.

beta-adrenergic receptor kinase (βARK) An enzyme responsible for the desensitization of the beta-adrenergic receptor as a result of continued stimulation by the receptor agonist (e.g. epinephrine). βARK causes inactivation of the receptor by phosphorylating serine residues on the cytosolic portion of the receptor. Inactivation of the beta-adrenergic receptor is due to elevated levels of βARK in cardiac muscle that occurs rapidly after a heart attack. The βARK gene is located on chromosome 11, centromeric to 11q13 (gene map locus 11q13).

beta-arrestin (βarr) A protein that binds to the cytosolic portion of the beta-adrenergic receptor following phosphorylation of the receptor by βARK. Binding of βarr effectively blocks the binding of the G-coupled receptor kinase, thereby inactivating all subsequent steps in the cascade of reactions that releases glucose from glycogen in muscle and liver.

beta-barrel A type of structure assumed by some transmembrane proteins in which the polypeptide(s) are arranged in a such a way as to give the appearance of a barrel. In a beta-barrel, 20 or more transmembrane polypeptide segments are aligned in a regular manner to form to a cylinder that acts as a channel to transport solutes across the cell membrane. Porins, which form channels in bacterial and mitochondrial membranes, are one of the best-known examples of beta-barrel structures.

beta-blockers A class of drugs used to treat high blood pressure (hypertension), congestive heart failure, abnormal heart rhythm (arrhythmia), and angina. Beta-blockers act by blocking beta-adrenergic receptors mostly on cardiac muscle tissue. Beta-blockers, particularly those specific for beta1 receptors, are selective for cardiac tissue. Some examples of beta-blockers are atenolol, meto-prolol, and propranolol.

beta-carotene A pigment that harvests light energy and transfers this energy to other photosensitive pigments, such as chlorophylls, in a photosystem. This pigment gives a red or orange color to carrots, tomatoes, and other plants.

beta-galactosidase An enzyme that catalyzes the hydrolysis of the disaccharide lactose to the monosaccharides glucose and galactose. The gene encoding this enzyme in *E. coli* is part of the *lac* operon.

beta-lactam ring A basic structure of penicillins and their synthetic derivatives.

beta-pleated sheet Rigid, extended sheetlike secondary structure of proteins that is held together by hydrogen bonds.

bFGF *B*asic *f*ibroblast *g*rowth *f*actor; one of a family of growth factors that induces ANGIOGENESIS and acts as a chemoattractant for fibroblasts and other cell types. It binds to heparan sulfate in the EXTRACELLULAR MATRIX.

BglII A restriction enzyme that recognizes a specific six base-pair sequence (AGATCT) and cuts the DNA in a staggered manner, creating single-stranded overhangs at the cut site.

bicoid genes A group of genes that code for proteins that play a determining role in the development of the head and the thorax in the embryo of the fruit fly, *Drosophila melanogaster.*

bidirectional replication Replication of a DNA molecule by two replication forks moving in opposite directions from a single initiation point.

binary fission Division of one cell into two after replication of the DNA.

biochemical oxygen demand (BOD) A measure of the amount of oxygen consumed in biological processes that breaks down organic matter in water. A measure of the organic pollutant load.

biochemistry The chemistry of biological systems and processes.

bioconversions The use of microbes to catalyze the production of economically valuable products. The process of fermentation is a bioconversion or a biotransformation.

biodegradation The field of study devoted to methods for removal of environmental pollutants using the degradative properties of microorganisms. Much of the work in this field centers on the creation of genetically engineered microorganisms designed to degrade organic compounds that are generated in industrial waste or that capture toxic metals present in toxic waste dump sites.

bioenergetics The field covering thermodynamic principles that are applied to biological systems.

biogel A matrix made out of a variety of materials such as dextran, polyacrylamide, agarose, and cellulose used in chromatography to purify proteins. See CHROMATOGRAPHIC TECHNIQUES.

bioinformatics Computational molecular biology and genetics. The use of computers to store and analyze data, usually nucleotide sequence data that can be analyzed for control regions of genes, amino acid sequence data that can be used to find functional domains of proteins, and both kinds of data used to find sequence similarities to other genes or parts of genes. Such gene sequence comparisons are being used to understand evolution of genes and organisms.

biomass The total mass of living matter present on Earth.

bioreactors Equipment designed to maximize product formation in a biocatalyzed reaction. Such equipment uses biocatalysts that are immobilized on a support and controls the contact between the catalyst and its substrate.

biosynthesis The synthesis of molecules in biological systems. These syntheses are carried out in small discrete steps, each step catalyzed by an enzyme, and are energy requiring, usually involving ATP or GTP as energy sources.

biotin A vitamin prosthetic group that carries activated carbon dioxide and that is bound to the enzyme pyruvate decarboxylase. This is an important enzyme because it replenishes one of the intermediates of the Krebs cycle.

biotin labeling (biotinylation) A nonradioactive labeling system in which biotin is covalently linked to a nucleic acid.

biotransformations See BIOCONVERSIONS.

biphasic growth curve The growth curve of a microorganism that is characterized by two exponential growth phases separated by a stationary phase. Such a growth curve is produced by culturing the organisms on two carbon sources, in which one carbon source is in a limiting concentration and must be used up before the second carbon source can be utilized.

bispecific antibodies Antibodies in which the two binding sites recognize different antigens. Such antibodies can have one binding site recognize the antigen of a tumor cell, and the other antigen recognize the antigen of a cytotoxic (T cell) lymphocyte, thus effecting the killing of the tumor cell by the cytotoxic T cell. Biospecific antibodies can be produced chemically, or biologically by fusion of two monoclonal antibodies, producing hybridomas or cells.

bithorax A genetic locus in the homeotic box defined by mutations that cause developmental defects in the thorax region of the fruit fly, *Drosophila melanogaster*. See HOMEOBOX.

bivalent A synapsed pair of homologous chromosomes found in prophase I and metaphase I of meiosis; also known as a tetrad.

BLAST *B*asic *l*ocal *a*lignment *s*earch *t*ool; a set of similarity search programs designed to look at all of the available protein or nucleic acid sequence databases. Access BLAST through the home page of the National Center for Biotechnology Information (http://www.ncbi. nih.gov).

blasticidin An antibiotic that inhibits protein synthesis in both prokaryotes and eukaryotes. A gene that confers resistance to blasticidin (BSD) has been incorporated into some plasmid vectors so that the antibiotic can be used to select stable cell lines that carry the vector.

blastoderm A stage in the development of insect embryos in which a layer of nuclei or cells around the embryo surround an internal mass of yolk.

blastula The stage in animal development in which a ball of cells is formed from the cleavage of cells of the zygote.

blood agar A culture medium in which animal blood, usually rabbit or horse, is added to provide nutrients or to be used diagnostically for hemolysins (enzymes that lyse red blood cells) secreted by certain strains of bacteria.

blood groups See ABO BLOOD GROUP.

Bloom syndrome (telangiectatic erythema) A rare autosomal recessive disease characterized by spider veins (telangiectases), sensitivity to light, impairment of growth and the immune system, and a predisposition to cancer. Bloom's syndrome is caused by a mutation in a gene called *BLM*, at gene map locus 15q26.1. The *BLM* protein is a DNA helicase.

blot A nylon or nitrocellulose membrane onto to which nucleic acids or proteins are transferred for the purpose of hybridization or interaction with antibodies. See NORTHERN BLOT and SOUTHERN BLOT HYBRIDIZATION.

blotting, capillary diffusion A procedure that transfers nucleic acid from a gel to a nylon or nitrocellulose membrane by capillary diffusion, that is, movement of water through the gel and through the membrane that results in depositing and trapping the nucleic acid on the membrane as the water moves.

blotting, electrophoretic A variant of capillary-diffusion-blotting procedure but using an electrical field to facilitate the transfer of the nucleic acid to the membrane.

blunt-end DNA Both strands of DNA at one end are even; that is, there are no single-stranded overhangs. This term is often used in reference to restriction enzymes that cut the DNA at the same position on both strands, as opposed to enzymes that make staggered cuts.

blunt-end ligation A cloning technique in which both the vector and the insert to be spliced into the vector have blunt ends that must be joined together by the enzyme ligase. Such a ligation is more difficult to achieve than one in which the vector and the insert have complementary single-stranded overhangs that first form hydrogen bonds before the ligation step.

blunt ends See BLUNT-END DNA.

B lymphocytes The antibody-producing cell of the humoral immune response. When stimulated with antibody, these cells divide and differentiate into plasma cells that secrete antibodies.

BMP Bone morphogenetic proteins; a family of proteins that can induce the formation of new bone (osteogenesis). The gene sequences of the BMPs place them in the TGF-b superfamily. Based on sequence homology, about 20 so-called growth/differentiation factors (GDFs) have been classified as BMPs. BMPs are being tested as a means of inducing bone growth at sites of extensive injury or after surgical procedures involving bone removal.

Bovine somatotrophin (BST) A growth hormone that has been manufactured in large quantities through recombinant DNA techniques used to enhance the production of milk. This is a very controversial product because of the public's concern over the long-term effects of recombinant DNA on the quality and safety of food.

Boyer, Herbert (b. 1936) A biochemist whose discovery of restriction enzymes and their application in creating recombinant DNAs initiated the field of genetic engineering. He and Stanley Cohen created the first recombinant DNAs that could be grown in bacteria. In 1976 he and Robert Swanson founded Genentech (for *gen*etic *en*gineering *tech*nology), the first biotechnology company. In 1985 Genentech was the first biotech company to produce a biopharmaceutical product, human growth hormone.

branched-chain alpha-ketoacid dehydrogenase (BCKDH) An enzyme that, if inactivated by mutation in any of the subunit genes (E1a, E1b, E2, E3), leads to a condition know as Maple Syrup Urine Disease (MSUD). BCKDH is necessary for the metabolism of the three branched-chain amino acids: leucine, valine, and isoleucine. Enzyme deficiency results in spillover of these amino acids and their corresponding alpha-ketoacids into the blood and urine, giving the urine a characteristic color and odor from which the name of the condition is derived. MSUD was first described in 1954 and leads to mental and physical disabilities and can be fatal if untreated. The genes for the different subunits are located on different chromosomes: E1a at gene map locus 19q13.1, E1b at gene map locus 6p21-22, E2 at gene map locus 1p21-31, and E3 at gene map locus 7q31-32.

branch migration A proposed step in the process of DNA recombination, or DNA crossing over, in which there is movement of the crossover point of the recombinant intermediate.

BRCA1/BRCA2 The first breast cancer genes identified. Both genes are tumor-suppressor genes. Mutations in these genes are believed to be responsible for about half of the inherited forms of breast cancer. Individuals inherit one copy of the mutated gene because if an embryo possesses two copies of the mutant gene, it does not survive.

breakage and reunion Physical breakage of DNA molecules and rejoining of parts of two different molecules, resulting in recombination or crossing over.

5-Bromouracil (5-BU) A chemical that causes mutations in DNA because it resembles thymine, a natural constituent of DNA. When incorporated into DNA in place of thymine in its enol-shifted form, it can readily pair with guanine. In its presence, an A-T base pair is replaced by a G-C base pair after two rounds of replication. This is called a transition mutation.

broth A liquid culture medium for microorganisms.

Bruton Agammaglobulinemia An X-linked immunodeficiency disease in males that is caused by mutations in a gene that codes for an enzyme known as the Bruton tyrosine kinase (BTK). The BTK enzyme is necessary for the maturation of antibody-producing B cells of the immune system, and therefore the disease is characterized by many types of systemic and pulmonary infections. Infections begin at about six months of age when the maternal antibodies begin to decline. In the absence of B-cell maturation, the organs where B-cell maturation normally takes place (the spleen, tonsils, adenoids, Peyer patches, and peripheral lymph nodes) are often reduced in size or completely absent. The *BTK* gene is found on the X chromosome at gene map locus Xq21.3.

buffer A substance in liquid that tends to resist changes in pH, by absorbing either hydrogen or hydroxyl ions.

Burkitt's lymphoma A tumor that is relatively common in East Africa and New Guinea but rare in other parts of the world. The Epstein-Barr virus (EBV), the etiological agent of infectious mononucleosis, is associated with this disease, but it is not currently known whether the relationship of the virus to the disease is casual or causal.

bursa of Fabricius A lymphoid organ of the chicken that is responsible for the maturation of B lymphocytes. The B cells were so named because of this organ. However, humans and other mammals do not possess a bursa, and its equivalent in these organisms is probably other lymphoid tissues, such as the tonsils, the appendix, Peyer's patches, and the lymphoid follicles.

burst number The number of viral particles that are produced per cell after infection.

C₃ cycle The Calvin cycle or that part of photosynthesis where CO_2 is fixed to form a three-carbon organic compound that is subsequently converted into a six-carbon sugar.

C₄ cycle The Hatch-Slack pathway, or an accessory very efficient pathway to fix CO_2 used by plants that grow in hot dry climates with low CO_2 levels.

C600 A strain of *E. coli* that is commonly used in genetic experiments and as a host for cloned plasmids.

C_0t value A parameter describing the rate at which complementary strands of DNA reassociate with one another to form double-stranded DNA. C_0t values are of significance historically because the theoretical relationship between C_0t and reassociation rate underlies the principle of DNA probe hybridization.

If DNA is denatured to a single-stranded state and then allowed to reassociate back to its native double-stranded form, the extent of reassociation for any particular DNA sample increases (1) with DNA concentration when allowed to renature for a given amount of time or (2) with time at a given DNA concentration.

C_0t is the product of the two variables:
$C_0t = (DNA\ concentration) \times (the\ time\ allowed\ for\ reassociation).$

The concentration (C) of double-stranded DNA formation as a function of C_0t is:
$$C/C_o = 1/(1 + kC_ot)$$
where k is the reaction rate constant and C_o is the initial concentration of unpaired DNA.

CAAT box A consensus nucleotide sequence in DNA that has homology to GGT(orC)AATCT and that is found in the promoter region of many eukaryotic genes and is required for efficient transcription.

cadherins A family of proteins found on the cell surface that mediate cell-cell adhesion and that play a central role in normal development. The cadherins are responsible for the selective cell-cell adhesion that accounts for the cell sorting by which cells are placed at their proper sites during development. The typical cadherin protein has five tandem repeated extracellular segments, a single membrane-spaning segment, and a cytosolic domain. Cadherins function as a signal transduction element in the Wnt signaling pathway. Cell-cell binding by E cadherins releases membrane-bound src, which acts to induce cyclin D through beta catenin. There is also recent evidence that altered expression of cadherins may be involved in invasion and metastasis of tumor cells.

caldesmon A CALMODULIN-binding protein involved in the regulation of contraction in smooth muscle. At high calcium concentrations caldesmon binds to the Ca^{++}-calmodulin complex. This leads to muscle contraction by allowing muscle actin to make contact with myosin.

calmodulin A ubiquitous calcium-binding protein that serves as an intracellular receptor of Ca^{++} and, in its active form, mediates an intracellular response to Ca^{++} as a second messenger. See SIGNAL TRANSDUCTION.

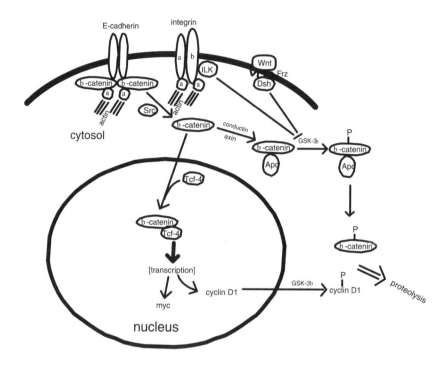

Cadherins function as a signal transduction element in the Wnt signaling pathway

calorie A unit of energy measurement: the amount of energy needed to raise the temperature of 1 gram of water by 1°C.

calpains A group of calcium-dependent proteases that act on various cytoskeletal, membrane, and certain proteins involved in regulatory processes. Calpains exist in two major isoforms—calpain I and calpain II—that have different calcium requirements for activation. Calpains has been linked to both neurodegenerative conditions, including ischemia and Alzheimer's disease. Calpain appears to play a role in apoptosis. Calpain-mediated proteolytic activation of apoptotic pathways leading to cell death is a late-stage event brought about by excitotoxic compounds, and therefore, therapeutic strategies aimed at limiting neuronal damage by selectively inhibiting calpains are currently under investigation.

calponin A protein of about 34 kDa involved in regulation of contraction in smooth muscle. Binding of calponin to muscle myosin head groups results in a state of relaxation.

cAMP See CYCLIC AMP.

cancer A class of diseases in which normal cell control is lost so that certain cells in the body proliferate uncontrollably, invade other tissues, and spread to distant sites (metastases) in the body.

CAP Catabolite activator protein, or catabolite repressor protein (CRP); a protein that when bound to cAMP will bind to the promoter region of some operons, encoding enzymes that metabolize sugars in bacteria to enhance transcription. Thus, the protein bound to cAMP acts as a positive regulator of transcription. See LAC OPERON.

capped 5′ ends A methylated guanosine residue attached to the 5′ end of a eukaryotic mRNA. The bond is made between the 5' phosphate group of the nucleotide and the 5' end of the RNA, so the nucleotide is referred to as inverted. This cap may give stability to the mRNA.

capping of mRNA The posttranscriptional process that adds a guanosine residue to the 5' end of a eukaryotic mRNA and then methylates it.

capsid The protein coat of a virus.

capsomere The protein subunits of the capsid.

capsule An envelope of CARBOHYDRATE, or a slime layer surrounding some microorganisms. Capsules contribute to the invasiveness of some bacteria because they enable the organisms to evade phagocytosis.

carbohydrate A sugar or the name for molecules that contain carbon hydrogen and oxygen in the ratio, $C_nH_{2n}O_n$ and that can be simple monomers, such as glucose or fructose; disaccharides; or two molecules joined together by a glycosidic bond (see GLYCOSIDE), such as sucrose (common table sugar) or lactose (milk sugar); or polymers containing as many as thousands of simple sugar molecules, such as starch, cellulose, and glycogen.

carbon dioxide cycle The flow of CO_2 from organisms that can photosynthesize (plants and algae) and convert it into organic foodstuffs to all other organisms that consume the organic molecules and give off CO_2 as waste product.

carbon fixation The process by which plants and algae (photosynthesizers) convert inorganic carbon, CO_2, into organic molecules, specifically carbohydrates, that are used as food for other organisms.

carbon source Any organic carbon-containing molecule that can be metabolized to produce energy in the form of ATP in an organism. In general, carbohydrates serve as carbon sources for most organ-

isms, but fats and some amino acids can also be utilized for energy production.

carbonyl group The atoms of carbon and oxygen, in which the oxygen is bonded to the carbon via two chemical bonds, C = O.

carboxyl group The atoms of carbon, oxygen, and hydrogen, in which one oxygen is bonded to the carbon via a double bond and the oxygen and hydrogen form a hydroxyl group (OH) and are bonded to the carbon via a single bond with the oxygen:

$$-C=0$$
$$|$$
$$OH$$

Carboxyl groups are found on organic acids.

carboxyl terminus The end of a molecule where a carboxyl group is found. Proteins that are made up of amino acids have a carboxyl terminus and an amino terminus.

carboxypeptidases Enzymes that remove successive amino acids from proteins starting at the carboxy terminal end (see above) by hydrolyzing the peptide bond between amino acids.

carcinogen A cancer-producing agent or any chemical or physical agent that can produce a tumor in an animal or cause normal cells in culture to become transformed.

carcinogenesis The process by which a normal cell transforms into a cancerous one. See TRANSFORMATION.

carcinoma A tumor derived from epthielial cells.

cardiac muscle The muscle tissue of the heart, thus responsible for pumping blood through the body's circulatory system. It is striated and looks very similar to skeletal muscle, but both muscles use different carbon sources for energy production.

carotene (beta) A specific CAROT-ENOID found in plants that assists in light absorption by the chloroplasts. This pigment gives the red color to such vegetables as carrots, tomatoes, and yellow squash.

carotenoids A family of pigments that can absorb a range of wavelengths of light and funnel the energy to chloroplyll a, the major photosynthetic pigment of plants. Thus, these accessory pigments greatly extend the range of wavelengths of light that can be used for photosynthesis.

carrier A substance involved with the transport of materials.

carrier protein A protein embedded within the cell membrane that binds to a specific compound or group of related compounds and aids in transporting it from the outside of the cell through the membrane lipid bilayer to the interior of the cell.

casamino acids A mixture of amino acids that results from the enzymatic breakdown of the milk protein, casein.

cascade A series of reactions that is triggered off by one reaction or compound.

casein hydrolysate The breakdown product of the milk protein casein to its constituent amino acids by either enzymes or acid hydrolyzing the peptide bonds between the amino acids.

caspase 3 (ced-3) A component of the apoptosis pathway that is activated by cleavage of procaspase 3 by caspase 9. Caspase 3 is an effector caspase that directly leads to the macro changes observed in apoptosis, such as membrane and DNA damage and large-scale proteolytic degradation. Ced-3 is the same protein characterized independently in the nematode *C. elegans.*

cassettes Genes or nucleotide sequences that can be easily spliced into chromosomes or plasmids. Some cassettes are naturally occurring and under appropri-ate conditions are moved behind promoters where they can be expressed. Cassettes can be constructed in the lab by flanking the gene or sequence to be made into a cassette with restriction enzyme cutting sites.

catabolic pathway A series of reactions that breaks down compounds to simpler ones, usually with the release of energy that is trapped into high-energy molecules such as ATP.

catabolism The degradation of complex substances to simple ones.

catabolite A substance that can be broken down by an organism to yield energy, usually in the form of ATP.

catabolite repression A process found in certain bacteria in which there is decreased synthesis of enzymes involved in catabolism when a preferred alternative catabolite is present. For example, the enzymes that metabolize the sugar lactose are not synthesized by bacteria when they are grown on the sugar glucose.

catalase An enzyme that breaks down hydrogen peroxide, a toxic waste product of metabolism, to water and oxygen; usually found in the microbody or peroxisome organelles.

catalysis (catalytic) The acceleration of a reaction by a catalyst.

catalyst A substance or physical agent that speeds up a reaction but is not consumed during the course of the reaction. Catalysts in biochemical reactions are enzymes. Catalysts change the rate at which reactions approach equilibrium but do not affect the position of equilibrium.

catalytic site The location on an enzyme where the active site or the place where substrates bind and the reaction proceeds. The catalytic site brings reactants of a reaction close together eliminating the need for random collisions, thus making it more efficient for the reaction to proceed.

CAT assay An assay for determining whether a given DNA fragment may contain promoter activity by ligating the fragment to the chloramphenicol acetyl transferase (CAT) gene in an expression vector and observing whether the CAT enzyme is made when the vector is transfected into animal cells. See APH.

catenanes Macromolecular rings that are mechanically interlinked with one another. Catenanes have been used to create molecular machines (nanomachines). Catenanes are being tested as nanoscale robotic devices that may be useful for the development of slow-release drug-delivery systems or to control chemical reactions in nanoscale laboratories on a chip.

catenation The linking together of multiple copies of a macromolecule.

CAT gene Chloramphenicol *a*cetyl *t*ransferase gene; a bacterial gene that catalyzes the transfer of an acetyl group to chloramphenicol. The CAT gene is commonly used as a reporter gene in experiments designed to demonstrate that certain DNA sequences can function as promoters.

cation A positively charged ion.

C-banding A technique for staining the highly repeated DNA sequences in the region of the chromosome surrounding the centromere. See SATELLITE DNA.

CBF See CCAAT-BINDING FACTOR.

CCAAT-binding factor (CBF) A transcription factor complex that binds to the CCAAT motif upstream of the promoters of many different genes, for example, type 1 collagen, albumin, and beta-actin genes. In the yeast *Saccharomyces*, CBF binding induces the transcription of genes required for growth based on a nonfermentable carbon source. CBF consists of four known subunits: HAP2, HAP3, HAP4,

and HAP5 and stimulates the transcription of various genes by recognizing and binding to a CCAAT motif in promoters. The human gene for CBF is at gene map locus 6p21.3.

CCBF transcription factor A TRANSCRIPTION FACTOR involved in regulation of genes (e.g., Cln1 and Cln2) required for progression through the G1 phase of the cell cycle in yeast. The CCBF transcription factor binds to a specific DNA sequence called the cell cycle box in the promoter region(s) of the critical cell cycle genes.

CD3 A complex of TRANSMEMBRANE PROTEINS in T cells that, in association with the T cell receptor, helps promote an interaction between the T cell and another cell containing an antigen on its surface. As a result of this interaction, there is a proliferation of T cell clones that recognize the antigen.

CD4 A TRANSMEMBRANE PROTEIN IN T CELLS that, like CD3, functions in the interaction of a T cell with an antigen-presenting cell to promote the proliferation of a T-cell clone specific for the antigen. During the interaction of the T cell and the antigen-presenting cell, CD4 (and CD8) activates a tyrosine kinase inside the T cell that leads to the proliferative response of the T cell.

CD8 A T-cell transmembrane protein that functions in essentially the same way as CD4.

cdc An acronym for cell division cycle.

cdks (cyclin-dependent kinases) The enzymatic subunit of the complexes that regulate progression of a cell through the cell cycle. Cyclin-dependent kinases are generally specific for tyrosine (tyrosine kinase) but are inactive unless they are complexed with a cyclin; the cyclin thus functions as the regulatory subunit of the cyclin-cdk complex. Different combinations of cyclins and cdks control passage through different phases of the cell cycle. In mammalian cells:

- Cyclin D-Cdk4 or 6 controls progression through G1 phase.
- Cyclin E-Cdk2 controls entry to S phase.
- Cyclin A-Cdk2 controls progression through S.
- Cyclin A-Cdk1 controls progression through G2.
- Cyclin B-Cdk1 initiates the onset of M phase.

cDNA (complementary DNA) The single-stranded complementary DNA that is copied from mRNA by the enzyme reverse transcriptase.

cDNA cloning A recombinant DNA technique in which double-standed cDNA is spliced into vectors so that the gene can be amplified or expressed.

cDNA library A collection of cDNA molecules spliced into vectors, made by using all the mRNA molecules in a cell or organism and copying it with reverse transcriptase. The library is subsequently screened with appropriate probes to pick out the clone of choice. See CLONE LIBRARY and LIBRARY.

cell The smallest membrane-bound unit capable of replication. Cells may either function independently, such as those of unicellular microorganisms, or function cooperatively as the cells of a tissue or organ.

cell abalation The selective destruction of cells. The technique of cell abalation is used in studies of the role of differentiating cells in the development of an organism. Genes encoding cytotoxins such as diphtheria toxin are introduced into the cells to be destroyed behind cell-specific promoters.

cell adhesion Any mechanical coupling of one cell to another or of a cell to a solid support (the substrate). Cell adhesion is mediated by specialized cell-cell or cell-substrate junctions or the EXTRACELLULAR MATRIX:

- circumferential belt—a ring of actin and myosin bundles that encircles the inner surface (cytoplasmic side) of the cell membrane just at the location of ADHERENS JUNCTIONS.
- hyaluronan (hyaluronic acid: HA) is a long acidic polysaccharide found in the extracellular matrix all over the cell surface. The presence of HA interferes with close cell-cell contacts and thus inhibits cell-cell junctions that mediate adhesion.
- stress fibers—bundles of ACTIN and MYOSIN that run internally along the ventral surface of cell. Stress fibers are attached at one end to adhesion plaques, and it is believed that they function in cell movement and cell-substrate attachment.
- proteins in cell adhesion—cell surface proteins that mediate cell-cell adhesion. These proteins fall into three major classes: cadherins, selectins, and immunoglobulins.

cell coat A layer of carbohydrates that protrude out into the extracellular space from the cell membrane in animal cells and is seen in the electron microscope as an electron dense coat on the surface of the cell.

cell culture A population of cells grown in a medium.

cell cycle A sequence of events involved in the replication of the genetic material of the cell and the orderly parceling of it out to two daughter cells. The cell cycle consists of the G1, S, and G2 phases that make up the interphase and the M phase or mitosis where chromosome division occurs.

cell-disruption techniques The release of intracellular proteins and nucleic acids from microorganisms. The techniques to be used depend on the sensitivity of the protein or nucleic acid to be isolated to each treatment and whether the extraction procedure is small scale or large scale:

- chemical—alkali conditions can be used to release certain proteins from microorganisms in both small-scale and large-scale preparations.

- detergents—both ionic such as sodium lauryl sulfate, or nonionic detergents such as Triton X-100 can be used to destroy the cell membrane and facilitate cell lysis.
- enzymatic—lysozyme, an enzyme prepared from hen egg whites, breaks down the peptidoglycan cell wall of bacteria.
- grinding—physical disruption of the bacterial cell wall by grinding with an abrasive material such as glass beads. This can be done either in small scale with a mortar and pestle or in large scale using a cell disrupter apparatus.
- osmotic shock—release of proteins from the periplasmic space of Gram-negative bacteria (see GRAM STAIN) by resuspending the cells first in a solution of 20 percent sucrose and then resuspending them in water.
- shearing—passage of cells through a narrow orifice at high pressure. Small-scale operations may use solid shear in which frozen cell are forced through the orifice. Large-scale preparations use liquid shear.
- sonication—disruption of cell walls by high-frequency sound waves.

cell-division-cycle genes Any of approximately 50 genes that control the cell cycle in yeast.

cell-division-cycle mutant Cell-division-cycle temperature-sensitive mutants of yeast that either become blocked or show aberrant behavior in various parts of the cell cycle at a temperature at which the mutation can be expressed (the restrictive temperature). See CELL CYCLE.

cell fractionation The process of preparing a cell-free extract and dividing the cell contents into fractions by centrifugation techniques.

cell-free extract The product of treating a suspension of cells with a substance(s) that destroys the cell wall (in the case of bacteria and plants) and/or the cell membrane, thus releasing the cytoplasm and cell organelles. Sometimes the cell-free extract refers to the soluble portion of the internal cellular contents after removal of the organelles and cell membrane debris.

cell-free protein synthesis The synthesis of proteins in a test tube using a cell-free extract to supply the necessary enzymes and components and dependent on addition of amino acids and mRNA.

cell fusion The process of fusing two different cells together, first creating a heterokaryon that contains both of the nuclei and then a fusion of the nuclei to create a synkaryon. The fusion occurs by reaction between the two cell membranes, brought about by treatment with Sendai virus or polyethylene glycol.

cell line A cell culture started from a particular type that can be cultured indefinitely in the laboratory and is thus characterized as "immortal."

cell lineage A complete set of ancestral cells and cell divisions that makes up a certain cell type during development.

cell-mediated immune system See CELL-MEDIATED IMMUNITY.

cell-mediated immunity A type of immunological response that is mediated by cytotoxic T lymphocytes or killer T cells and is used by the body to destroy cells that carry foreign antigens, such as virally infected cells, tumor cells, and nonmatching tissue grafts.

cell membrane The boundary that limits the cell contents from its environment. It is composed of a phospholipid bilayer that is associated with proteins either embedded in the bilayer (intrinsic or transmembrane) or external to it (extrinsic). The cell membrane provides a selectively permeable barrier to the cell, allowing entrance by substances that are needed by the cell and preventing leakage of important substances out.

cell plate The boundary between two newly formed nuclei in a plant cell that is about to divide into two daughter cells; these daughter cells consist of cell wall material and cell membrane that grows and eventually becomes contiguous with the existing cell wall and cell membrane. It is also called the phragmoplast.

cell sorter An instrument used to separate and analyze different classes of cells from mixed populations. The fluorescence activated cell sorter (FACS) separates different cell types in a population based on external antigens that bind to antibodies labeled with fluorescent dyes.

cell sorting The process of sorting out different cell types in a heterogeneous population. See FLUOURESCENCE-ACTIVATED CELL SORTING.

cell synchronization The process by which all cells in a population come to be in the same phase of growth and consequently undergo cell division simultaneously.

cell theory The theory that states that the cell is the basic structural unit of all organisms and that all cells arise from preexisting cells.

cellulase An enzyme that hydrolyzes long polymers of cellulose to cellobiose, which is a disaccharide of glucose units.

cell wall The rigid or semirigid layer peripheral to the cell membrane of bacteria, algae, fungi, and plants. In plants the cell wall is composed of microfibrils of cellulose embedded in a matrix. The bacterial cell wall, the peptidoglycan layer, is a complex structure of chains of alternating residues of N-acetylmuramic acid and N-acetyl glucosamine held together by peptide bridges.

Center for Inherited Disease Research (CIDR) A center of the National Institutes of Health (NIH) supported by nine of the NIH institutes to provide genotyping and statistical genetics services for researchers identifying genes that cause human disease.

centimorgan (cm) One hundredth of a morgan, the unit of genetic distance or a map unit distance, named in honor of Thomas Morgan's contribution to mapping genes in *Drosophila*. The centimorgan is defined by recombination frequency between two genetic markers expressed as a percentage.

central dogma The concept that all genetic information flows from DNA. The information in the DNA is passed on to progeny cells by the process of DNA replication, and the information stored in DNA is first transcribed into RNA, which is then translated to synthesize proteins.

central nervous system The sensory and motor cells (neurons) of the brain and spinal cord.

centrifugal force The force that tends to impel substances outward from a center of rotation.

centrifuge An instrument that separates substances from liquids by centrifugal force and separates substances from other substances based on how each moves in a centrifugal field.

centriole A structure composed of microtubules that is found in the nucleus and is involved in the formation of the spindle apparatus that aids in the orderly parceling out of duplicated chromosomes to daughter cells in the process of cell division. The centriole is also identical in appearence to the basal body, the organelle that is embedded at the periphery of the cell and serves as the base of the cell's locomotive appendages, the flagella and cilia.

centromere The point along the chromosome to which duplicated sister chromatids are joined before the chromosomes are divided into the two daughter cells. It also serves as the site of attachment of the kinetochore, the structure on which microtubules of the spindle apparatus attach to pull the duplicated chromosomes to opposite ends of the cell during mitosis.

centromere binding factor A complex of proteins that binds to a particular DNA sequence in the centromeric region of the yeast chromosome (the CEN sequence) and also to one of a microtubule in the spindle fiber. In this way, the centromere binding factor forms a physical connection between the chromosome and the spindle apparatus during mitosis.

centromeric sequences Special nucleotide sequences in the DNA of chromosomes that serve as sites where the spindle apparatus attaches to chromosomes during mitosis. See YEAST ARTIFICIAL CHROMOSOMES.

centrosome The cell center or a microtubule organizing center consisting of granular material surrounding two centrioles (see CENTRIOLE).

cephalosporin-C One of a group of antibiotics, the cephalosporins, that is produced by the fungus *Cephalosporium* and that resembles penicillin in structure and mode of action.

cerebellum Part of the hindbrain consisting of two hemispheres and a small central portion.

cerebroside A class of membrane lipids derived from sphingosine similar in structure to gangliosides but differing in that cerebrosides have only a single sugar molecule. Both gangliosides and cerebrosides are widely found in the cell membranes of neural cells, where they play essential roles in neural functioning.

cerebrospinal fluid (CSF) The fluid that is produced in the ventricles of the brain and fills the ventricles and the central canal of the spinal cord. It serves to cushion the brain and protect it from blows to the skull or bruises resulting from sudden movements of the head.

cerebrum That part of the brain, located above and in front of the hind-

brain, that consists of a pair of hollow, convoluted lobes.

cesium chloride gradient centrifugation A method used to separate and/or purify molecules, usually nucleic acids. The nucleic acids to be separated are mixed with an appropriate density of the dense chemically inert salt, cesium chloride, and centrifuged at high speeds for hours to days. The cesium chloride establishes a density gradient during the centrifugation, and the molecules of nucleic acid move up or down the gradient to reach their position of bouyant density in the gradient.

CFTR See CYSTIC FIBROSIS TRANS-MEMBRANE CONDUCTANCE REGULATOR.

CH$_3$ choline A small alcohol of the structure, $HO-CH_2-CH_2-N(CH_3)_3$, that is found in membrane phospholipids and is part of the important neurotransmitter, acetylcholine.

channel protein A cell-membrane-embedded protein, part of a channel structure that allows substances of appropriate size and charge to pass through the membrane by diffusion.

chaotrophic The ability of an agent to disrupt the structure of water. Such substances weaken hydophobic interactions.

chaperones Proteins responsible for the proper folding of proteins once they are translated (see HEAT-SHOCK GENES).

Chargaff's rules The discovery that in DNA the concentration of adenine always equals the concentration of thymine and that the concentration of guanine always equals the concentration of cytosine.

Charon phage A vector used for cloning DNA constructed from bacteriophage lambda. The name *Charon* is taken from the ancient Greek myth of the ferryman Charon, who transported the spirits of the dead across the river Styx.

checkpoint Places in the cell cycle where specialized processes can arrest progression through the cell cycle. Cell-cycle arrest at a checkpoint is generally caused by DNA damage or other type of injury at an early stage, which would lead to malfunction.

chelator An organic compound in which atoms form bonds with metals, thus removing free metal ions from solution.

chemiluminescence The emission of light as the result of a chemical reaction. Chemiluminescence is widely used as a means of detecting DNA and protein probes in various analytic techniques, particularly in Southern, northern, and western blotting.

chemiosmotic theory A model proposed by Peter Mitchell that couples electron transport to oxidative phosphorylation (ATP synthesis during respiration) or photophosphorylation (ATP synthesis during photosynthesis). It postulates that the energy needed to drive the synthesis ATP is stored in a proton gradient across the inner membrane of the mitochondrion or the thylakoid membrane of the chloroplast and that this gradient forms during electron transport. When the gradient is relieved by the transport of protons across the membrane, the stored energy is used to drive the synthesis of ATP.

chemoautotroph An organism that obtains its energy from the oxidation of chemical bonds, usually the oxidation of inorganic metal ions, and can use inorganic carbon (C) or carbon dioxide (CO_2) to make biological molecules.

chemolithotroph A synonym for chemoautotroph.

chemoorganotroph An organism that obtains its energy from the oxidation of chemical bonds and requires organic carbon compounds for growth. A heterotroph.

chemostat An apparatus used to maintain a bacterial culture in continuous culture or exponential growth. This is done by coordinating the rate of addition of some limiting nutrient to the removal of spent medium and cells.

chemotaxis The movement of an organism to an attractant and away from a repellant.

chemotherapy The treatment of a disease with chemicals, but the term is usually used to define the treatment of cancer with drugs that selectively kill faster growing tumor cells.

chemotroph An organism that obtains its energy from the oxidation of chemical bonds. See CHEMOAUTOTROPH and CHEMOORGANOTROPH.

chiasma (chiasmata, pl.) The location of a crossover event between two chromatids in the tetrad structure of synapsed duplicated pairs of chromosomes that occurs during prophase I of meiosis.

chimera An animal formed from aggregates of genetically different groups of cells. Chimeras are made by combining early stage embryos that arise from fertilized eggs of two different sets of parents or by injecting cells from an early embryo of one genotype into the blastocyst of another genotype. The term *chimera* is derived from the name of the mythological creature with the head of a lion, the body of a goat, and the tail of a serpent.

chimeric DNA A recombinant DNA molecule, or one in which a fragment of DNA from a some source is spliced into a vector from another source.

chiral compound A compound, usually a carbon compound, that is optically active, that is, has the ability to rotate the plane of polarized light to the left or to the right, due to its ability to exist in one of two mirror images. See ENANTIOMERS.

chirality The nonidentity of a compound with its mirror image. See ENANTIOMERS.

chi sequence A sequence of bases on the genome of the bacterium *E. coli* that signals a nuclease to cut at that site for recombination or crossing over to occur. It serves as a hot spot of recombination as it is used preferentially as a site where recombination occurs.

chi structure The structure generated when the figure eight–shaped molecule, which is an intermediate form in the process of recombination between two cir-

cular DNA molecules, is cut by a restriction enzyme that cuts each circular DNA once. The name is derived from the fact that the four-armed structure, as seen by electron microscopy, resembles the Greek letter *chi*.

chitin The structural polysaccharide present in the exoskeleton of insects, in the cell

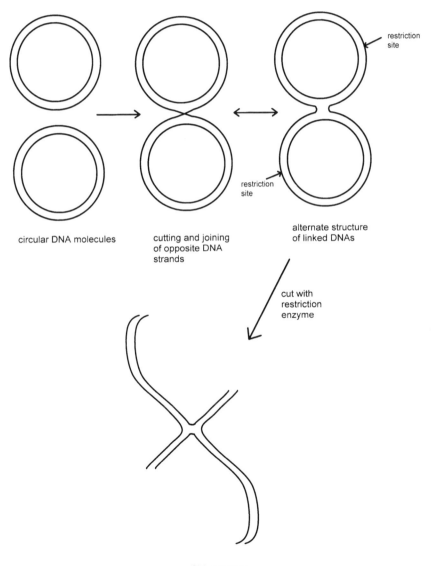

circular DNA molecules

cutting and joining
of opposite DNA
strands

restriction
site

alternate structure
of linked DNAs

restriction
site

cut with
restriction
enzyme

Chi structure

walls of fungi, and in crustacean cells composed of units of N-acetylglucosamine.

chk kinases A group of critical intermediates (Chk1, Chk2, and Chk3) in an important type of checkpoint control that operates in the G2 phase of the cell cycle. Chk kinases block mitosis in cells that have undergone DNA damage and other types of injury by phosphorylating cdc25, which then becomes inactive. Inactivation of cdc25 stops the dephosphorylation of cdc2, which is normally carried out by unphosphorylated cdc25.

chloramphenicol An antibiotic produced by *Streptomyces venezuela* that inhibits protein synthesis in bacteria, mitochodria, and chloroplasts but not in higher organisms. It is used to amplify recombinant DNA molecules when a plasmid vector is used to make the recombinant. Chloramphenicol will specifically inhibit the host cell's replication because it is dependent on new protein synthesis but will not inhibit certain plasmid replication. Thus, in the presence of the antibiotic, the recombinant molecule preferentially replicates as many as 200 copies per cell.

chlorophyll A light-absorbing pigment found in the chloroplasts of plants and algae that is essential as an electron donor in the process of photosynthesis. It is the pigment that gives the green color to plants.

chloroplast The organelle in plant cells and algae that is responsible for photosynthesis. It contains the chlorophyll and the proteins used to carry out the reactions of photosynthesis.

cholecystokinin A small polypeptide hormone secreted by the mucosal epithelial cells of the duodenum and also by the nerve cells of the enteric nervous system that stimulates the release of digestive enzymes stored in the pancreas and bile stored in the gallbladder into the small intestines. The secretion of cholecystokinin is stimulated by the presence of partially digested fats and proteins in the duodenum.

cholera toxin A toxic protein produced by the bacterium *Vibrio cholerae* that causes the disease cholera. Cholera toxin acts by binding to a receptor (GM1 ganglioside) present on intestinal mucosal cells. This activates adenylate cyclase, which stimulates the formation of cAMP, which, in turn, causes rapid loss of H_2O, Na^+, K^+, Cl^-, and HCO_3^- into the small intestine. The lost H_2O and electrolytes are replaced from the blood, which causes the diarrhea, loss of electrolytes, and dehydration that are characteristic of cholera. If untreated, the disease ultimately progresses to shock, kidney failure, and death. The toxin contains five binding (B) subunits, an active (A1) subunit, and a bridging piece (A2) that links A1 to the five B subunits. The A1 subunit is an enzyme that transfers ADP ribose from NAD to a G protein that more or less irreversibly maintains adenyl cyclase in an active state.

cholesterol A nonpolar lipid molecule containing four conjugated rings that is a major component of cell membranes and which is the precursor molecule for a variety of important biomolecules, including steroid hormones, vitamin D, and bile salts. Cholesterol is carried by lipoproteins in the blood that are responsible for carrying cholesterol to sites where it can be metabolized. Inappropriate deposition of cholesterol in the arterial wall is a factor in the formation of plaques that leads to atherosclerosis.

cholinergic The term applied to all neurons that utilize acetylcholine as a neurotransmitter.

cholinergic neuron Pertaining to the general class of neurons that utilizes acetylcholine as a NEUROTRANSMITTER.

chondroitin sulfate A sugar acid that is commonly a component of the fuzzy layer, an extracellular layer of collagen and glycosaminoglycans peripheral to the CELL COAT of some animal cells.

chromatid One of the daughter-duplicated chromosomes that is joined at the centromere to its sister chromatid and is seen at the prophase and metaphases stages of mitosis and meiosis.

chromatin The material of the chromosome, consisting of DNA associated with histone proteins.

chromatin remodeling The process of altering the DNA-protein complexes in chromatin to allow access of transcription factors to the promoter regions of genes. Chromatin remodeling is carried out by protein complexes known as chromatin remodeling factors (or "machines") that temporarily remove nucleosomes from the DNA in regions of DNA where the transcription factors bind. The remodeling complex in yeast is called the SWI/SNF complex, which contains an ATP-binding protein that provides energy for the movement of the complex along the DNA.

Derivatives of cholesterol

chromatographic techniques

- affinity chromatography—a technique for separating a substance from a mixture based on its natural tendency to bind to another substance (a ligand) for which it is said have an affinity. In affinity chromatography, a liquid suspension or solution containing the mixture is generally poured through a hollow column containing the ligand that is bound to some inert supporting substance. The substance to be separated remains bound to the ligand in the column while the unbound molecules in the mixture pass through the column.
- column chromatography—the separation or purification of substances, generally proteins, based on their specific binding to a column, prepared by attaching a ligand to some support that will specifically bind to the substance and washing the bound column with a solution that will compete with the bound substance for the column.
- gas-liquid chromatography (GLC)—a type of chromatography in which substances in a sample are evaporated into a stream of an inert gas such as argon, helium, or nitrogen and then separated at a high temperature by passing the evaporated material through a column containing a liquid, such as silicone oil or polyethylene glycol, on an inert matrix material such as firebrick.
- gel-exclusion chromatography—a variant of gel filtration in which separation of substances in a mixture is achieved by collecting the fraction of the sample containing the molecules whose size is greater than the exclusion size of the gel beads.
- gel-filtration chromatography—the separation of molecules, generally proteins, based on their sizes, as seen by their flow through a column prepared with porous beads of a carbohydrate polymer that will trap smaller molecules impeding their flow but permit larger molecules to flow faster. This procedure is also called molecular sieving. See GEL FILTRATION.
- high-performance liquid chromatography (HPLC)—a chromatographic method in which a high resolution of separation is achieved by improvements in the packing of columns and the flow of solvents through the columns under high pressure. Such a method yields very sharp peaks of substances eluted from the column.
- hydroxyapatite chromatography—the separation of molecules based on their relative binding to a column prepared with calcium phosphate. Such a column is used to separate double-stranded DNA that will bind to it from single-stranded DNA, which will pass through.
- ion-exchange chromatography—the separation of substances based upon their charge and thus their affinity to a column prepared with a charged support material. Substances are eluted from the column with a solution of ions that compete with the substances binding to the column.
- paper chromatography—the separation of substances based on their relative solubilities in a mixture of solvents. Substances to be separated are applied to paper support, and the solvents travel up or down the paper via capillary action, dissolving and transporting the substances on the paper.
- reverse-phase chromatography—the separation of substances based on their relative hydrophobicities. The support matrix is prepared to contain large hydrophobic carbon chains that will bind hydrophobic proteins more strongly and are thus eluted from the column more slowly than hydrophilic ones.
- thin-layer chromatography—the same as paper chromatography, but the support is a glass plate that is coated with a silica gel.

chromatography An analytical technique used to separate molecules from each other based on differences in their affinities and/or migration on some support resulting from the flow of a solvent (see specific types in CHROMATOGRAPHIC

TECHNIQUES). Chromatographic methods differ with respect to the nature of the solid support and the type of mobile phase (solvent).

chromogenic label Any molecule attached to a biological probe molecule that generates a colored compound(s) as a means of visualizing the location and amount of probe bound to a particular target.

chromomeres The beadlike structures on lampbrush chromosomes that are seen during meiosis when the chromosomes become extended. See LAMPBRUSH CHROMOSOME.

chromosomal mutation A change in the sequence of base pairs of the DNA encoding a gene that results in a change on the protein. The change can be as simple as a change in one base (missense mutation involving one amino acid on the protein) or the addition or deletion of one base (resulting in a change in the reading frame, thus affecting many amino acids on the protein), to more complex changes, such as the addition or deletion of many bases or the transposition of part of one chromosome to the other.

chromosome The structure in the nucleus that contains the genetic information composed of DNA and the histone proteins associated with the DNA. The term *chromosome* is also used to refer to the genes-containing unit of bacteria, viruses, mitochondria, and chloroplasts, although these do not resemble the chromosomes of higher organisms in structure or histone content.

chromosome cycle A term first coined by Barbara McClintock in 1942 to describe the cyclical series of changes in chromosome structure that takes place during the cell cycle.
The chromosome cycle:

- G1 phase — Chromosomes become dispersed as a result of changes in the way chromatin fibers are coiled.

- S phase — Relaxation of the chromatin and unwinding of the DNA helices during DNA replication

- G2 phase — Condensation of chromatin begins.

- Mitosis

PROPHASE — Sister chromatids become detectable.
Assembly of the mitotic spindle
Breakdown of the nuclear envelope

METAPHASE — Alignment of centromeres

ANAPHASE — Disjunction of sister chromatids

TELOPHASE — Disassembly of chromosomes

chromosome jumping See CHROMOSOME WALKING.

chromosome mapping The techniques used to assign specific genes locations on the chromosome, based upon crossover frequencies and linkage frequencies between genes.

chromosome puffs The uncoiled regions of DNA found on the giant polytene chromosomes of the salivary glands of certain members of the *Diptera* group (e.g., fruit fly) with the appearance of puffs when observed by conventional light microscopy, which have been shown to be sites where the chromosomal DNA is actively in the process of transcription.

chromosome walking (jumping, crawling) A procedure used to locate a gene by using cloned genes close to the target, preparing probes from these genes, and using them to isolate members of a genomic library that hybridize to the probe but contain other genetically linked material. If each member of the library contains an insert of 10,000 base pairs (bp), a probe that can hybridize to the first 100 bp can be used to isolate a gene that is located 9,000 bp away. If the gene is, in fact, located more than 10,000 bp away, then the first probe is used to isolate a clone to make a second probe, which can be used in turn to isolate a third probe, until the specific gene is found. This procedure is called chromosome walking because probes are isolated and then used to identify portions of the chromosome that are contiguous to each other.

chymotrypsin A digestive enzyme that hydrolyzes peptide bonds, thus cleaving proteins to their component amino acids, found in the small intestine.

cilium (cilia, pl.) Short, hairlike membrane-bounded appendage composed of microtubules used in the locomotion of cells.

circadian clock A biological timing mechanism that controls a type of natural synchrony (see CELL SYNCHRONIZATION) by controlling cell division.

cis A term used in genetics to define an event or gene whose action occurs on the same chromosome.

cis acting Pertaining to a genetic element that exerts an effect on a target located within the same unit. For example, a promoter element is said to be cis acting with respect to the genes it controls because both are on the same strand.

cis-acting gene A regulatory gene that controls transcription of genes that lie near it on the same chromosome by binding protein factors needed to turn transcription on or off. See CIS-TRANS TEST.

cis face The portion of the Golgi complex stack of vesicles that has just formed, also called the forming face, which is oriented toward the rough endoplasmic reticulum. See GOLGI APPARATUS.

cisterna A flattened membrane-bound sac, such as found in the endoplasmic reticulum.

cis-trans test A test to determine a functional genetic unit or the gene (see CISTRON) by genetic complementation (the ability to make functional gene prod-

ucts in a cross that does not involve a recombination event or crossing over) when crosses are made between genes carrying two mutations. If two mutations are found on separate genes, they are said to be in the trans configuration; if they are on the same gene, they are in the cis configuration. Complementation will only occur between transmutations in different genes, not in the same gene.

cistron A genetic unit or gene as defined by the cis-trans test.

citric acid An organic acid containing three CARBOXYL GROUPS and an important intermediate in a cyclic pathway called the Krebs cycle, tricarboxylic acid (TCA) cycle, or citric acid cycle that is responsible for the metabolism of glucose to water and carbon dioxide in the presence of oxygen.

c-jun N-terminal kinase (JNK1, JNK2) An enzyme that activates the TRANSCRIPTION FACTOR AP-1 (AP-1 is identical to the oncogene product jun) by addition of phosphate groups to certain amino acids at the N-terminal end.

clathrate A semisolid structure in which water molecules assume a cagelike structure around a "guest" molecule. In the most common clathrates, the guest molecule is methane (CH_4). These types of clathrates (also known as hydrates) were discovered by Sir Humphrey Davy in 1810. In the natural environment, methane clathrates are formed by bacterial or thermal degradation of organic materials in oceans. Clathrates are under consideration as a possible source of renewable energy.

clathrin A large protein that forms a basketlike structure around vesicles that transport molecules into or through cells or at sites (coated pits) where endocytosis will occur.

cleavage 1. The breaking of bonds of between units of macromolecules, such as the enzymatic cleavage of amino acids from protein.

2. The furrowing that occurs in animal cells to form two daughter cells from a parent cell after mitosis when the chromosomes have divided.
3. A series of cell divisions that occur during early animal embryogenesis.

cleavage divisions   See CLEAVAGE.

cleavage furrows   See CLEAVAGE.

clinical trials   Testing of new drugs or therapies on humans in a rigorous, controlled setting.

CLN1, CLN2, CLN3   The genes that code for the three G1 phase cyclins—cln1, cln2 and cln3—in yeast. These cyclins function to drive the initiation of S phase in the yeast cell cycle by forming complexes with cdc28, the yeast homologue of the mammalian cyclin dependent kinase, cdc2. cln3 forms a complex with cdc28, which stimulates transcription of the CLN1 and CLN2 genes. This results in accumulation of the cln1 and cln2 cyclins, which then form complexes with cdc28. The cln1-cdc28 and cln2-cdc28 complexes induce the transition from G1 into S phase.

clonal deletion   The selective loss, early in development, of B and T cells of the immune system that produce antibodies or have receptors for antigens that are an integral part of the organism (self antigens). This process is necessary to prevent the immune system from attacking the cells and tissues of the organism later in life (autoimmunity).

clonal selection   The theory that the stimulation of an immune response specific caused by the introduction of a foreign antigen results from proliferation of a single preexisting antibody-producing cell of the immune system such that a clone of cells bearing antibodies specific for the antigen is produced. Clonal selection was originally put forth as a counterhypothesis to the instructional theory that stated that an antibody-producing cell altered the type of antibody it pro-

duced after being "instructed" to do so by contact with the antigen.

clone (cellular)   A population of cells that have been derived from the divisions of one cell, so the population is genetically identical.

clone (DNA)   Recombinant DNA molecule or recombinant molecule. A gene or fragment of DNA that has been spliced into a vector, so that the DNA can be amplified many times by transferring the recombinant molecule into a host organism (usually a bacterium or yeast) that can be grown in large quantities.

clone bank   A collection of recombinant DNA molecules of the genomic material of a particular organism, prepared by fragmenting the DNA of the organism and splicing each of the fragments into vector molecules. Also known as a library.

clone library   See CLONE BANK.

cloning   The process of creating a recombinant DNA molecule, isolating it and amplifying it. See GENE CLONING.

cloning vector   The molecule of DNA that is used to house the DNA fragment to be cloned. Vectors are small chromosomes, either plasmid or bacteriophage, capable of self-replication in a host cell and producing many copies of itself per host cell, thus amplifying the number of copies of the cloned fragment.

Clostridium   The genus of organisms that are obligate anaerobes and produce spores. Members of this group produce powerful toxins and are responsible for diseases such as botulism, gas gangrene, and tetanus.

clustal   A computer program for aligning multiple nucleotide or peptide sequences. Carrying out alignments on multiple sequences simultaneously allows the delineation of similar segments in genes from different sources. This information can be used to group sequences into gene families or to define regulatory elements or other motifs or to study molecular evolution.

coated pit An invaginated site on the cell membrane that is lined with clathrin facing the interior of the cell and containing specific receptors at the exterior where molecules interact with the receptors for transport into the cell via receptor mediated endocytosis.

coated vesicle Small, membrane-bound droplets coated with a basket of clathrin transporting molecules either from the outside of the cell via receptor mediated endocytosis, having arisen from coated pits, or transporting newly made proteins to be sorted to either organelles or secreted to the outside of the cell.

coat protein(s) The proteins that make up the outer layer, or coat of a virus.

coccus (cocci, pl.) The name for a type bacterial cell with a round morphology.

Cockayne syndrome A rare hereditary disease first described by Edward Alfred Cockayne that is characterized by sensitivity to sunlight, short stature, and an aged appearance. The molecular basis of the disease is the inability to perform a certain type of rapid DNA repair called "transcription-coupled DNA repair" after exposure to ultraviolet light. Defects in at least two genes have been identified in Cockayne syndrome: CSA (also called ERCC8 for Excision-Repair Cross Complementing rodent repair deficiency), located on chromosome 5, and CSB (also called ERCC6), at gene map locus 10q11-21.

code Refers to the way the genetic information is stored in the DNA. See CODON.

coding strand The strand of DNA that is used as a template to make mRNA. It contains the complement of the code to be translated.

coenzyme Also called cofactor, a small nonprotein organic molecule associated with an enzyme (apoenzyme) and is required for catalytic activity. The coenzyme plus the apoenzyme is called a holoenzyme. Although the apoenzyme does not change during the course of catalysis, the coenzyme may be chemically altered, but it is regenerated and reused in subsequent reactions. A number of vitamins serve as components of coenzymes. For example, two common coenzymes involved in energy metabolism are nicotinamide adenine dinucleotide (NAD^+) and flavin adenine dinucleotide (FAD). Both the nicotinamide and flavin portions of the molecules are derivatives of the B vitamins, nicotinic acid and riboflavin.

coenzyme A (CoA or CoASH) A small organic molecule composed of adenosine diphosphate that is linked to the vitamin pantheteine phosphate, which serves as a carrier of acyl groups. CoA is particularly important as an acyl carrier during the oxidation of sugars for energy production.

cofactor A metal ion, such as Mg^{++}, Fe^{+++}, or Mn^+, or coenzyme that functions in association with enzyme proteins and that are necessary for complete enzymatic activity.

Cohen, Stanley (b. 1922) A molecular biologist who carried out the first cloning experiments by splicing the gene encoding resistance to the antibiotic tetracyclin from one strain of bacteria *(Staphylococcus aureus)* into a plasmid from another strain *(Escherichia coli)* in a test tube. The recombinant molecules were transferred into cells of *E. coli,* and transformed cells with tetracyclin resistance grew into colonies. These experiments demonstrated that genes isolated from one organism, spliced into a vector, and transferred into a host organism are intact and capable of producing functional proteins. However it was the discovery of growth factors, including epidermal growth factor, that won him the Nobel Prize in physiology and medicine, which he shared with Rita Levi-Montalcini in 1986.

cohesions See CONDENSINS.

cohesive ends Also known as sticky ends. The single-stranded extentions of a double-stranded DNA molecule that show complementarity to other single-stranded extensions of DNA molecules. Such sticky ends are generated by restriction endonucleases.

coiled-coil A type of higher-order protein structure in which two helices are wrapped around each other. Because the wrapping causes each helix to be shaped into another helix form, the helix "coils" are thought of as being a "coiled-coil" in this configuration.

coincidental evolution (concerted evolution) In genes that have become duplicated, the tendency for mutations occurring in one copy to appear in the other with the result that the effects of evolution appear in both copies at the same time.

cointegrate structure A molecule of DNA in which a transposon has mediated the joining of two plasmids, with copies of the transposon occurring at the joints between the two plasmids. This is the first step in the transposition of the transposon from one plasmid to another.

colcemid A drug that blocks microtubule formation and thus disrupts events, such as chromosome separation during mitosis, which depend upon microtubule function.

colchicine A drug that disrupts microtubule function as does colcemid.

ColE1 A naturally occurring plasmid that is carried by some strains of *E. coli* and has been used as a basis for constructing a number of cloning vectors for making recombinant DNA molecules. It is one of a family of plasmids that encodes genes for a colicin and immunity proteins that protect Col-harboring cells from the bacteriocidal effects of the colicin it produces.

colicin An antibiotic that is encoded by certain *E. coli* plasmids, such as ColE1. Colicins kill bacteria by a number of different mechanisms, including inhibition of protein synthesis, inhibition of active transport, and DNA degradation.

coliforms A group of bacteria that includes the genera *Escherichia, Klebsiella, Enterobacter,* and *Citrobacter,* which are small rod-shaped facultative anaerobes, stain Gram negative, and ferment lactose with gas production within 48 hours of growth. They are used to assess fecal pollution of water.

colinear Having the linear array or sequence of one molecule correspond to that of another. The sequence of bases found on the mRNA corresponds to the sequence of amino acids found on the protein that it encodes. This relationship extends to the sequence of bases on the DNA for bacteria, but in higher organisms, the DNA also contains some intervening sequences (see INTRONS) that must be eliminated before obtaining colinearity.

colinearity The condition of being colinear. See COLINEAR.

coliphage A bacterial virus that infects and reproduces in coliforms.

collagen A fibrous protein that is a major component of connective tissue and is found in the fuzzy layer that envelops animal cells.

Collins, Francis (b. 1950) A researcher in the field of genetic disease who gained fame as the head of the scientific team that succeeded in cloning the gene for cystic fibrosis through chromosome walking in 1989. He is currently director of the Human Genome Project at the National Institutes of Health.

colloid A suspension of microscopic particles ranging in size from 1 nm to 1 μm that is dispersed in some medium. Hydrophilic colloids are composed of macromolecules that remain dispersed in an aqueous solution because of the particles' affinity for water. Hydrophobic colloids are less stable and are composed of insolubles particles suspended in water and remaining in a suspended state due to repulsive forces among particles.

colony A group of cells that grow from a single cell on some solid medium, such as an agar plate.

colony counter An instrument used to count the number of colonies on an agar plate. There are two types. The manual type has an electronic stylus that creates a signal that is counted when touched to a colony. Automatic colony counters have scanners that detect density differences and can read an entire plate for the total number of colonies.

colony-forming unit (CFU) A viable cell that gives rise to a colony.

colony hybridization A method used to identify colonies harboring a particular gene or DNA sequence. Colonies on an agar plate are partially transferred to a membrane, generally nitrocellulose or nylon, by gently pressing the membrane on top of the colonies. The membrane is treated with alkalai to denature the DNA in the cells, heated to fix the DNA onto it, then washed with a labeled probe to identify those colonies that carry the sequence. Once the colony is identified on the membrane, it can be picked from the original plate and cultured to study or to further isolate the gene or sequence.

colony-stimulating factors (CSFs) A group of hormonelike substances that stimulates the production of various types of white blood cells. Colony-stimulating factors include: granulocyte-macrophage colony-stimulating factors (GM-CSF or sargramostim), granulocyte colony-stimulating factors (G-CSF or filgrastim), and promegapoietin. CSFs are used to treat conditions of low white-blood-cell counts, such as in patients receiving radiation or chemotherapy, patients with AIDS, and those with white-blood-cell diseases. Commercially available CSFs are generally made by recombinant DNA techniques.

colorimeter An instrument that quantitates amount of substance in solution by measuring the amount of light, at a given wavelength, that is absorbed by the solution. Colorimetry is based on Beer's and Lambert's laws defining extinction coefficient of a substance and the relationship of light absorbed by a substance to its concentration. Colorimeters are also used to measure the turbidity of solutions, which is an indication of the number of particles in suspension or the number of bacterial cells in a culture.

combinatorial library A mixture of polymeric chains in which individual chains differ from one another by virtue of the linear arrangement of the monomers that make up the polymer. For example, a combinatorial peptide library would contain polypeptides that are created from the same amino acids, but the amino acids in different polypeptides in the library would be arranged in a unique order, for example, Ala-Leu-Tyr-Ser- . . . v. Leu-Ser-Tyr-Ala-. . . .

combining site The site on an antibody molecule where the antigen interacts.

commensalism A relationship between members of different species living within the same cultural environment with one organism benefiting from the relationship, but the other not being affected.

comparative genomics A branch of bioinformatics concerned with the analysis of gene structure and function by comparing similarities and differences in DNA and protein sequences from different organisms. Comparative genomics uses computer programs to compare DNA and protein sequences in genetic databases such as the GenBank and Swiss Protein Database. These programs align multiple sequences to look for regions of similarity. This type of analysis is used to predict the function of genes in higher organisms from the known functions of similar genes in lower organisms such as bacteria and yeast. Differences among similar genes from different organisms can be used to create models of evolutionary relationships (molecular evolution).

compatibility group Defining a group of plasmids on their ability to coexist in the same cell with another plasmid from a different group.

competence The state of a bacterial cell that has the ability to take up DNA from the environment. Some species of bacteria develop natural competence by synthesizing competence factors and DNA receptor proteins that aid in the uptake of DNA into the cell. Other species, such as *E. coli* can be made competent by treatment of cells with high concentrations of $CaCl_2$ in the cold.

competition hybridization A technique for determining the degree of similarity between two nucleic acids by measuring the degree to which the two nucleic acids hybrize to one another in the presence of a third nucleic acid that acts as a standard.

competitive inhibition The inhibition of an enzyme by a substance that reversibly binds to the active site of the enzyme and thus competes with the substrate for the site.

complement A group of serum proteins that are activated by reaction with antigen-antibody complexes. Once activated, they aid in the killing of pathogenic bacteria and/or facilitate phagocytosis.

complementary base pairing The formation of hydrogen bonds between adenine and thymine and between guanine and cytosine. See COMPLEMENTATION.

complementary base sequence A sequence of bases that can form hydrogen bonds with another sequence. See COMPLEMENTARY BASE PAIRING.

complementary DNA See cDNA.

complementation 1. The ability of one chain of polynucleotides (either DNA or RNA) to form hydrogen bonds with another chain because of a coincidence of adenine/thymine pairs of bases and guanine/cytosine pairs of bases on each strand.
2. Genetic complementation is the ability of one mutant to supply a required function to another mutant. See CIS-TRANS TEST.
3. Also, cloning by complementation is a technique in which a mutant host cell (e.g., lacks the ability to synthesize some nutrient) is infected with a library and a clone is picked that has the ability to synthesize the nutrient. This clone is derived from a mutant cell that picked up a recombinant molecule containing a functional gene that has the ability to replace its own faulty one.

complementation test See CIS-TRANS TEST.

complement-fixation test (CF) A serological test for antibodies based on the ability of complement to lyse red blood cells. Serum to be tested is mixed with antigen and complement. An indicator system of sheep red blood cells (RBCs) and antibody against the sheep RBCs is added. If specific antibody for the antigen is present in the serum, it will combine with the antigen and bind the complement, and no complement will be available to lyse the sheep RBCs. Thus no lysis of sheep RBCs indicates the presence of antibody in the serum in a complement-fixation test.

complete medium A culture medium that supplies all the nutrients (amino acids, vitamins, and bases found in nucleic acids) that an organism needs for growth.

complexity A measure of the number of different base-pair sequences on a given genome.

composite transposon A transposable element made up of insertion sequence (IS) elements flanking a central portion of DNA sequence that usually contains a gene or genes encoding antibiotic resistance determinants. Common composite transposons are Tn5 and Tn10.

compost A mixture of decaying organic material used for fertilization or rejuvenation of soil.

concanavilin A (con A) A lectin isolated from the jack bean *(Canavalia ensiformis)* that binds to certain sugar residues. It is used in affinity CHROMATOGRAPHY to purify glycoproteins and is also used to agglutinate cells by cross-linking glycoproteins found at the cell surfaces. In addition, con A induces resting lymphocytes to divide.

concatamer A series of the same DNA molecules linked in tandem, thus creating a dimer, a trimer, or a multimer.

c-oncogene A normal cellular gene that has a viral oncogene, or tumor-producing homologue. Such genes are also called proto-oncogenes and can be activated by mutation, amplification, or overepression to become a cancer-producing cell.

condensation The chemical reaction that results in the joining of two molecules with the elimination of a water molecule. An example is the formation of the peptide bond between two amino acids.

condensing vacuole A membrane-bound vacuole arising from the Golgi complex and developing into a secretory granule by the progressive loss of water.

condensins Families of proteins that mediate the condensation of chromosomes during mitosis. The process of chromosome condensation involves an ordered coiling of the DNA-chromatin that uses energy derived from hydrolysis of ATP. The cohesions are highly conserved in evolution and are found in eukaryotes and prokaryotes.

conditional mutation A mutation that is expressed only under certain conditions. An example, is a temperature-sensitive mutation that encodes a protein functional at certain permissive temperatures (e.g., 32°C), but not functional at higher nonpermissive temperatures (e.g., 42°C). Such mutations can define essential genes because mutations in essential functions are lethal events, but conditional mutations allow the organisms to survive at permissive temperatures.

conformation The three-dimensional structure of a macromolecule, such as a protein.

congenital Aquired during development in the uterus.

conidiophore A specialized fungal structure that bears the spores of conidia.

conjugation A means of gene transmission between any coliform bacterial strain that carries an F factor (carried either on the bacterial chromosome or extrachromosomally) and another strain that lacks the F factor. During conjugation, neighboring bacteria come into direct contact with one another and transfer DNA from one (the donor) to another (the recipient) by means of a mating tube formed at the point of contact. Because the genes from the donor are always transferred in a given order, conjugation has been used to map genes on the bacterial genome by observing which genes are transferred to recipients following controlled interruption of the mating process. See HFR STRAIN.

conjugation, bacterial A means of gene transmission between bacteria by cell-to-cell contact or bacterial mating.

conjugation, chemical The covalent attachment of a molecular group to a molecule for the purpose of enhancing or altering its function. The attachment of fluorescent or chromogenic groups to antibodies or nucleic acids to create probes is an example of conjugation.

connective tissue Fibroblast cells that secrete collagen. Collagen gives cells adhesive strength that is needed to maintain form. Some examples of connective tissue are bone, cartilage, tendons, and ligaments.

connexon A structure of the gap junction composed of six protein subunits around a hollow center. Two aligned connexons of two cells provide a means of communication between the two cells.

consensus sequence An order of bases that has the most common nucleotide at each position when different examples are compared. Consensus sequences are found in promoters and are responsible for binding RNA polymerase and other proteins needed for transcription. Consensus sequences also signal other events, such as splicing of introns out of primary transcripts.

conservative replication A mechanism of DNA replication in which each strand of a parental molecule remains together after replication. See SEMICONSERVATIVE REPLICATION.

constant region The carboxy terminal regions of the heavy chain or light chain of the antibody molecule, which has the same or nearly identical amino acid composition as each member of the same class.

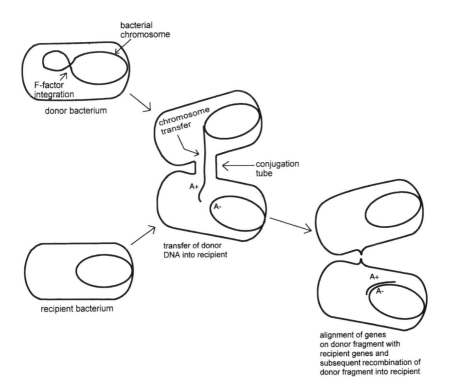

Bacterial conjugation

constitutive gene Genes that are expressed continuously and are not subject to either induction or repression. Such genes encode housekeeping functions and are expressed in all cells at a low level.

constitutive heterochromatin Chromosomal regions that remain in a permanently condensed state during interphase in every cell in the organism and are never genetically active in any cell, such as the centromere.

constitutive mutant An organism that has a mutation in some regulatory gene so that the expression of the gene(s) it controls is constitutively expressed. See CONSTITUTIVE GENE.

contact inhibition The property of normal animal cells in culture to stop dividing once they have formed a contiguous monolayer over the surface of the medium on which they are growing.

contamination Growth of undesirable organisms in some culture or material.

contig A term that describes the assembly of sequence data from multiple overlapping DNA fragments into a larger segment. The term reflects the practical size limitations of the sequencing techniques used for sequencing large stretches of genomic DNAs such as those employed in the Human Genome Project. Because only relatively short fragments can be reliably sequenced in a single run, sequence data representing a large DNA segment, such as may be present in a yeast artificial chromosome (YAC), must be derived by assembling data from many smaller pieces.

continuous culture A system that maintains a cell culture at a steady growth rate. This is achieved by use of a chemostat or turbidostat.

continuously fed stirred tank reactor (CSTR) A BIOREACTOR apparatus in which fresh substrate is continuously fed in and a corresponding volume of contents is removed. Such procedure quickly

dilutes the incoming material and is useful in waste treatment or any other procedure in which any components of the incoming substrate would inhibit the enzymatic bioconversion process.

contractile ring A beltlike structure composed of actin microfilaments and found under the plasma membrane that functions to divide an animal cell into two daughter cells after mitosis.

controlling element Transposable elements of maize that were found to cause mutations and chromosomal breakage when they transposed into or excised out of genes.

Coombs reaction An immunological test for identifying blood groups using specific antibodies to antigens of red blood cells, red blood cells, and antiantibodies to overcome a natural replusion by red blood cells, which can mask a positive test.

coordinated enzyme synthesis The regulation by the same event or signal of the synthesis enzymes involved in the same metabolic process. In bacteria this is accomplished generally by the organization of the genes that encode the enzymes into operons with a single regulatory element. In higher organisms, the genes are usually scattered but have common regulatory elements that respond to the same signal.

coordinate regulation The expression of multiple genes in unison, for example, the genes of an operon.

copia **elements** A family of transposable elements found in *Drosophila*. A typical *Drosophila* genome carries about 50 of these elements in widely scattered regions.

copolymer A synthetic polymer of two deoxyribonucleotides in random order (e.g., ACACCACCCAA), or two ribonucleotides in alternating order (e.g., AUAUAUAUAU). These were used to elucidate the genetic code by using them in an in vitro protein synthesizing system and analyzing the amino acid composition of the resulting polypeptides.

copy choice A mechanism of genetic recombination in which the recombinant molecule is formed by selectively replicating parts of the parental DNA molecules.

copy number The number of plasmid molecules per bacterial cell. Some plasmids are said to be relaxed in the control of their replication and are defined as high-copy-number plasmids, e.g., more than 20 to 100 copies per cell. These are used as cloning vectors and result in high yields of recombinant DNA or the protein encoded by the recombinant. Stringently controlled plasmids exist in cells in low copy number, or one to a few copies per cell. These plasmids are used to clone genes that produce proteins that are toxic to bacterial cells when produced in high concentrations.

cordycepin An antibiotic that acts by blocking transcription. Cordycepin is a derivative of the normal nucleoside, adenosine, in which the hydroxyl group on the 3′ carbon is missing. When cordycepin becomes incorporated into newly synthesized RNA in place of the normal adenine nucleoside, the RNA strand terminates.

core particle An octamer of histones (H2A, H2B, H3, and H4) with 146 base pairs of DNA wrapped $1^3/_4$ times around in a nucleosome.

corepressor The effector molecule that binds to a repressor to form a complex. The effector–corepressor complex functions to repress or prevent transcription of a bacterial operon.

Cornybacteria A genus of small Gram-positive straight to slightly curved rod-shaped bacteria, frequently club shaped, with aerobic to facultative chemoorganic (see CHEMOORGANOTROPH) metabolism. A pathogenic member of this group is *C. diphtheriae*, the causative agent of diphtheria.

corpus luteum The temporary endocrine gland formed from a ruptured ovarian follicle after release of an egg.

cortical cytoplasm A region of the egg cytoplasm just under the cell membrane that undergoes rearrangements after fertilization and has profound consequences in the development of the embryo.

cortical reaction Release of enzymes from the cortical vesicles after fertilization of an animal egg that results in a hardening of the vitrelline membrane to prevent additional sperm penetration.

cortical vesicles Membrane-bound structures of the egg cell that release proteases and other enzymes during the process of fertilization.

corticosteroid The steroid hormone, a derivative of cholesterol, that is synthesized in the adrenal cortex.

COS cells A derivative of CV-1 monkey cells that are infected with, but do not produce SV40 virus. COS cells express the SV40 early genes (T antigens) needed for viral replication. COS cells are used as host cells in cloning experiments, when derivatives of SV40, lacking the early genes, are used as cloning vectors for eukaryotic DNA. The early genes expressed by the COS cells allows the recombinant molecules to replicate and thus amplify its cloned inserts.

cosmid A cloning vector constructed of a plasmid origin of replication, an antibiotic resistance gene, and the cos sites of lambda DNA. These molecules can be packaged in vitro into a lambda phage coat and can then be transferred into host cells by viral infection. Colonies containing the cosmid are selected on medium supplemented with the antibiotic of the resistance gene on the cosmid. Such cloning vectors can be used to clone fragments up to about 47 kilobases long. This is about three times the amount of DNA that can be cloned into lambda-phage-derived vectors.

cos site The 5′ 12 base-pair (bp) overhang termini of bacteriophage lambda DNA that are base-pair complementary to each other: When a cell is infected with this bacteriophage, the DNA, injected in as a linear molecule, circularizes through hydrogen bonding between nucleotides of the overhang termini. Lambda is replicated in long tandem repeats of its genome. The cos sites serve as packaging markers for an endonuclease that cuts in a staggered fashion, creating unit headfuls of DNA with 5′ 12 bp overhangs that are packaged into phage coats.

cotransduction The introduction of two linked genes into a bacterial cell by the genetic transmission process of transduction.

cotransformation The introduction of two linked genes into a bacterial cell on the same fragment of DNA by the genetic transmission process of transformation.

cotranslational transfer The insertion of one end of a polypeptide into the membrane of the ENDOPLASMIC RETICULUM before synthesis of the whole polypeptide is completed. See LEADER SEQUENCE.

cotransport The simultaneous movement across a membrane of two different substances in a coupled manner. The two substances may move in the same direction (symport) or in the opposite direction (antiport).

Coulter counter An instrument that automatically counts cells by measuring the changes in resistance that occur when cells in suspension are passed through a small slit.

coupled reactions Two enzymatically controlled chemical reactions that must occur simultaneously. For example, many reactions that require an input of energy to proceed are coupled to the hydrolysis of ATP (ATP ADP + Pi), which releases 7 kcal of energy.

coupled transport The obligatory simultaneous transport of two solutes across the cell membrane. See COTRANSPORT.

covalent bonds Strong chemical bonds formed between atoms in which there is a sharing of two or more electrons.

covalently closed circular DNA A circular double-stranded molecule of DNA, such as a plasmid, in which there are no nicks or breaks in the sugar phosphate backbone. Usually, covalently closed circular (ccc) DNA exists as supercoiled; that is, the molecule folds in on itself due to strain in the molecule. If a nick is introduced into the backbone, the molecule relaxes and is referred to as an open circle (oc). See SUPERCOILED DNA.

Coxsackie viruses An antigenically distinct group of viruses of the enterovirus genus (viruses that are found in the intestines and excreted in the feces), including some human pathogens.

CpG rich islands Regions of DNA believed to be regulatory elements of gene activity that are characterized by an unusually high content of cytosine and guanine nucleotides arranged in the repeating sequence: CGCGCGC. . . . CpG islands are located in a segment 5′ to the coding region in many genes and are often sites of methylation.

C-reactive protein A protein whose levels increase during systemic inflammation and which functions to activate the complement pathway and prepares foreign substances for phagocytosis. C-reactive protein is a member of the pentraxin protein family, was discovered by Tillet and Frances in 1930, and was named for the fact that it reacts with the C polysaccharide of *Streptococcus pneumoniae*. C-reactive protein is made in the liver, and blood levels increase within six hours of an acute inflammatory stimulus. These protein levels in the blood are under investigation as a means of assessing cardiovascular disease risk.

creatine phosphate A high-energy compound in muscle cells that is used to regenerate the ATP needed for muscle contraction.

CREB Cyclic-AMP *r*esponse *e*lement *b*inding; CREB proteins are transcription factors that are activated by stimuli that increase cAMP levels. The binding of CREBs to sequences called CRE elements in the promoters of a number of genes regulates their transcription. cAMP production is controlled by the binding of various ligands to certain cell surface receptors linked to adenylyl cyclase. cAMP activates a protein kinase which, in turn, activates a protein kinase that migrates into the nucleus and activates a CREB protein. CREB proteins have been shown to be involved in the process of long-term potentiation in the neurons of lower organisms, including snails, fruit flies, and rats. In humans abnormalities in the gene coding for the CREB protein CBP is associated with Rubenstein-Taybi syndrome.

Crick, Francis (1918–2004) A British scientist who with James Watson won the Nobel Prize in physiology and medicine in 1962 for postulating a double-stranded helical structure for DNA, using the X-ray diffraction data of Maurice Wilkins, also a Nobel Prize winner in 1962. The double helix accounted for the known physical and chemical properties of DNA, but also suggested a mechanism for its replication.

crista The infolding of the inner membrane of the mitochondrion, which increases the surface area of the membrane responsible for electron transport and production of ATP via oxidative phosphorylation.

critical concentration The minimal concentration of subunits required for formation of a polymer.

critical dissolved oxygen concentration (Ccrit) The concentration of dissolved oxygen in a submerged culture when oxygen is the limiting substrate. The air supply to a fermentor is adjusted to maintain an oxygen level above its Ccrit.

cro protein A protein that interferes with the synthesis and action of the C repressor of a lambda bacteriophage and is necessary in the lytic response of the phage after infection into a cell.

crossing over The physical exchange of genetic information between a pair of homologous DNA molecules.

crosslink A covalent bond between strands of DNA. See CROSS-LINKING.

cross-linking A reaction in which two strands of DNA are covalently bonded together. Certain mutagenic agents, such as X-rays, cause cross-linking, and the DNA must be repaired if it is to replicate and function properly.

crossover fixation An alternative model to saltatory replication to explain the occurrence of highly repeated sequences. In crossover fixation, additional copies of a certain sequence are created on one DNA strand by unequal crossing over.

cross-reactive antibodies Nonspecific antibodies that will bind to antigens and give a false positive response in an antigen-antibody test.

crown gall plasmids The Ti (tumor-inducing) plasmid of *Agrobacterium tumefaciens* that is responsible for the malignant transformation of dicotyledonous plants infected with this organism. Part of the plasmid DNA incorporates into the plant chromosome to cause the production of a tumor. These plasmids lacking the tumor-producing genes have been constructed as potential vectors for recombinant DNA molecules for plant genetic engineering.

crown gall tumor See CROWN GALL PLASMIDS.

cruciform structures DNA structure in which strands separate and self-anneal through complementary base pairing

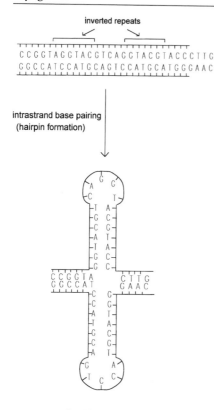

inverted repeats

C C G G T A G G T A C G T C A G G T A C G T A C C C T T G
G G C C A T C C A T G C A G T C C A T G C A T G G G A A C

intrastrand base pairing
(hairpin formation)

Cruciform structures

to form cruciforms or crosslike structures. Cruciforms can arise at regions of inverted base-pair repeats.

cryogenics The science of freezing, especially with reference to methods for producing very low temperatures.

cryopreservation The preservation of cells, organs, tissues, or other biological materials at very low temperatures, in freezers (−80°C), over dry ice (−79°C), or in liquid nitrogen (−196°C). At low temperatures, preserved biological materials remain genetically stable and metabolically inert.

cryoprotectants Chemicals that reduce the formation of ice crystals during freezing so that survival of cryopreserved cells is enhanced. Common cryoprotectants are dimethyl sulfoxide (DMSO), glycerol, and sucrose.

cryptic plasmid A plasmid that contains no genes or apparent phenotypic markers other than those needed for replication and transfer.

crystallography (X-ray) A technique used to analyze the structure of molecules by analysis of diffraction patterns of X-rays that pass through crystal specimens.

CsCI centrifugation See CESIUM CHLORIDE GRADIENT CENTRIFUGATION.

ctDNA The DNA found in a chloroplast.

culture A population of cells cultivated in a medium.

curing Any action that causes the loss of a plasmid or lysogenic bacteriophage from a culture of bacteria.

cutaneous Pertaining to, existing on, or affecting the skin.

cuvette The container for samples for a spectrophotometer, or other instruments that are used to make measurements on liquid samples.

C value A value representing the total amount of DNA, given in base pairs, in the haploid genome of a particular species.

cyanine dyes Water-soluble fluorescent dyes widely used as labeling molecules in a variety of probe applications, including probe hybridizations for microarrays and analytic techniques based on antigen-antibody reactions. Cyanine dye-labeled molecules can be detected with sensitivities close to that for radiolabels. The most commonly used cyanine dyes are Cyanine 3 (or 5) bihexanoic Acid; Cy3 or Cy5.

cyanobacteria Blue-green algae. Procaryotic, photosynthetic, oxygen-evolving organisms.

cyanogen bromide (CNBr) A chemical that recognizes methionine residues and cleaves polypeptide chains at these residues. CNBr is used to cleave genetically engineered proteins that have been constructed as a composite of cloned material and vector material. CNBr is also used to cross-link proteins to various support materials for affinity chromatography purposes.

cyclic AMP (cAMP) A molecule of adenosine monophosphate in which there is a covalent bond between the 3' hydroxyl (OH) and the 5' phosphate group. It is an important molecule in controlling metabolic processes in higher organisms (see CYCLIC AMP–DEPENDENT PROTEIN KINASES). Because its intracellular concentration is often controlled by hormonal action and the metabolic activities it controls is in response to the hormone, it is called a second message. cAMP also plays a role in the control of bacterial metabolism by binding with the catabolite activator protein and regulating transcription of some genes.

cyclic AMP–dependent protein kinases Enzymes that add phosphate

NH$_2$

Cyclic AMP (cAMP)

groups to proteins, either at a serine or tyrosine residue. Phosphorylation of proteins is an important mechanism of regulation of metabolism in higher organisms because the added phosphate group either activates or inactivates the protein and thus stimulates or inhibits the metabolic reaction.

cyclic GMP A molecule of guanosine monophosphate in which there is a covalent bond between the 3' hydroxyl (OH) and the 5' phosphate group.

cyclin(s) The regulatory subunits of cyclin-dependent kinases (cdks); cdks are activated upon binding a specific cyclin. Once activated, cdks induce cells to transit through a certain stage of the cell cycle. Originally discovered as a component of MPF (mitosis-promoting factor), a substance isolated from cells of embryos of the frog *Xenopus laevis* that were found to induce entry of cells at any stage of the cell cycle into mitosis (M phase).

cyclohexamide An antibiotic that inhibits yeasts and other fungi but does not inhibit bacteria. It is used as an agricultural fungicide.

cyclooxygenase enzymes Cyclooxygenases (COXs) are mixed-function oxidases that catalyze the addition of oxygen atoms to carbons 9, 11, and 15 as well as form a covalent bond between carbons 8 and 12 of arachidonic acid to create prostaglandin H$_2$ (PGH$_2$), the precursor of other prostaglandins. The COX enzymes are also known as prostaglandin H2 synthases. COXs have two isozymic forms termed COX-1 and COX-2. Aspirin acts to inhibit both COX isozymes by acetylating a serine residue in the active site.

cycloserine An antibiotic from *Streptomycetes* that acts by blocking two steps in the biochemical pathway by which the bacterial cell wall is synthesized. Because cycloserine is structurally similar to the amino acid D-alanine, it

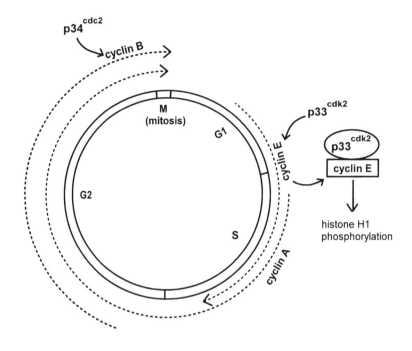

Regulation of cell-cycle kinetics by cyclins. The transition from the G2 phase to mitosis (M) is regulated by a complex between cyclin B and a cyclin-dependent kinase (cdk). p34cdc2 contains three phosphorylation sites at Thr 14, Tyr 15, and Thr 161; in the active form, Thr 161 is phosphorylated, and Thr 14 and Tyr 15 are not (left inset). p34cdc2 is phosphorylated at Thr 161 during G1 and becomes additionally phosphorylated at Thr 14 and Tyr 15 on binding to cyclin B that begins to accumulate at the end of S phase. Dephosphorylation of the Thr 14 and Tyr 15 sites mediated by the phosphatase p80cdc25 occurs at the G2/M boundary, resulting in the active complex. Cyclin B is then rapidly degraded by ubiquitin just after the start of mitosis. A-type cyclins begin to accumulate during S phase and appear to function at the G2/M boundary. Net synthesis of E-type cyclins occurs in G1; E-type cyclins activate a second cdk, p33cdk2, which acts at the G1/S boundary. p33cdk2 acts to induce phosphorylation of histones and certain cellular proteins involved in mitosis.

competitively inhibits the incorporation of D-alanine into a pentapeptide that is used to construct the bacterial cell wall.

cysteine An amino acid with a sulfhydryl in its side chain:

$$NH_2$$
$$|$$
$$HS-CH_2-CH-COOH$$

Disulfide bonds between two cysteine residues on the same polypeptide chain or between chains contributes to the overall shape of the protein.

cystic fibrosis An inherited disease that afflicts almost one in 2,000 children in the United States. In 1989 the gene, whose mutant allele accounts for a majority of the cases, was cloned.

Cystic Fibrosis Transmembrane Conductance Regulator (CFTR) A protein that functions as a channel for the transport of chloride ions across the cell membrane. Mutations in CFTR are responsible for the diseases cystic fibrosis and congenital bilateral aplasia of the vas deferens. The CFTR gene is located on chromosome 7q31.2.

cytidine monophosphate (CMP)   The nitogenous base CYTOSINE attached to a ribose sugar molecule with a phosphate residue at the 5' end of the ribose.

cytidine triphosphate (CTP)   The same as CMP above, but with three phosphate residues.

cytochalasins   A family of drugs produced by certain fungi that interferes with polymerization of actin microfilaments and hence inhibits cell movements that depend on actin polymerization-depolymerization reactions.

cytochrome c oxidase   An enzyme complex of the electron transport chain that reduces molecular oxygen to water.

cytochrome P-450 (also P450)   A cytochrome found in the smooth endoplasmic reticulum that is important in drug detoxification, especially in the liver. Cytochrome P-450 is a type of mixed-function oxidase that carries out hydroxylation reactions (addition of OH groups) to molecules, thus aiding in solubilizing them so that they can be flushed out of the body.

cytochromes   Heme-containing proteins of the electron transport chain involved in cellular respiration that carry out oxidation-reductions reactions, thus passing electrons down the chain from iron atom of one cytochrome to iron atom of another until it is passed to the final electron acceptor, molecular oxygen.

cytogenetics   That area of study of chromosomes and their behavior.

cytokeratins   A class of proteins that make up the intermediate filaments char-acteristic of epithelial cells previously identified as tonofilaments by electron microscopy. The cytokeratin family contains a number of different proteins, and keratin filaments found in different types of epithelia are comprised of different keratin proteins. Expression of keratins is used clinically as a diagnostic criterion in tumor diagnosis.

cytokinesis   The process of dividing the cytoplasm of a cell into two daughter cells following mitosis.

cytokinins   Substances that promote cell division and cell-and-shoot differentiation in plant tissue cultures. Some common cytokinins are benzylaminopurine (BAP) and 2-isopentenyladenine.

cytology   The study of cells based on microscopic observations.

cytomegalovirus (CMV)   A usually nonpathogenic human virus that can be pathogenic in immunocompromised hosts. Following infection, the virus persists in the host but the carrier is protected from disease by its immune system. Both (T-cell immunity) and humoral (antibody) activities are believed to be involved in the defense against such CMV-induced disease, and thus the organism is a good model to study immunological processes. Many plasmid vectors have incorporated a promoter from CMV so that mammalian genes cloned behind the promoter will be expressed in cell culture.

cytoplasm   The liquid colloidal substance between the cell membrane and nucleus of the cell.

cytoplasmic inheritance   Patterns of inheritance carried by genes not contained in the chromosomal DNA; i.e., the genes carried in mitochondria or chloroplasts. Since the sperm cytoplasm does not contain mitochondria, patterns of cytoplasmic inheritance involving mitochondria are always maternal. One example of cytoplasmic inheritance in humans is LHON (Lebers Hereditary

Optic Neuropathy), which is caused by defects in the electron transport Complex I in mitochondria. Other examples include: MELAS syndrome (mitochondrial myopathy, encephalopathy, lactic acidosis, and stroke-like episodes), caused in most cases by an A to G transition in a mitochondrial leu-tRNA; Kearns-Sayre syndrome (KSS), which leads to loss of vision, hearing, and heart problems due to the accumulation of defective mitochondria; and MMC (maternally inherited myopathy and cardiomyopathy), where energy-demanding muscle cells are compromised.

cytoplasmic streaming The back-and-forth movement of cytoplasm in some algae and the circular flow of cytoplasm around a central vacuole in plant cell, also known as cyclosis.

cytosine One of the nitrogenous bases found in nucleic acids. Cytosine is a pyrimidine that forms hydrogen bonds with the purine guanine.

cytoskeleton A complex network of microtubules, microfilaments, and intermediate filaments extending throughout the cytoplasm that gives shape to a eukaryotic cell and is involved in cellular movement.

cytosol The cytoplasm that contains the organelles of a eukaryotic cell.

cytotoxic T cell An activated T lymphocyte, also known as a killer T cell, that lyses cells that are recognized as a combination of self and foreign, such as virally infected cells, tumor cells, and foreign tissue graphs.

D

dalton A unit of mass used generally for macromolecules that is equal to 1.000 on the atomic mass scale, or almost the same as the mass of a hydrogen atom. The term *dalton* can be used interchangeably with the term *molecular weight.* Thus a 100,000 dalton (or 100 kilodalton protein) can be described as having a molecular weight of 100,000.

dansyl chloride A compound that reacts with the amino group of an amino acid to produce a fluorescent derivative that can be easily detected and identified. It is used in procedures to identify the amino terminal residue of peptides.

dark reactions A series of enzymatically catalyzed reactions in which organisms that carry out photosynthesis synthesize organic compounds in the form of sugars from inorganic carbon dioxide. These reactions use energy, in the form of ATP, and reducing power, in the form of NADPH made during the light-phase reactions of photosynthesis.

databases Information stored in computers to be used in the sequence analysis of genes and proteins. The National Institutes of Health maintains such databases (GenBank). EMBL is a European database established in 1980 to collect and store nucleotide sequence data. Its counterpart, SwissProt, translates the sequence data into protein data. EMBL, GenBank, and the DNA database of Japan, DDBJ, collaborate to collect and exchange data on a daily basis, as sequence data are being deposited at a rate of one sequence per minute.

DCC Deleted in *c*olon *c*arcinoma; a tumor-suppressor gene associated with colon cancer. The name DCC derives from the observation that segments of chromosome 18 now known to contain DCC are frequently deleted in colon carcinoma tumors. DCC is expressed on neural membranes, where it appears to serve as a receptor for the protein netrin. Its tumor-suppressive activity appears to stem from its ability to induce apoptosis in tumor cells. The apoptotic activity of DCC is upregulated by caspase 3 and downregulated by netrin. DCC gene map locus is 18q21.3.

DDBJ See DATABASES.

ddNTPs D*i*deoxyribo*n*ucleotide*tri*phosphates. See SANGER SEQUENCING.

deaminase An enzyme that removes amino groups from molecules.

deamination The process by which a deaminase removes amino groups from molecules. Deamination of bases in DNA results in mutations, and cytosine is the most susceptible base.

death phase The final phase in the growth curve of a population of cells in which the cells die exponentially; that is, for each time increment, a certain percentage of cells die.

decline phase See DEATH PHASE.

degrees of freedom The number of independent variables in an experiment.

defective virus A virus that is missing some essential genetic information so that it cannot reproduce itself. Such viruses can be propagated in a host cell only if a helper virus that supplies the missing proteins coinfects the same host cell.

defensins Part of the innate host defense system against invading microbes. These small peptides, produced by many different organisms, have a broad spectrum of activity against bacteria, fungi, and some enveloped viruses. Defensins have been found in great abundance in the phagocytic white cells of mammals and birds, but they have also been found in cells of the intestine and in skin cells. Their main mechanism of activity is to insert into the membranes of microbes and destroy them.

defined medium A medium used to grow organisms in which all the components are known. For heterotrophs, that would be a medium with a known carbon source, nitrogen source, metals, and any amino acids, vitamins, or other growth factors required by the organism.

degenerate code Referring to the fact that in the GENETIC CODE many amino acids are specified by more than one codon or sequence of three bases (triplet). The degeneracy of the code accounts for 20 different amino acids encoded by 64 possible triplet sequences of four different bases (see NUCLEIC ACID). For example the amino acid leucine has six different codons, UUA, UUG, CUU, CUC, CUA, CUG. See WOBBLE.

degradation The process by which substances are broken down. A degradative pathway is one in which molecules are enzymatically cleaved into smaller molecules.

dehydration-condensation reaction The joining of two molecules together with the elimination of a molecule of water. See HYDROLYSIS.

dehydrogenation The process by which hydrogen ions or protons are removed from an organic molecule. Such a process is also called oxidation. It is carried out by enzymes called dehydrogenases.

delayed hypersensitivity An allergic reaction that takes 24 to 48 hours to appear. An example is the skin test for exposure to tuberculosis. After injection with the ALLERGEN, a positive response (a swelling at the injection site) does not appear before 48 hours.

deletion mutation A change on the DNA due to the elimination of one or more nucleotides. A deletion can alter the genetic information in a very profound way. The deletion of one base pair results in a frameshift, where every codon is changed following the deletion; a deletion of many bases results in a message with fewer codons. See FRAMESHIFT.

delivery system An artificial system to deliver a drug to a specific target, such as inclusion of a drug in a LIPOSOME or conjugating a drug to an antibody. See FUSOGENIC VESICLE.

demyelination The loss of the MYELIN sheath, layers of membrane surrounding segments of nerves that provide rapid transmission of nerve impulses down such nerves. Demyelination occurs in some degenerative nerve diseases, such as multiple sclerosis and polio, resulting in loss of function of those demyelinated nerves.

denaturation Change in the three-dimensional shape or structure of a protein or nucleic acid by a physical or chemical agent, such as heat or strong acid (the denaturant), such that normal functioning is altered.

denaturation of DNA The splitting apart of the double-stranded structure into single strands by heating the molecule or treating it with acid, alkali, salts, or urea.

denaturation of proteins See DENATURATION.

dendrite A branch of a nerve cell the receives signals and transmits them inward toward the nerve cell body.

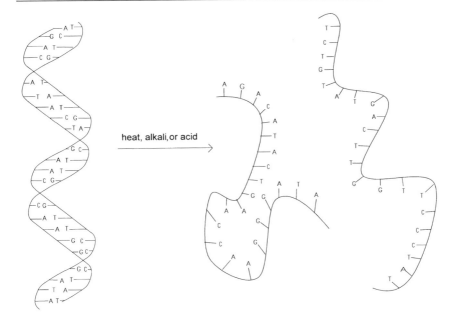

Denaturation of DNA

Denhardt's solution A commonly used solution for carrying out probe hybridizations on filters, for example, Southern and northern blots. Denhardt's solution contains polyvinylpyrrolidone (PVP), ficoll, bovine serum albumin, and a nonspecific DNA at high concentration to prevent nonspecific probe hybridization. See SOUTHERN BLOT HYBRIDIZATION.

denitrification The process of reducing nitrogen compounds to a lower oxidation level, e.g., nitro (NO_3) to nitrate (NO_2) or nitrate to nitrite (NO).

density gradient A solution in which there is a range of densities with the solute being more concentrated at the bottom and less concentrated at the top. These gradients can be stepwise, formed by discrete layers of different density solutions, or continuous, formed by small incremental changes in density. The gradients are generally made from solutions of sucrose or the heavy salts, cesium chloride or cesium sulfate.

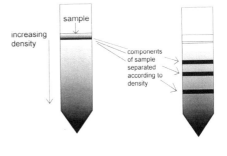

Density gradient centrifugation

density gradient centrifugation Technique used to separate macromolecules according to their buoyant densities by either layering the macromolecules on top of a preformed gradient and subjecting the gradient to centrifugation, or mixing the molecules in a solution of some support that will form a gradient on centrifugation. See CENTRIFUGE, CESIUM CHLORIDE DENSITY CENTRIFUGATION, and DENSITY GRADIENT.

deoxynucleoside Any of the nitrogenous bases found in nucleic acids (adenine, guanine, thymine, uracil, or cytosine) attached to deoxyribose, a five-carbon sugar. See NUCLEOSIDE.

deoxyribonuclease (DNase) An enzyme that breaks the chemical bond between the phosphate and sugar groups (the backbone) of DNA molecules. These enzymes can be exonucleases, removing deoxynucleotides from the ends of the molecule, or endonucleases, that cleave bonds of internal nucleotides.

deoxyribonucleic acid (DNA) A macromolecule consisting of two complementary (see COMPLEMENTARY BASE PAIRS) chains of deoxyribonucleotides. The chains are formed by chemical bonds between the sugar and phosphate portions of the deoxyribonucleotides, and the two chains are held together by hydrogen bonds between the bases (A pairs with T and G pairs with C). This molecule contains the genetic information of the cell because the sequence of nucleotides of the chains specify the sequence of amino acids of proteins made by the cell.

Deoxyribose

deoxyribonucleotide A deoxyribonucleoside with phosphate groups attached to the sugar, usually at the $5'$ position. This is the basic building block of DNA. See NUCLEOTIDE.

depolarization The change in electrical charge across a membrane. When a nerve cell receives an impulse, it becomes momentarily depolarized; its interior becomes more positivily charged with respect to its exterior. Repolarization restores its interior to a negative charge. A nerve impulse is propagated down a nerve fiber by waves of depolarization-repolarization events. See ACTION POTENTIAL.

depurination Removal of purines (guanine or cytosine) from DNA with the sugar-phosphate backbone remaining intact. Such a process occurs either enzymatically, as in a DNA repair process, or nonenzymatically, when a chemical interacts with the base and weakens its bond to its sugar residue.

desalting Removal of salt. This can be accomplished for preparations of macromolecules by DIALYSIS or by column chromatography. See GEL FILTRATION.

desmin A protein component of the INTERMEDIATE FILAMENTS found in muscle cells.

desmoplaquin One of the protein components of the DESMOSOME.

desmosome Regions of tight contact between adjacent epithelial cells; this contact gives strength to tissues and enables cells of a tissue to function together. There are two types of desmosomes: Belt desmosomes are bands of attachment that encircle the cell, and spot desmosomes are small local points of contact.

desmotubule A tubular structure that lies in a channel of the plasmadesmata, a means of communication between two plant cells. Such a channel is made up of the fusion of cell membranes from two adacent cells through the pores of the cell wall.

desulfurization The removal of sulfur from a molecule.

detergent A molecule with a hydrophobic part and a hyrophilic portion. The detergent can dissolve lipids (fats and oils).

determination The irreversible committment of a cell to a particular developmental pathway. If a determined cell is transplanted, it will develop into the structure it would have if it had not been transplanted.

deuteromycetes One of the four major classes of fungi, also called *Fungi imperfecti*. The deuteromycetes are important because this group contains the majority of human pathogens.

dextran A storage polysaccharide in yeasts and bacteria that is made up of glucose.

dextranase An enzyme that catalyzes the breakdown of dextran.

dextrin A molecule made up of several glucose residues and one of the products resulting from alpha-amylase hydrolysis of starch.

dextrose Another name for D-glucose, the most common sugar found in living organisms.

dextrotatory isomer An isomer of a sugar that rotates polarized light to the right. Dextrose is the dextrotatory isomer of glucose.

diacylglycerol (DAG) A molecule of glycerol with two fatty acids attached to it by ester linkages. It is formed along with INOSITOL TRIPHOSPHATE (InsP3) by hydrolysis of a membrane lipid (phosphatidylinositol-4,5- bisphosphate) by the enzyme PHOSPHOLIPASE C. Both DAG and InsP3 serve as second messengers in the cell. DAG activates protein kinase C, which in turn activates other enzymes. See PROTEIN KINASE.

diagnostic 1. n. A test that is used to determine the source of a problem.
2. The method of determining the nature of a disease by analyzing the symptoms. adj. A specific characteristic that allows one to determine the source of a problem or the nature of a disease.

diagnostic test A procedure that gives the ability to determine the nature of a problem. See DIAGNOSTIC.

dialysis 1. A technique used to separate molecules from each other through a semipermeable membrane that allows water and small molecules such as salts to pass through. The separation is based on the permeability of the molecules. Large molecules are retained by the membrane and smaller ones pass through. Thus proteins can be desalted by dialysis.
2. A medical procedure used to clear the blood of impurities after kidney failure. See DESALTING.

dialyzable The ability to be dialyzed. See DIALYSIS.

diatomaceous earth A finely pulverized mixture of earth composed largely of the silicon shells of the microorganisms diatoms used as a filtering substance or as an absorbant.

dibasic An acid that has two hydrogen atoms that may be replaced by basic molecules or metal ions to form a salt.

dicentric chromosome A chromosomal aberration involving breakage and then fusion of chromosomal fragments resulting in the formation of a hybrid chromosome with two centromeres.

dicotyledon Any plant characterized by flower parts in fours and fives, net-veined leaves, a cambium, and an embryo with two cotyledons, or two seed leaves.

dictysome A stack of flattened membranous sacs found in plant cells located adjacent to the ENDOPLASMIC RETICULUM. Its function is to mediate the secretion of proteins outside the cell or to target newly synthesized proteins to organelles such as the lysosome. This organelle is called the Golgi (see GOLGI APPARATUS) in animal cells.

dideoxynucleotide A nucleotide with a ribose having a hydrogen atom at the 3' position instead of an OH group. Such a nucleotide cannot form a 3' to 5' phosphodiester linkage (the linkage of the sugar phosphate backbone of DNA) with another nucleotide. Thus, adding a dideoxynucleotide to a growing DNA chain will terminate further synthesis of the chain. See DIDEOXY SEQUENCING.

dideoxy sequencing An enzymatic method of sequencing DNA using dideoxynucleotides to stop the synthesis of DNA chains at specific points prematurely, developed by Fred Sanger. DNA to be sequenced is divided into four tubes containing DNA polymerase, the four deoxynucleotides (dA,dT,dG,dC) and one of the dideoxynucleotides, either ddA, or ddT, or ddG, or ddC. The ratio of dideoxynucleotide to regular nucleotide is fixed so that during DNA synthesis the DNA polymerase has the option of incorporating a regular or dideoxynucleotide. Because incorporation of a DIDEOXYNUCLEOTIDE into a growing DNA chain stops further synthesis of that chain, each tube will contain a series of fragments, each ending with the dideoxynucleotide of that tube, for example, the tube containing ddA will have fragments that end in A. The size of each of the fragments can be determined by gel electrophoresis, and the sequence can be read up the gel. For example, if the tube containing ddA produces three fragments, with sizes of four bases, seven bases, and 10 bases, the sequence of the DNA synthesized has an A residue at the fourth, seventh, and 10th position. See SANGER SEQUENCING.

differential centrifugation A technique used to separate cells, organelles, or molecules that differ in size or density by using successively higher centrifugal forces.

differentiation The process during the development of an embryo in which cells become specialized in structure and function and go on to form different tissues of the adult.

differentiation antigen Any biomolecule that is detectable by an immunologic assay only in a specific cell subtype in an organism and that may therefore be used as a marker of that subtype.

diffusion The free movement of molecules from an area of greater concentration to an area of lower concentration.

diffusion coefficient or constant (k_D) The measure of the ability of a solute to diffuse through a concentration gradient. The factors that affect the value of the k_D include the size of the particle, its degree of polarity, and temperature. The rate of diffusion depends upon the k_D in the following way: $v = k_D$ ([X]outside-[X]inside); where v is the rate of diffusion and ([X]outside-[X]inside) represents the concentrations of solute [X] of the concentration gradient.

digitalis A cardiac drug derived from the plant foxglove that is used to treat congestive heart failure and arrhythmias. Digitalis was discovered by the Scottish doctor William Withering in 1785 and was found to contain a mixture of steroids including digoxigenin and digitoxigenin that act to slow the heart rate while increasing the intensity of the contraction.

digoxigenin A plant-derived steroid that, when covalently bound to a biological probe molecule, has been used as a hapten in some antibody-hapten-based probe systems. See HAPTEN.

dihybrid cross A cross between two individuals who are heterozygous for two different genes; for example, a cross between pea plants that carry heterozygous alleles for short/tall and for red/white flower phenotypes.

dihydrouridine An unusual pyrimidine base that is found only in tRNA. Dihydrouridine is derived from uridine by the addition of two hydrogen atoms.

dimer 1. A molecule that has two subunits. The subunits may or may not be identical.
2. Denoting two units, for example, a dimer of nucleotides, dCdG.

dimorphism The state of having two different forms. In botony this can be seen in a plant or a species of plant that has two distinct leaf types, flowers, or some other structure. In zoology this can be seen in two individuals of the same species exhibiting coloring, size, or other characteristics.

dioxin A group of heterocyclic hydrocarbons or any of a number of isomers of the chlorinated teratogen, TCDD, which is highly toxic and is found as impurities in some defoliants and herbicides.

diploid Having two sets of chromosomes so that each gene is represented twice in a cell or an organism. Describing a cell or an organism that contains two copies of each chromosome. See HAPLOID.

diplotene One of the stages of prophase I during meiosis I in the formation of germ cells. During diplotene, the chiasmata, or region where crossing over took place, can be visualized.

dipole A polar molecule in which the centers of positive and negative charge are separated. A molecule of water has a triangular shape and exists as a dipole. The oxygen at the head of the triangle is electronegative, and the two hydrogen tails are electropositive. See POLARITY.

diphtheria toxin The toxin secreted by the bacterium *Corynebacterium diphtheriae,* the pathogen that causes the disease diphtheria, which is characterized by

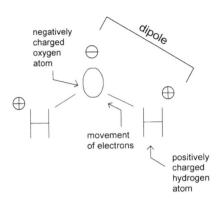

Dipole

skin lesions, swollen lymph glands, sore throat, and fever. Diphtheria toxin consists of two subunits, A and B. The toxin binds to a receptor (called the HB-EGF receptor) and is taken into the cytosol in an endosome where proteolytic cleavage releases the A subunit. The A subunit has enzymatic activity that causes ADP-ribose from NAD to ribosylate the eukaryotic Elongation Factor 2, which then blocks the function of this factor in protein synthesis and ultimately causes cell death.

direct terminal repeats Sequences of nucleotides that are duplicated on each end of a polynucleotide molecule. See LONG TERMINAL REPEAT and TERMINAL REDUNDANCY.

disaccharide A molecule consisting of any two sugar units. Maltose is a disaccharide consisting of two glucose molecules linked together by a beta-glycosidic bond. Sucrose is composed of fructose

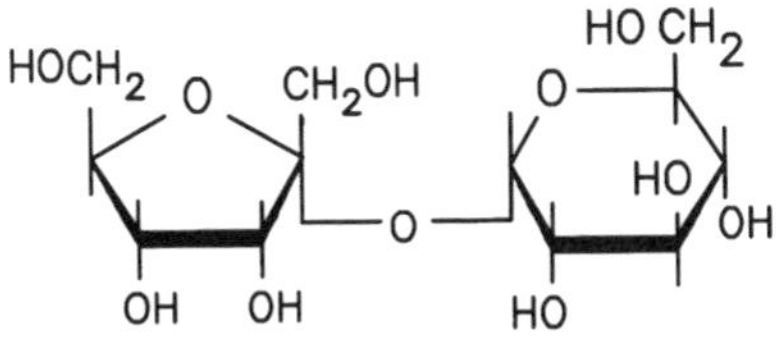

Disaccharide

and glucose linked together by a beta-glycosidic bond.

disc electrophoresis Shortened term for discontinuous electrophoresis, a refinement of polyacrylamide gel electrophoresis in which the sample is electrophoresed through two polyacrylamide phases: a low percentage (stacking) gel that sits on top of a higher percentage (resolving) gel. The two-phase approach produces higher resolution between closely migrating bands. See GEL ELECTROPHORESIS.

disinfectant Any chemical that can kill bacteria and viruses.

disjunction The separation of chromosomes during anaphase of mitosis or meiosis.

dissociation constant The constant that relates the dissociation of two atoms, molecules, or even large particles from one another. For the dissociation of substance A from substance B, where AB is a complex of A and B:

$$AB \rightarrow A + B$$
$$K_{dissociation} = [A][B]/[AB]$$

where
[A] is the molar concentration of A
[B] is the molar concentration of B
[AB] is the molar concentration of AB

distillation The process of separating and purifying liquids from a mixture based on each liquid's boiling temperature. The more volatile substance will boil at a lower temperature from the others in a mixture. The vapor is then collected, cooled, and condensed, thus extracting and refining it from the mixture.

disulfide bond A covalent bond between two sulfhydryl (SH) groups by oxidation to form an S-S linkage. Such bonds occur in proteins between cysteine residues and stabilize the tertiary structure of the protein.

divalent An atom or radical group having two valences or the ability to combine with two different atoms or molecules.

D loop A structure of DNA in which there is a localized denaturation of the duplex or displacement of a single strand from the duplex resulting in a shape that resembles the letter *D*. This structure is usually stabilized with proteins called single-stranded binding proteins.

DMSO Dimethylsulfoxide [(CH3)2SO]; a reagent used for the cryopreservation of cultured animal cells. DMSO is also used to increase the efficiency of transfection of DNA. See CRYOPROTECTANTS.

DMT Dimethoxy trityl; a molecule that is used as a blocking group to prevent unwanted reactions in automated oligonucleotide synthesis.

DNA, repetitive Sequences repeated many times on the genome. These sequences vary in length from three to five base pairs to 300 base pairs and are found on the genome in hundreds to thousands of copies. Some of the repetitive DNA makes up the satellite DNA, a distinct band from the bulk of chromosomal DNA found after cesium chloride density centrifugation. See ALU ELEMENTS.

DNA cloning Any procedure that generates many copies of a particular DNA sequence. The sequence can be inserted into a plasmid or bacteriophage, which will be duplicated manyfold in a bacterial cell, or the sequence can be copied manyfold by polymerase chain reaction, or PCR.

DNA fingerprinting A process of identifying trace evidence such as blood, semen, saliva, and hair found at crime scenes. The most common procedure is to amplify specific regions of DNA found in the evidence by PCR and then analyze them using specific probes to Southern blots (see SOUTHERN BLOT HYBRIDIZATION) of the amplified DNA. Because certain regions of human DNA are very variable, comparisons between blots of

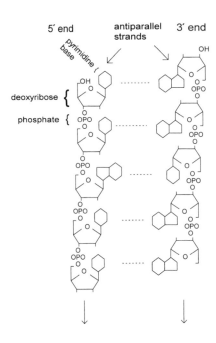

5' end   antiparallel   3' end
          strands

pyrimidine base

deoxyribose {

phosphate {

Double-stranded DNA

DNA of suspects, victims, and evidence can be used to ascertain with great likelihood whether the trace evidence came from a particular individual.

***dna* genes** The genes encoding the DNA proteins *(dnaA, dnaB, dnaC)* that function in the initiation of DNA replication at the replication origin in prokaryotic genomes and plasmid DNAs.

DNA glycosylase An enzyme that recognizes a deaminated base and catalyzes its removal from the DNA molecule, creating an apurninic or apyrimidinic site in the DNA molecule.

DnaG primase The enzyme responsible for catalyzing the formation of the short RNA primers in Okazaki fragments.

DNA gyrase An enzyme that catalyzes the introduction of negative supercoils or relaxes positively supercoiled DNA by unwinding one strand of duplex DNA around the other so that each strand is wrapped around the other less than one turn per every 10 bases.

DNA linkers Short stretches of nucleotide-sequence-carrying restriction-enzyme (see RESTRICTION ENDONUCLE-ASE) cutting sites that can be added by LIGATION to the ends of the sequence of a gene to facilitate the cloning of the gene into a vector.

DNA photolyase An enzyme that catalyzes the repair of pyrimidine dimers formed as the result of ultraviolet irradiation. The first step in the repair involves the excision of the dimer that takes place only in the presence of visible light. DNA photolyase is also known as photoreactivating enzyme. See ULTRAVIOLET REPAIR.

DNA polymerase(s) Any enzyme that can use a chain or a strand of deoxynucleotides as a template, a primer that is a short fragment of deoxy nucleotides and that can synthesize a complementary strand. All DNA polymerases synthesize DNA from the 5' phosphorylated end to the 3' hydroxyl end.

DNA polymerase I A specific DNA polymerase that has not only the 5' to 3' polymerizing activity but also has two nucleolytic or degradative activities, a 3' to 5' exonuclease, an editing function, and a 5' to 3' exonuclease. This enzyme, with all of its activities, is used by the cell during different steps of DNA replication. It is also used during DNA repair processes. In addition, purified DNA polymerase I with or without its nuclease activities is used in various in vitro procedures, such as preparing of labeled DNA probes and DNA sequencing via the dideoxy method. See KORNBERG ENZYME.

DNA probe A sequence of deoxynucleotides used to identify or isolate specific genes or RNA transcripts that have

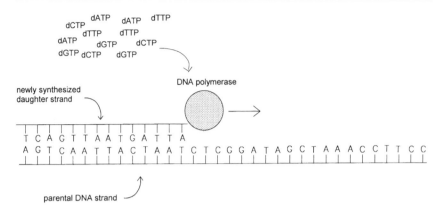

DNA polymerase

complementary sequences. Such probes are used in hybridization procedures (Southern, northern, slot, and dot blots and colony or plaque hybridizations) and are labeled with either a radioactive atom of ^{32}P or ^{35}S, which allows for detection using autoradiography, or nonradioactive materials such as biotin or digoxigenin, which are detected via specific reactions. See PROBE.

DNA profiling See DNA FINGER-PRINTING.

DNA repair Any process that restores damaged DNA. Generally, these are multistep processes, requiring an enzyme to remove the damaged nucleotide (see DNA GLYCOSYLASE and/or ENDONUCLE-ASE) alone or with other nucleotides, a polymerase (see DNA POLYMERASE I) to replace the removed nucleotides, and an enzyme to seal the sugar phosphate backbone. See EXCISION REPAIR.

DNA rescue, in positional cloning A technique for cloning genes in bacteria or yeast by transforming (see TRANSFOR-MATION, BACTERIAL) plasmids containing normal genes into a host containing a mutation that inactivates a particular function, for example, the ability to synthesize an amino acid. Addition of the plasmid that leads to reactivation of the function indicates that the plasmid contains the gene of interest.

DNA-RNA hybrid A DNA-RNA duplex molecule composed of a single chain of deoxyribonucleotides (DNA) and a chain of complementary ribonucleotides (RNA). Such molecules may be created experimentally from purified DNA and RNA and are also formed when chromosomal DNA is fragmented, heated, and mixed with RNA transcripts.

DNase I A DNA-degrading enzyme that catalyzes the cleavage of phosphodiester bonds of DNA; DNase I is isolated in large quantities from pancreas. See ENDONUCLEASE.

DNase I hypersensitivity sites Regions on the chromosome that are extremely sensitive to digestion by DNase I. These sites are generally found near active genes where transcription factors or other regulatory elements bind to the DNA. See HYPERSENSITIVE SITE.

DNase I sensitivity Increased susceptibility to digestion by DNase I, which correlates with genes that are actively transcribing RNA. This shows that the chromatin of genes being expressed has an open conformation that is accessible to DNase I and that inactive chromatin is condensed.

DNA sequencing A method to determine the order of nucleotides on a DNA fragment or molecule. Methods include use of chemicals to break the DNA chains at specific bases (see MAXAM-GILBERT SEQUENCING) and enzymatic incorporation of dideoxynucleotides that result in chain termination. See DIDEOXY SEQUENCING.

docking protein (DP) A receptor protein located on the membrane of the rough endoplasmic reticulum (RER) that binds the signal recognition particle (SRP). This protein-RNA complex bound to the initial sequence of a protein is in the process of being synthesized and is destined for secretion outside the cell. The docking protein anchors the partially synthesized protein to the membrane of the RER so that as its synthesis is completed, it is deposited into the RER, where it will be targeted for secretion. See ROUGH ER.

dolichol phosphate A group of phosphorylated long-chain hydrocarbons comprised of cis-isoprene repeat units that acts to collect sugars for glycosylation of proteins being processed in the endoplasmic reticulum for export to the cell surface. Branched sugar chains attached to the phosphate moiety on dolichol phosphate are transferred to proteins in the lumen of the rough ER.

domain A compact globular unit of protein structure. Many large proteins have a number of domains usually connected by flexible regions of polypeptide chain. In the antibody molecule, there are variable domains that recognize different antigens, as well as constant domains that characterize each class of antibody molecule. See EPITOPE.

Dolichol phosphate

dominance The ability of a genetic trait to be phenotypically or physically expressed, whether it occurs as heterozygous or homozygous. See RECESSIVE.

dominant control regions (DCRs) An ENHANCER element found in the human beta-globin gene cluster that is being incorporated into viral vectors to stimulate and regulate the expression of genes cloned into such vectors.

dominant negative Any mutant that codes for an inactive protein that, in the heterozygous state, is expressed in dominant manner over the functional, or wild type protein.

dopamine A monamine neurotransmitter derived from the amino acid tyrosine. Dopamine is involved in several clinical syndromes including schizophrenia and amphetamine-induced psychoses. A deficiency of dopamine is responsible for Parkinson's disease.

dosage effect The ability of a phenotype to be altered by an increase in the amount of gene product.

dot blot A hybridization technique used to quantitate rapidly the amount of DNA or RNA in a crude preparation that is placed directly onto a hybridization membrane. See BLOT.

double crossover Two recombination events on the same chromosome. See CROSSING OVER.

double digestion The treatment of a preparation of DNA with two restriction enzymes. This technique is used to map DNA and to isolate fragments of DNA with two distinct sticky ends to clone into a vector in a particular orientation. See RESTRICTION ENZYMES and CLONING.

double helix Another name for a molecule of DNA, consisting of two antiparallel, complementary strands of deoxypolynucleotides held together by hydrogen bonds between the complementary pairs. The molecule has a right-handed twist resulting in one strand wrapped about the other to form a helix conformation. See DNA.

double minutes Small pieces of a chromosome that contain many copies of a particular gene. The amplification of the dihydrofolate reductase (DHFR) gene following exposure to methotrexate may be manifest either in terms of the formation of double minutes or as a homogeneously staining region of a giant chromosome. See HOMOGENEOUSLY STAINING REGION.

double reciprocal plot A method for analyzing the kinetic parameters of an enzyme (Km and Vmax), by plotting 1/v versus 1/[S], where v = rate of product formation and [S] = substrate concentration.

double thymidine block A technique used to synchronize cells in culture. A high concentration of thymidine added to the culture will block DNA replication, so all treated cells proceed through their cell cycle and stop at the same point. See CELL SYNCHRONIZATION.

doubling time The same as a generation time, or the time it takes for a population of cells to double in number.

down-promoter mutation A mutation or change in the sequence of the promoter of a gene that results in less expression or transcription of that gene. See PROMOTER.

Down's syndrome The most frequent genetic cause of mental retardation. The disease also includes variety of phenotypic abnormalities: broad skull, short stature, epicauthal fold (a fold of skin around the eye), stubby hands and feet. The disease is the result of an individual inheriting three copies (trisomy) of chromosome number 21.

downstream Denoting the region of a gene that is located away from the gene in the direction of the 5′ end.

downstream processing An industrial term referring to the process of protein extraction and purification that occurs after recombinant DNA technology has cloned a gene and expressed its protein product.

Drosophila A genus of small flies that includes *Drosophila melanogaster,* the common fruit fly. A well-defined genetic organism, it is used as a model system to study and understand cell processes, development, and genetics of higher organisms.

Drosophila **heat-shock proteins** Several proteins that are immediately synthesized after the organism is subjected to a short treatment of heat above its lethal limit. Some of these proteins are highly conserved in evolution in that they are very similar to hsps found in bacteria and other higher organisms. Synthesis of these proteins can also be induced by exposure to certain toxic chemicals, alcohol, and other types of stress. See HEAT-SHOCK PROTEINS.

drug-delivery systems See DELIVERY SYSTEMS.

duplex Another name for double-stranded helical DNA. See DOUBLE HELIX.

duplex melting The process of denaturing double-stranded DNA by heating so that the hydrogen bonds between complementary bases are disrupted. See DNA DENATURATION.

dyad Two units, or a pair.

dyad symmetry of DNA Two regions of the DNA that have inverted, repeated, or palindromic base-pair sequences. Restriction-enzyme cutting sites exhibit dyad symmetry.

dystrophin One of a number of proteins that serve to anchor the muscle myofibril to the plasma membrane. The protein derives its name from the finding that a defect in the structure of dystrophin, or absence of the protein, is found in patients with muscular dystrophy.

E

E1A An adenovirus early gene that is responsible for the oncogenic properties of the virus.

E2A A TRANSCRIPTION FACTOR that acts in a complex with a second transcription factor, MyoD, to bring about the developmental process that causes cells to become myoblasts.

early development A stage in the growth cycle of a bacteriophage that precedes DNA synthesis.

early genes Viral genes that are the first to be expressed after the virus infects its host. The early genes are generally responsible for replication of the virus DNA and for inducing expression of the late genes at some specific point in the viral life cycle.

E-cadherin A TRANSMEMBRANE PROTEIN that anchors cells to one another at specialized junctions (ADHERENS JUNCTIONS) and DESMOSOMES where the membranes of two adjacent cells make contact with one another. The extracellular DOMAINS of two opposing E-cadherins make contact with one another on the extracellular side of the cell membrane. The intracellular domains of E-cadherins are embedded in a plaque that also anchors cytoskeletal filaments.

ecdysone A hormone that induces expression of critical genes during larval development in insects. Ecdysone is known to be responsible for gene transcription seen in chromosome puffs.

E. coli *Escherichia coli,* a bacterium normally found in the intestinal tract. Because of its ability to grow rapidly under minimal nutritional conditions, *E. coli* is widely used as a vehicle for carrying recombinant DNAs and as material for studying bacterial genetics.

ecology The field of study that deals with the interrelationship between a population of organisms and the environment, including physical factors and populations of organisms.

EcoRI A restriction enzyme derived from the bacterium *E. coli* that has the recognition sequence:

$$5'\text{-GAATTC-}3'$$
$$3'\text{-CTTAAG-}5'$$

EcoRI methylase An enzyme that catalyzes the transfer of methyl groups from the compound s-adenosylmethionine to an adenine nucleotide in the restriction site of the enzyme EcoRI:

```
                EcoRI          me
              methylase         |
···GAATTC··· – – – – – →  ···GAATTC···
···CTTAAG··· s-adenosyl  ···CTTAAG···
              methionine         |
                                me
```

Edman degradation A procedure for determining the sequence of amino acids in a polypeptide. The procedure is based on reaction of each amino acid in the peptide chain, in order, with the Edman reagent, phenyl isothiocyate (PITC). The Edman degradation is used in devices for automated polypeptide sequencing.

EDTA Ethylene *d*iamine *t*etra *a*cetate; a chemical that binds tightly to magnesium and calcium and that is used to remove even trace amounts of these metals effectively from a solution. EDTA is used to control unwanted magnesium- and calcium-dependent side reactions in a biochemical mixture.

effector A regulatory molecule; a chemical that brings about an increase or a decrease in the rate of reaction in a specific biochemical pathway.

efferent Running in the direction away from a certain structure. For example, efferent nerve fibers carry nerve impulses away from the brain to an effector such as a motor neuron.

effluent Waste fluid such as buffer emerging from a chromatographic column either before or after the actual chromatography.

EGFR Epidermal growth factor receptor; a family of transmembrane proteins whose extracellular domains are the receptors for epidermal growth factor and whose cytosolic domains are receptor tyrosine kinases. The EGFR family consists of four members, EGF-R (ErbB1), ErbB2 (Neu), ErbB3, and ErbB4. In addition to the normal ligand (EGF), the tyrosine kinase portion of EGFR can be activated by various chemical stimuli and ultraviolet radiation. The EGFR tyrosine kinase activates the MAP kinase signaling pathway, which in turn activates the transcription factors fos, AP-1, and Elk-1 that stimulate gene expression related to cell proliferation. Abnormal stimulation of the EGFRs or mutations in an EGFR gene have been implicated in the development of cancers of the lung, breast, prostate, colon, and ovary.

egg The common term for an oocyte.

egg coat A specialized extracellular matrix, comprised of glycoproteins, that surrounds the oocyte plasma membrane. In mammalian eggs, the egg coat is called the zona pellucida; in sea urchins, it is referred to as the vitelline layer. In addition to protecting the egg, the egg coat sometimes functions as a selective barrier to fertilization by sperm from different species.

eicosanoids A group of paracrine hormones derived from arachidonic acid through a series of enzymatic pathways, involving cyclooxygenases (COX1, COX2) and lipoxygenases. The eicosanoids derived from COX reactions are prostaglandins and thromboxane, while lipoxygenases transform arachidonic acid to the leukotrienes and lipoxins.

electroblotting A technique that utilizes an electric field to transfer protein or nucleic acids from a gel to a blotting membrane, generally for the purpose of carrying out northern, Southern, or western blot hybridizations.

electrodialysis The technique of accelerating the process of dialysis by applying an electric field across the dialysis membrane.

electrodiffusion The induction of movement of a charged substance by an electric field.

electroendosmosis The diffusion of water into or out of a gel or membrane in the presence of an electric field. Electroendosmosis resulting in the shrinkage or swelling of an agarose gel is a factor that influences the migration of nucleic acids during agarose gel electrophoresis.

electroimmunodiffusion A method of quantifying antigen-antibody reactions in which antisera is incorporated into a layer of support medium such as agarose and the antigen that reacts with the antibody is induced to migrate through the gel electrophoretically (see ELECTROPHORESIS). Interactions of antigen and antibody produce rocket-shaped precipitin lines, the heights of which are proportional to the antigen concentration.

electrolyte A charged atom or molecule in solution.

electron carrier In the biochemical context, a molecule that accepts electrons or hydrogen atoms from a specific donor molecule and then transfers them to a specific electron acceptor. FAD, NAD$^+$, and

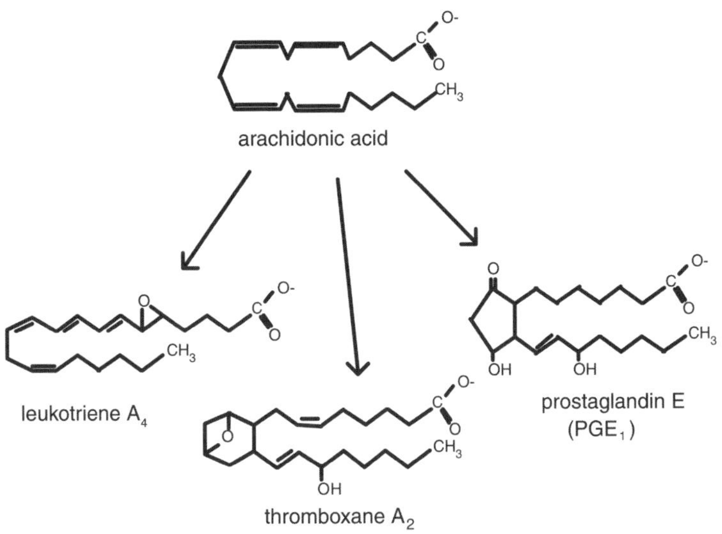

arachidonic acid
leukotriene A4
thromboxane A2
prostaglandin E
(PGE1)

Eicosanoids

ubiquinone are examples of important biochemical electron carriers.

electronegativity The affinity that an atom or molecule has for electrons.

electronic potential The measure of electron pressure in volts; the relative difference in the concentration of electrons in two compartments, such as the inside of a cell membrane versus the outside of the membrane.

electron microscope A device that utilizes a beam of electrons passing through a specimen, instead of light, to visualize and magnify the features of the specimen. In an electron microscope, a powerful magnet that is used to bend the electron beam is the equivalent of the glass lens that, in a conventional microscope, is used to bend the light beam as a means of achieving magnification.

electron microscopy A procedure for using the electron microscope to achieve high levels of magnification. Because electron microscopy must be carried out in a vacuum, biological specimens are gener-

electron gun
anode
electron beam
specimen
magnet
lenses
fluorescent screen
specimen image

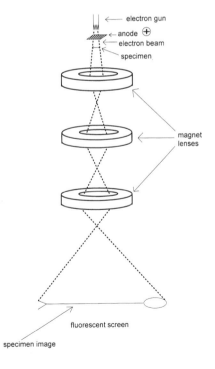

Electron microscope

ally first coated with a thin layer of metal that conveys the outlines of the structural features of interest.

electron transport The process of passing electrons among electron carriers according to a defined sequence.

electron transport chain A series of large complexes located in the inner mitochondrial membrane that uses the energy contained in electrons derived from the metabolism of fats and carbohydrates to generate ATP. The transport chain consists of five complexes designated by the roman numerals I, II, III, IV, and V. Complexes I and II accept electrons from metabolites, and these complexes transfer the electrons to complex III. Complex III transfers electrons to complex IV, which then reduces molecular oxygen to form water. During the process, protons are pumped out of the mitochondrion to create a proton gradient. The energy stored in the proton gradient is used by complex V to create ATP.

electrophoresis The movement of substances through a medium induced by an electric field.

electroporation A technique for introducing substances into cells by using a pulsed electric field to cause the target substance to be electrophoresed across the cell membrane.

ELISA *E*nzyme *l*inked *i*mmuno*s*orbant *a*ssay; a sensitive technique for detection of a substance by allowing the substance of interest, if present in a sample, to attach to an immobilized antibody on some solid substrate such as plastic. The presence of the substance is visualized and quantitated using a second, labeled antibody.

elongation factor Any of several protein factors that are necessary to carry out the part of the process of translation in which amino acids are added to the growing polypeptide chain (elongation). See TRANSLATION.

elongation factors A group comprised of at least three proteins (EF-G, EF-Ts, EF-Tu) that are required for the elongation of a polypeptide that is in the process of being synthesized on ribosomes (translation).

eluant In column chromatography, the fluid, such as a buffer solution, that runs through a column and in which separated substances appear as they are washed through the column.

elution profile In column chromatography, a graph showing the amount of material appearing in the eluant of a column over time. The elution profile is generally seen as a series of peaks rep-

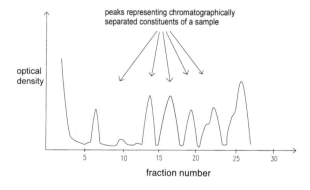

Elution profile

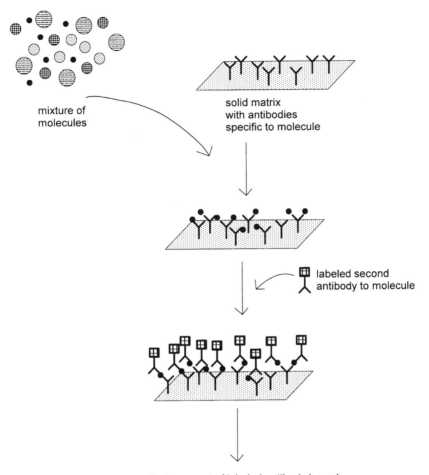

ELISA

resenting the optical density or biological activity of the eluant at various times during elution of the material that is undergoing separation.

elution volume In column chromatography, the amount of eluant that passes through a column before a particular peak in the elution profile is observed.

EMBL See DATABASES.

embryo In vertebrates the organism that develops from the fertilized egg at any stage prior to birth.

embryology The field of study devoted to the development of the embryo.

emulsifier A chemical, such as a detergent, that is capable of breaking up a mass of insoluble material into small particles that then form an emulsion. The most common biochemical emulsifiers produce emulsions from otherwise water-insoluble fatty substances.

emulsion For two unmixable liquids, a colloid of one of the liquids suspended in the other (e.g., emulsified oil and water).

enantiomers A pair of optical isomers that are direct mirror images of one another.

encapsulation The process by which particles become engulfed in or coated by a continuous matrix.

3'-end The terminal nucleotide at one end of a polynucleotide chain, whose 3' carbon carries an unreacted hydroxyl group (–OH group). The other end is the 5'-end.

5'-end The terminal nucleotide at one end of a polynucleotide chain, whose 5' carbon carries an unreacted phosphate group. The other end is the 3'-end.

endergonic reaction A chemical reaction that requires the input of energy, such as heat, or mechanical agitation.

end-filling Creating a blunt end from a ragged-ended or staggered-ended double-stranded DNA through the use of a DNA polymerase.

endocrinology The field of study devoted to the function and pathology of the endocrine glands, for example, the thyroid and pituitary glands.

endocytic vesicle The membrane enclosed vesicle that forms in the cytoplasm of a cell during the process of endocytosis.

endocytosis A process in which cells take up small particles or large molecules by an invagination of the cell membrane, which leads to the formation of a membrane-enclosed vesicle in the cell cytoplasm. A number of important effectors influence cell behavior including the induction of gene transcription after being delivered to the cytoplasm via endocytosis.

endogenous Originating from within; as from within a cell or a tissue.

endogenous virus A virus present, usually in a latent form, inside a cell. This term applies to various viruses that are found in an inactive state in the cells they infect but that may become activated following exposure of the infected cells to various chemical and physical agents. This is true of bacteriophage proviruses and of some forms of herpes virus.

endonuclease A type of enzyme that produces nucleic-acid-strand breaks in the interior of the nucleic acid strand.

endoplasm The inner part of the cell cytoplasm, that is, the portion closest to the nucleus.

endoplasmic reticulum (ER) A complex cytoplasm membrane network to which ribosomes engaged in the synthesis of proteins destined to be exported outside the cell are attached. Portions of the endoplasmic reticulum containing completed proteins for export are transported to the Golgi apparatus to which they fuse.

endorphin Any of a group of short peptides that bind to receptors on neurons in the brain with the effect of reducing the sensation of pain. The term is derived from *endo-* or *endogenous morphine* because endorphins are seen as naturally produced opiates. See METHIONINE-ENKEPHALIN.

endosome The structure formed by the fusion of several endocytic vesicles in the cytoplasm following endocytosis. See CLATHRIN, COATED PIT, and COATED VESICLE.

endospore A tough, resistant, membrane-enclosed cell that is formed by some Gram-positive bacteria and actinomycetes under conditions of limited food supply. The endospore is highly dehydrated and metabolically inactive and can survive harsh environmental conditions such as prolonged heat, drying, and exposure to toxic chemicals.

endosymbiont A symbiotic organism that lives inside the body of its symbiotic partner.

endothelial cells The cell type of which blood vessels are comprised.

endothelial growth factors A class of growth factors released by various tissues under conditions of oxygen deprivation that stimulates angiogenesis. The production of angiogenic factors including endothelial growth factors by tumors accounts for the vascularization of large tumor masses that would otherwise become internally necrotic. For this reason, these factors are studied as a target of cancer therapies designed to kill tumors by starvation.

endothermic A term used to describe a chemical reaction that requires heat in order to proceed.

endotoxins Toxic lipopolysaccharides associated with the cell wall of Gram-negative (see GRAM STAIN) bacteria. These cell-bound toxins cause a variety of physiological effects, including fever, hemorrhagic shock, and diarrhea.

end product The final chemical product of the series of enzymatic reactions in a particular biochemical pathway.

end-product repression See FEED-BACK INHIBITION.

engrailed gene One of a class of genes known as segment polarity genes in *Drosophila melanogaster* (fruit fly). The engrailed gene is a major gene responsible for dividing the segments of the *Drosophila* embryo into posterior and anterior halves.

enhancers Certain DNA nucleotide base sequences that act over distances as great as several kilobases to stimulate transcriptional activity of a particular gene or group of genes.

enriched medium A supplemented nutrient broth for the culture of cells or microorganisms that require unusual nutrients or unusually high levels of normal nutrients. Enriched medium is required for the culture of auxotrophic mutants.

enteric organism A microorganism that inhabits the intestinal tract.

Enterobacteriacae Any of a large group of bacteria that inhabit the intestinal tract.

enterotoxins Toxins that affect the intestine, or those that cause food poisoning. These toxins are secreted by the bacteria that produce them and are ingested in foods contaminated with these enterotoxigenic (toxin-producing) bacteria. Enterotoxin A is produced by the Gram-positive (see GRAM STAIN) bacterium, *Staphylococcus aureus*, and is most frequently associated with outbreaks of food poisoning. Recently epidemics of food poisoning caused by ingestion of water, meat, or fruit con-

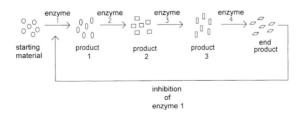

End-product inhibition

taminated with animal fecal matter are due to the potent toxin produced by the enterotoxigenic *Escherichia coli* (ETEC), strain *E. coli* O157:H7.

entomology The field of study that deals with insects.

entropy The variable that measures the degree of disorder in a molecule. Changes in entropy that occur in molecules undergoing chemical reaction are one component of the free energy change that determines whether a reaction will occur under a given set of conditions.

env gene(s) One of the three genes contained in most retroviruses that codes for the ENV glycoprotein(s).

ENV glycoproteins The protein product of the retrovirus env gene(s) which forms a major component of the virus envelope in the mature virus particle.

enzyme A polypeptide or protein that acts as a catalyst for biochemical reactions. Enzymes do not actually cause a reaction to occur, but rather speed up the rate at which an ongoing reaction takes place. Virtually all significant biochemical reactions in living systems are catalyzed by enzymes.

enzyme derepression The induction of enzyme activity by removing or inactivating an inhibitor such as the induction of a galactosidase activity by lactose. See LAC OPERON.

enzyme engineering Modification of enzymes through recombinant DNA techniques and SITE-DIRECTED MUTAGENESIS so that they can be used for industrial purposes. Some of these modifications include increasing protein stability, enhancing catalytic activity and/or substrate specificity, changing optimal requirements for catalysis so that the engineered enzymes will function under nonphysiological conditions, and/or become resistant to feedback regulation.

enzyme immobilization The chemical bonding of an enzyme to some solid matrix in a manner that preserves the enzymatic activity. The attachment of enzymes to solid matrices is an essential step in the development of many enzyme-based biochemical assays.

enzyme inactivation The loss of the activity of an enzyme under conditions other than that found in the intact cell. Enzyme inactivation is an important consideration when purified enzymes are employed in an environment where they may be subject to conditions of temperature, salt, pH, and so on that are not found in their native environment. Spontaneous inactivation of enzymes that occurs for unknown reasons is also often observed in enzyme preparations, particularly in dilute solutions.

enzyme replacement therapy The method used to treat disease states caused by enzyme deficiencies by direct injection of the missing enzyme. Enzyme replacement therapy has been used successfully for treating patients with Gaucher's disease.

enzyme stabilization Inhibition of enzyme inactivation. Enzyme stabilization is often achieved by altering the salt concentration pH or lowering the temperature of an enzyme solution. Recently, modification of the enzyme by attachment of organic groups or altering the amino acid compostion of the enzyme polypeptide have been used to achieve enzyme stabilization.

eosinophil One of the three subclasses of leucocytes. Eosinophils are named for their characteristically intense staining with eosin. Eosinophils are amoeboid scavenger cells similar to macrophages and are found in greatly increased numbers in the blood of individuals carrying parasitic infections.

ephrins A family of proteins implicated in guiding axons and patterning the nervous system during neural development.

Ephrins act as ligands for receptors (designated EPH-related receptors) that are protein-tyrosine kinases. There are two classes of ephrins: the A-subclass (ephrin A1–ephrinA5) and the B-subclass (ephrinB1–ephrinB3).

epidemiology The field of study devoted to analysis of the occurrence of disease in a population including the distribution, incidence, and factors that control the spread of a disease.

epidermal growth factor (EGF) A small polypeptide growth factor discovered by Stanley Cohen as a factor that caused premature eyelid opening in newborn mice. EGF has since been shown to be active in stimulating the growth of epithelial as well as some nonepithelial cell types. A portion of the gene that codes for the EGF cell receptor has been found to be virtually identical to the Erb-B oncogene.

epigenetic The term applied to any factor that influences cell behavior by means other than via a direct effect on the genetic machinery, that is, the DNA.

epimerase A type of enzyme that catalyzes the conversion of one epimer into its opposite epimer.

epimers Optical isomers that differ from one another at only a single carbon atom. The sugars glucose and galactose are examples of epimers.

epinephrine (adrenaline) The biochemical secreted by the adrenal glands and by the synaptic vesicles of certain types of neurons. Epinephrine serves as both a hormone that stimulates the breakdown of glycogen into glucose and a neurotransmitter.

Epinephrine

episome Bacterial DNA that is not integrated into the bulk of the chromosomal DNA and therefore replicates separately, and in different copy number from, chromosomal DNA.

epistasis A term coined by William Bateson in 1909 to describe the control of a certain phenotypic trait by two or more genes. A gene is considered epistatic when it suppresses the effect of another gene. Epistatic genes are also called inhibiting genes because of their suppressive, hypostatic effects on other genes. Pleiotropy, in which a single gene controls the expression of more than one phenotypic trait, is the opposite of epistasis.

epistatic gene A gene that suppresses the effect of another, nonallelic, gene. See ALLELE.

epithelial Of or pertaining to the cell layers that interface between the tissue and the external enviornment, such as the cells of the skin and the lining of the gut and lung airway passages.

epitope The segment on a polypeptide that constitutes the actual site of antibody binding by a specific antibody molecule. The antigenic determinant.

epitope tag A technique by which the function of a protein can be studied by inserting a short nucleotide segment that codes for a known epitope (an antigenic oligopeptide) into the gene for the protein to be studied. When the protein is made, it will contain the epitope. Antibodies to the epitope can then be used to obtain various kinds of information on the protein, such as where it is located in the cell, what other proteins it interacts with, if there are changes in location within the cell in response to stimuli, if there is a subunit structure, etc.

Epstein-Barr virus A member of the herpes family of DNA viruses that has been associated with Burkitt's lymphoma in West Africa and New Guinea.

equatorial plate The early stage of the formation of the membrane that divides two daughter cells at the end of the process of mitosis; the metaphase plate.

equilibrium centrifugation A technique for separation of proteins or nucleic acids from a mixture by subjecting the mixture to density gradient centrifugation for a period of time sufficient for each component of the mixture to form a band at a point equal to its density.

equilibrium potential The membrane potential at which there is no net diffusion of a particular type of ion across the membrane. Equilibrium potentials are important determinants of nerve-impulse generation.

erb-A An oncogene carried by the avian erythroblastosis virus. There are two distinct human erb-A proto-oncogenes, erb-Aα and erb-Aβ. Both forms encode proteins that are thyroid hormone receptors, but they are located on different chromosomes: erb-Aα on chromosome 17q21 and erb-Aβ on chromosome 3p24.

erb-B An oncogene carried by the avian erythroblastosis virus. The human proto-oncogene of erb-B (c-erb-B) encodes the protein for the EGF receptor of which as many as five variant forms may exist. These are designated erbB-1, erbB-2, etc. The erb-B proteins are receptor tyrosine kinases that stimulate cell division via the MAP kinase signaling pathway. erbB-2 which is also known as HER-2 or neu, has been implicated in a number of highly malignant breast cancers. The erb-B genes are located within the region of chromosome 7p12.3-p12.1. See EGFR and HEREGULIN.

ERCC1 Excision repair cross complementing; a polypeptide that is required for nucleotide excision repair (NER) in DNA that has been damaged. In the nucleotide excision repair mechanism, single-stranded cuts are made on either side of the damage, after which the segment of DNA between the two cuts is excised. ERCC1 is the homologue of the RAD10 gene in *Saccharomyces cerevisiae*, which functions in repair and recombination between chromosomes. ERCC1 levels are elevated in cancer cells that have become resistant to the chemotherapeutic agent, cisplatin. In human syndromes in which NER is defective, such as xeroderma pigmentosum (XP), there is a greatly increased incidence of skin cancer.

ERK Extracellular *r*eceptor tyrosine *k*inase; a group of TRANSMEMBRANE PROTEINS that function as signal transducers for signals in the form of biochemicals (ligands) that bind to the ERK extracellular domains. Ligand binding activates a tyrosine kinase function of the intracellular domain. Phosphorylation of a tyrosine residue(s) on an intracellular protein(s) initiates a series of subsequent biochemical changes.

erlotinib (Tarceva) An anticancer drug that acts by blocking the human epidermal growth factor receptor (EGFR) that inhibits the tyrosine kinase activity of the receptor. Tarceva is a quinazolinamine, with the chemical name N-(3-ethynylphenyl)-6,7-bis(2-methoxyethoxy)-4-quinazolinamine.

error-prone repair Another term for SOS repair. The terminology is derived from the observation that repair of pyrimidine dimer damage is often inaccurate. See EXCISION REPAIR and SOS REPAIR SYSTEM.

erythroblast A bone-marrow stem cell that gives rise to erythrocytes.

erythrocyte A red blood cell.

erythrocyte ghosts Red blood cells whose contents have been removed. Erythrocyte ghosts are used as vehicles to deliver drugs and other bioactive compounds to cells. See DELIVERY SYSTEM.

erythromycin An antibiotic that acts by binding to bacterial ribosomes and inhibiting the process of translocation during protein sysnthesis.

erythropoesis The process by which erythrocytes are generated from stem cells in the bone marrow.

erythropoetin A glycoprotein, produced by the kidney, that stimulates erythropoesis.

Escherichia coli See E. COLI.

essential amino acid Any amino acid that cannot be synthesized by an organism from other components. In humans about half of the 20 amino acids are essential; in most bacteria none are.

essential gene Any gene whose malfunction is lethal to an organism. A number of classical experiments on bacterial molecular genetics, such as fluctuation analysis, depended on the use of mutations in essential genes.

established cell line Cells that have become immortalized during the process of maintaining them in cell culture.

establishment of cell lines The process by which cells in tissue culture become immortalized so that they can be maintained indefinitely. Establishment is believed to involve some genetic change that occurs spontaneously during the course of culture. Because cells derived from cancerous tissue are more readily established than cells from normal tissue, the genetic changes involved in the process of established cell lines are also believed to be related to the process by which cells become cancerous.

esterase A type of enzyme that catalyzes the breakage of ester linkages. Esterases are important in the breakdown of many lipids and in the metabolism of nucleic acids.

estrogen A steroid hormone, produced by the ovaries, that cause changes in the lining of the uterus in preparation for implantation of the embryo during estrus.

estrus cycle A set of changes occurring periodically in female primates that prepare the reproductive tract for pregnancy and are governed by changes in the levels of the female hormones. The peak of the cycle (called estrus) coincides with ovulation.

ethanol Ethyl alcohol (drinking alcohol); the alcohol produced from the fermentation of sugar by certain strains of anaerobic yeast.

ethidium bromide A widely used fluorescent stain for visualizing DNA under ultraviolet light. Ethidium bromide is called an intercalating dye because it has a multi-ring structure that allows it to insert between the nucleotide bases. (See figure on next page.)

ethylene A simple two-carbon hydrocarbon with the formula $H_2C=CH_2$.

etiology The study of the cause of a disease or pathological condition.

ets oncogene An oncogene that is carried by Avian leukemia virus E26 (v-ets) that causes leukemias in chickens. The product of the ets proto-oncogene (c-ets) is a nuclear protein that has been found to have DNA binding activity and is believed to play a role in the activation and proliferation of T cells.

euchromatin One of the two classes of chromatin seen in interphase cells that is distinguished from the other class (heterochromatin) by being much less condensed and transcriptionally active.

eugenics The science of selective breeding to achieve a predetermined set of genetic characteristics.

Euglena A primitive single-celled organism classified as belonging to the algae in the plant kingdom. Euglena exhibits the properties attributed to both plants and animals, being photosynthetic in the presence of light and a motile, food-seeking organism in the absence of light. Euglena

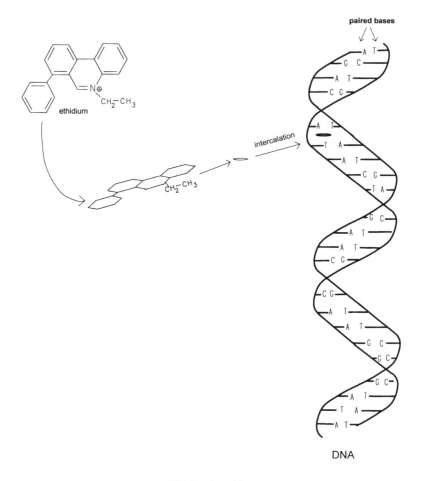

Ethidium bromide

is believed to represent or be related to ancestral organisms that gave rise to both plants and animals.

eukaryote The general term for any higher plant or animal distinguished by the presence of a true nucleus that contains the DNA. Bacteria and viruses are the organisms that comprise the noneukaryotes (i.e., prokaryotes).

eukaryotic cell Any cell in which the cellular genome is contained in a membrane-enclosed nucleus. In general, eukaryotic cells include all cells in the plant and animal kingdoms other than bacteria and viruses.

European Bioinformatics Institute (EBI) The European Bioinformatics Institute (EBI) is a molecular genetics facility that forms part of the public domain European Molecular Biology Laboratory (EMBL) that was established in Heidelberg, Germany, in 1980. The EBI is organized into three subprograms:

1. Service program: biological databases and information services. The EBI provides access to the EMBL Nucle-

otide Sequence Database, SWISS-PROT protein-sequence databases, the Macromolecular Structure Database (EBI-MSD), and the RHdb database of radiation-hybrid maps

2. Research program: tools and information for the study of molecular structure, gene comparison, metabolic pathways, three-dimensional structure, and database searching

3. Industry program: molecular biological resources geared to the needs of the biotechnology, chemical, and pharmaceutical industries

European Molecular Biology Lab (EMBL) The EMBL has a large DNA sequence database containing sequence data compiled from international sources

and maintained in Heidelberg, Germany; it is the European equivalent of the GenBank DNA sequence databank.

evolution In the biological context, the term *evolution* is generally equated with natural selection as proposed by Darwin: the process of change in the composition of a population resulting from the selection of a subpopulation that is better fit than the population as a whole for survival under a particular set of environmental conditions.

excision repair The process of repairing damaged regions of DNA that involves the removal of the damaged region by excision followed by recopying of the excised region by DNA polymerase

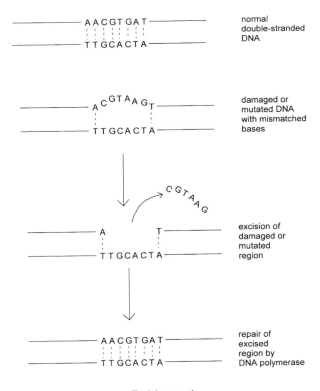

Excision repair

and ligation of recopied region by DNA ligase. See ULTRAVIOLET REPAIR.

exergonic A chemical reaction that releases energy in various forms such as heat or light.

exit domain One of the two major classes of binding sites on the ribosome. The finished polypeptide that is the product of the process of TRANSLATION leaves the ribosome at this site.

exocellular Pertaining to processes or reactions that originate within the cell but take place outside the cell. For example, the digestion of extracellular proteins by proteolytic enzymes secreted by a cell is an exocellular event.

exocrine Secretion of a glandular-produced substance via a duct or canal that leads to the exterior. Exocrine glands are distinguished from endocrine glands, which secrete their products into the bloodstream. Sweat produced by eccrine glands or milk produced by mammary glands are examples of exocrine secretion.

exocytosis The process in which substances contained in a specialized vesicle within the cytoplasm of a cell are secreted to the outside by fusion of the vesicle with cell plasma membrane. The secretion of neurotransmitters in synaptic vesicles by neurons is a common example of exocytosis.

exon The regions of a gene in eukaryotic cells that, as opposed to the introns, actually contain the coding sequences for a polypeptide. See INTRON and SPLICING.

exon shuffling The evolutionary process of creating new genes by duplication and recombination of preexisting EXONS.

exonuclease A class of enzymes that catalyze the cleavage of nucleotides from the end(s) of a nucleic acid.

exonuclease III An exonuclease that catalyzes the cleavage of single nucleotides one at a time from the 5′ end of double-stranded DNA that has a non-phosphorylated 3′ end.

exonuclease VII An exonuclease that catalyzes the cleavage of short oligonucleotides from both the 3′ and 5′ ends of single-stranded DNA.

exotoxin Any of a variety of toxic substances produced by a microorganism and released to the surrounding fluid.

expansins A class of plant cell wall proteins required for cell growth. Under conditions of low pH, expansins induce the breakdown of the hydrogen bonds that bind cellulose microfibrils to one another. This makes the plant cell wall less rigid, which, in turn, allows cell growth to occur.

explant The growth of a portion of a tissue outside of its normal location.

exponential growth phase The phase of growth of a population of cells or organisms during which the overall population number is seen to double at a regular interval. See BIPHASIC GROWTH and GROWTH PHASES.

export The transport of substances across the cell membrane from the interior of a cell to the exterior via specialized systems.

expressed sequence tags (EST) Short cDNA sequences used to link physical maps of genomes.

expression library A library of DNA fragments that has been created using a vector designed to express any genes that are present in the library. See EXPRESSION SYSTEM and EXPRESSION VECTOR.

expression-linked copy (ELC) The particular variable surface glycoprotein gene that is being expressed at any one time during the developmentmental cycle of the trypanosome. See VARIABLE SURFACE GLYCOPROTEIN.

expression site The term for the genetic location of an EXPRESSION-LINKED COPY of a variable surface glycoprotein. Expression sites are all located near the telomere of a chromosome.

expression system An expression vector that contains the cloned DNA it is designed to express, together with the host with which the vector is to be used.

expression vector A specialized cloning vector that contains the elements needed to transcribe a cloned DNA. Expression vectors contain sequences required for DNA replication and promoter elements adjacent to the cloned DNA to initiate transcription.

extinction coefficient The constant of proportionality relating the molar concentration of a substance and the absorbance of its solution. See BEER-LAMBERT LAW.

extracellular Outside the cell.

extracellular fluid The liquid outside the cells in a tissue.

extracellular matrix A complex mixture of proteins (such as FIBRONECTIN, LAMININ, collagen) that is deposited on the outside of a cell and plays a crucial role in the attachments of cells to the surfaces on which they grow. The extracellular matrix is believed to play an important role in regulating the growth and differentiation of a cell partly because the composition of the extracellular matrix is often dramatically altered in cancerous tissue.

extrinsic protein A protein present in a cell or tissue but which originated elsewhere.

extrusion The energy-requiring process by which cells export large particles or organelles.

ezrin A cytoskeletal element that links the transmembrane adhesion molecule, ICAM-1 (intercellular adhesion molecule 1) to the actin cytoskeleton.

F

F(ab)₂ fragment A portion of the IgG antibody molecule containing the two antigen binding domains but not the Fc portion. Such fragments are generally produced by treating antibodies with certain proteases that specifically cleave the molecule near the end of the Fc segment.

F-1 generation First filial generation; the first generation offspring of a genetic cross. The term is generally applied to the immediate offspring of higher plants and mammals.

F-2 generation Second filial generation; the offspring of a mating between members of the F-1 generation.

Fabry's disease One of the lysosomal storage genetic diseases characterized by the lack of alpha-galactosidase. When this enzyme is missing in the lysosome, long-chain carbohydrates that need to be degraded are not. These charbohydrates then accumulate in the bloodstream and deposit in the capillaries and other organs, eventually leading to stroke, heat attack, and fatality in the young adult. This is a disease that is a promising candidate for enzyme replacement therapy because cloned alpha-galactosidase can be administered to individuals and directed to the lysosome, where it needs to function.

facilitated diffusion The process of passive diffusion through a membrane via membrane channels or with the aid of carrier proteins in the membrane.

faciogenital dysplasia (Aarskog-Scott syndrome) A disorder characterized by wide-spaced eyes (ocular hypertelorism), anteverted nostrils, a malformed scrotum, and excessive looseness of the liga-

ments. The genetic form of the disease is an X-linked autosomal recessive. The gene for the disease (called FGD1), of as yet unknown function, has been mapped to chromosome band Xp11.21. The disease was described by the Norwegian pediatrician D. J. Aarskog in 1970 and American C. I. Scott Jr. in 1971.

f-actin (filamentous actin) The functional actin filament composed of G-actin subunits.

factor, blood clotting Any of a group of protein factors in the blood serum that act according to a defined pathway to produce a blood clot. In blood clotting, the breakdown of platelets at the wound site is the first step. Factors VII, VIII, IX, and XI become activated by tissue factor and, in the presence of calcium, convert factor X to activated thromboplastin. Thromboplastin then converts prothrombin into thrombin, the clotting enzyme. Thrombin catalyzes the conversion of soluble fibrinogen in the serum into the insoluble protein fibrin. Fibers made of fibrin are the basic structure of the final clot and are made firm by factor XIII, the fibrin-stabilizing factor. The genetic disease, hemophilia, is the result of a defect in the gene that codes for one of these factors (factor VIII).

facultative anaerobe A microbe that lives under anaerobic conditions but that can adjust its metabolism to utilize oxygen when placed in an aerobic environment.

facultative heterochromatin A highly condensed form of heterochromatin that is believed to not be transcribed.

facultative microoganisms See AEROBE.

FAD (flavin adenine dinucleotide)

FAD, FADH$_2$ Flavin *a*denine *d*inucleotide; a cofactor for enzymes involved in oxidoreductions and electron transfer in numerous biochemical reactions, but particularly those involved in the oxidative metabolism of sugars for energy production. FAD is a combination of two nucleotides, one of which is derived from the B vitamin riboflavin (vitamin B$_2$).

FADD *F*as *a*ssociated via *d*eath *d*omain; a component of the classical apoptosis pathway mediated by FAS.

familial hypercholesterolemia A genetic disease characterized by abnormally high serum cholesterol levels resulting in early onset atherosclerosis and heart attack. The genetic defect results in low tissue levels of the receptor for low-density lipoproteins (LDLs). Cells lacking these receptors cannot take up LDLs from the blood to digest them.

familial Mediterranean fever A genetic disease characterized by recurrent fevers and inflammation of the abdominal cavity resulting in severe abdominal pain. Other symptoms include arthritis, chest pains, and skin rashes.

Fas A transmembrane receptor protein involved in the "classical" apoptosis pathway. In this pathway, activation of the Fas receptor (caused by binding of ligand to the extracellular portion of Fas) results in activation of caspase-8 (also known as FLICE) by FADD, which is bound to the cytosolic domain of the Fas receptor.

fast component That portion of a preparation of eukaryotic DNA that reassociates first when the strands of a native DNA helix are dissociated from one another, for example, when heated and then allowed to reassociate. The fast component was shown to contain highly repeated DNA sequences.

fastidious Pertaining to microrganisms with complex nutritional requirements; requiring enriched media. See ENRICHED MEDIUM.

fatty acid Any of a group of long-chain carboxylic acids that are found mostly in animal fat. The most common fatty acids found in animal tissues are oleic, palmitic, stearic, and palmitoleic acid.

FBJ An acronym derived from the names of the discoverers (*F*inkel, *B*iskis, and *J*inkins) of the FBJ murine osteosarcoma virus. See FOS.

Fc The portion of the immunoglobulin heavy-chain molecule that does not contain the antigen-binding region; the antibody molecule constant region.

fecundity The measure of fertility; for example, sperm count or the production of viable eggs.

feedback control The general term for regulation of an enzyme's activity by one of its own metabolites.

feedback inhibition Inhibition of enzyme activity by a product (generally the final product) of the metabolic pathway of which the enzyme is part.

feeder layer In tissue culture, a layer of cells that produces a product that supports the growth of another cell type in co-culture. Feeder layers are used as a means of growing cells that will not grow in purely synthetic culture medium.

feedforward control Stimulation of enzyme activity by a product of the metabolic pathway of which the enzyme is part.

feline sarcoma virus (FSV) A retrovirus that causes sarcoma tumors in cats.

fermentation The process in microorganisms in which the metabolism of sugars for energy is accompanied by the formation of alcohol or lactic acid.

fermentor A device for growing large bacterial cultures. A fermentor consists of a large vessel (usually containing more than 10 liters of culture) that is mechanically shaken or rapidly rotated for aeration of the culture. There is also a heater in contact with the culture vessel that maintains the culture at the proper temperature, usually 37°C.

ferritin A protein that forms a complex with iron. Ferritin, which normally functions as an iron storage protein, is used as a noradioactive label for visualizing antibodies bound to a specific antigen, such as in a western blot.

fertilization The fusion of two gametes and of their respective nuclei to create a diploid or polyploid zygote.

fes *F*eline *s*arcoma; the oncogene carried by Snyder-Theilen strain of FELINE SARCOMA VIRUS (FSV). fes is believed to be a protein kinase that catalyzes the phosphorylation of tyrosine residues.

fetal calf serum (FCS) The serum from the blood of embryonic calves; an essential component of most tissue culture media. The factors in FCS that promote the growth of cells in tissue culture are largely unknown but are believed to include growth factors and hormones.

Feulgen reagent A DNA-specific stain (fuchsin sulfite) that, because it was found to stain chromatin in the nucleus strongly, was cited as evidence that DNA was the hereditary material in experiments carried out by Robert Feulgen in 1914.

F factor A bacterial plasmid containing fertility genes that establish the donor characteristics for CONJUGATION (see HIGH-FREQUENCY RECOMBINATION STRAIN). These genes are responsible for the ability of the donor cell to establish contact with and transfer its chromosome to the recipient. F′ plasmids are F factors that have picked up a region of the bacterial chromosome after a faulty recombination event in an Hfr strain.

FGP *F*luorescent *g*reen *p*rotein; A naturally occurring fluorescent protein isolated from certain coelenterates such as the Pacific jellyfish. Because the gene for FGP has been cloned and found to be active when fused to other genes, it has found wide application as a molecular tag for determining the cellular location of various proteins to which it can be fused by various molecular genetic techniques.

fgr The oncogene carried by the Garden-Rasheed strain of the FELINE SARCOMA VIRUS.

fibrin The protein formed from fibrinogen that polymerizes to form the fibers that comprise a blot clot at the site of a wound. See FACTOR, BLOOD CLOTTING.

fibrinogen The protein that is released by platelets at the site of a wound and that gives rise to fibrin when thrombin is present. See FACTOR, BLOOD CLOTTING.

fibroblast A cell type that comprises the bulk of the living cells in connective tissue and in the supporting matrix (the stroma) of skin and other epithelial tissues. Fibroblasts are embedded in a complex extracellular matrix, much of which they secrete and that is responsible for the strength and flexibility of the stroma.

fibroblast growth factor receptor 3 (FGFR3) The cell surface receptor for fibroblast growth factors (FGFs); a family of polypeptide growth factors involved in cell division, angiogenesis, and wound healing. The FGF receptor is comprised of an immunoglobulin-like extracellular domain, a transmembrane domain, and an intracellular tyrosine kinase domain. Mutations in the FGF receptor cause achondroplasia, a disease of bone development characterized by stunted bone growth. The FGFR3 gene is located on the HD region on chromosome 4 (gene map locus 4p16.3).

fibronectin A ubiquitous glycoprotein found in blood and in virtually all tissues of the body that is thought to play a key role in cell adhesion and in the control of cell growth and differentiation. Fibronectin is particularly prominent in the fibroblast-containing tissues where it is complexed with collagen.

Fick's law of diffusion The premise that a substance in solution will diffuse in a direction that will tend to eliminate any concentration gradient, that is, make the solution homogenous with respect to concentration.

ficoll A synthetic polymer of the sugar sucrose. Ficoll is biochemically inert and is used primarily to increase the density of solutions for purposes of density gradient centrifugation and nucleic acid hybridization.

ficoll gradient A solution of ficoll created in such a way that the concentration of ficoll varies continuously along an axis through the solution. Ficoll gradients are often used to separate different cell types from one another by sedimentation. See DENSITY GRADIENT.

figure eight An intermediate stage in the process of recombination in which two circular DNAs are covalently bound to one another.

filamentous bacteriophage A subclass of single-stranded DNA bacteriophage in which the bacteriophage genome is encapsidated by an elongated viral coat resembling a filament. F1 and M13 are the most common members of this class of bacteriophage.

filamin A dimeric protein that crosslinks actin filaments to produce a viscous aggregate with the properties of a gel.

filopodia Long microspikes (50 μm) that extend out of the growing tip of the axon of a developing neuron.

filter sterilize A technique for rendering a solution sterile (i.e., free of microbes the size of bacteria) by passing it through a fine filter.

fine-structure mapping A mapping technique that can detect changes in nucleotide base sequence covering a few nucleotides based on very rare recombination events between strands of DNA carrying different forms of the same gene (alleles).

fingerprinting A general term for techniques that define a unique identity for a given protein or nucleic acid molecule by breaking the molecule into a pattern of fragments based on its amino acid or nucleotide base sequence by using various proteases or restriction enzymes. Fingerprinting has been developed as a tool in forensic medicine primarily in the form of DNA fingerprinting in which an individual's unique pattern of DNA fragments is visualized by Southern blot hybridization using a probe for a gene that is known to vary widely.

finger protein A protein that contains segments of regularly spaced cysteine amino acids that appear to be involved in binding zinc atoms. This type of structure is characteristic of nucleic acid binding proteins.

first-order kinetics Any chemical reaction in which the rate at which the reaction occurs is proportional to the molar concentration of only one reactant. For example, for the reaction $A \dashrightarrow B + C$, the rate of reaction = $k[A]$, where k is the reaction rate constant.

flagella Long, external, flexible filaments that are used to propel cells in a liquid medium. Bacterial flagella, which differ in structure from the flagella in eukaryotic cells, also serve as a chemotactic organ that guides the cell to sources of food.

flagellin The protein that makes up the bacterial flagellum.

flash evaporator A device for removing solvent from large volumes of a solution by evaporation to concentrate the solute. A flash evaporator consists of a heated, rotating glass sphere with a tube to allow the evaporating solvent to exhaust.

flavin A compound that is derived from riboflavin (vitamin B$_2$). Important flavin biomolecules are FAD and FMN. See COENZYME.

A flavin molecule

FLICE Fadd-like ICE; an alternative name for caspase 8. FLICE-2 is an alternative name for caspase 10.

flippases A group of enzymes that catalyzes the movement of membrane lipids from one membrane leaflet to the other (flipping). Flippases are believed to serve important functions in the trafficking of vesicles, particularly for the formation of secretory and endocytic vesicles. Some flippases do not require an energy source, while others require energy derived from the hydrolysis of ATP.

flocculation The rapid precipitation out of solution of large amounts of material.

flora Plant life.

flow cytoenzymology A technique for separation and analysis of cells by fluorescence-activated cell sorting (FACS) based on the presence of certain enzymes that generate colored compounds from synthetic substrates.

flow cytometry A technique based on automated measurement of fluorescence

emitted by individual cells. Flow cytometry is carried out in an instrument in which individual cells are illuminated by a laser beam as they pass by a window where a sensitive photocell records the quantity of light emitted at a given wavelength. Because antibodies can be labeled with fluorescent compounds, this technique has been widely used as an automated procedure for quantitating amounts of various antigens present in a population of cells.

fluctuation analysis A method developed by Salvatore Luria and Max Delbruck in 1943 that used statistical analysis of the rate of mutation occurring in bacterial cultures containing small numbers of cells to demonstrate that mutations occur spontaneously.

fluid mosaic model A model of the eukaryotic cell membrane proposed by S. J. Singer and G. L. Nicolson in 1972. The model is based on the idea of a semisolid lipid bilayer into which transmembrane and integral membrane proteins are embedded.

fluorescence The property of certain molecules whereby they emit light at a specific wavelength (emission wavelength) when illuminated by a light beam at another specific wavelength (excitation wavelength).

fluorescence-activated cell sorting (FACS) A variation of flow cytometry in which fluorescently labeled cells are physically sorted into different compartments based on the amount of fluorescence emitted at a given wavelength.

fluorescence in situ hybridization (FISH) A process like AUTORADIOGRAPHY, but instead of using radioactively labeled DNA, the DNA probe is tagged with a fluorescent dye that will vividly show up chromosomes or parts of chromosomes.

fluorescence resonance energy transfer (FRET) The direct transfer of an excited photon from one fluorescent molecule to another located within 1-50Å from one another. Since the ability to transfer the photon from one molecule to the next varies as a function of the sixth power of the distance, the transfer of the photon (which can be measured from the fluorescence of the recipient molecule) is a sensitive indicator of the distance between the two molecules.

fluorescent-antibody techniques Techniques for visualization of the location of a certain antigen in a tissue section or other cell preparation that is based on the binding of an antibody with a fluorescent label to the antigen of interest.

fluorescent label Any molecule that fluoresces and can be attached to another, nonfluorescing probe molecule such as an antibody or a DNA hybridization probe.

fluorimetry Quantitative measurement of fluorescence.

5-fluorodeoxyuridine The nucleotide derivative of 5-fluorouracil that is formed in cells treated with 5-fluorouracil. 5-fluorodeoxyuridine and 5-fluorouracil are both used as anticancer agents.

5-fluorouracil (5-FU) An analog of thymine used as an anticancer agent. 5-FU is an inhibitor of the enzyme, dihydrofolate reductase (dHFR) and therefore is an inhibitor of nucleotide synthesis, which is particularly harmful to the rapidly growing cells in tumors.

flush ends Termini of a double-stranded DNA molecule that have no single-stranded overhanging regions. See BLUNT-END DNA.

flux-control coefficient A measure of the relative change in flux (i.e., change in the rate of substrate conversion to product) caused by some specific modulation of an enzyme (e.g., chemically altering its activity) when the system is in a steady state. If the activity of an enzyme increases by 5 percent in response to a change in substrate concentration when the system is isolated, but increases only by 2 percent when the enzyme is in the intact cell, the enzyme would have a flux-control coefficient of 2/5, or 0.4.

FMN Flavin mononucleotide; a nucleotide derived from the vitamin riboflavin (vitamin B_2). FMN, one of the two nucleotides that comprise FAD, also functions in the transport of electrons during the oxidative metabolism of sugars for energy production. See COENZYME and FLAVIN.

Fmoc A molecule (9-fluorenymethoxycarbonyl) used to protect the free amino end of a growing peptide chain against unwanted side reactions during the chemical synthesis of a peptide.

fms The oncogene carried by the McDonough strain of feline sarcoma virus. The human proto-oncogene (c-fms) encodes a protein that is the membrane receptor for macrophage colony stimulating factor (MCSF). The chromosomal location of the c-fms gene is 5q33.2-5q33.3.

focus-forming assay A test for the presence of DNA that contains oncogenic activity. In a focus-forming assay, test DNA is transfected into animal cells that ordinarily show contact inhibition. If the test DNA contains oncogenic activity, the recipient cells lose contact inhibition, begin to divide, and then form areas of dense packing (foci). The appearance of foci is taken as an indication of oncogenic activity.

focus-forming units (FFU) A measure of the concentration of live virus in a given volume of fluid. Focus-forming units are determined by spreading a known amount of virus-containing fluid over a layer of cells that the virus infects and then observing the number of areas in the cell layer that show evidence of viral infection. See TITER.

folate antagonist A type of compound, for example, methotrexate, that blocks certain critical reactions in the synthesis of nucleotides and that requires the B vitamin, folic acid. Folate antagonists are widely used as chemotherapeutic agents for treatment of cancer because the rapidly growing cells of malignant tumors are more dependent on these reactions than normal cells.

foldback DNA A segment of DNA that contains palindromic repeat sequences that may base pair to one another during reassociation. See PALINDROME.

follicle cells A layer of cells found in both vertebrates and invertebrates that surround the oocyte and supply it with certain low molecular weight nutrients.

footprint The region on a DNA molecule to which some particular regulatory protein binds. The footprint can be visualized by partially digesting the protein-bound DNA with DNase 1 and then separating the digested DNA fragments by electrophoresis. The region bound by the protein will not be cut by the DNase and will appear as a blank area on the gel.

forensic science The science developed by Edmund Locard in the early part of the 20th century to establish whether there has been a transfer of trace evidence between the criminal and crime scene or between the crime scene and the criminal. Forensic scientists focus on trace evidence such as blood, semen, saliva, and hairs found at the crime scene. Before the advent of DNA technology, trace evidence was analyzed by a series of blood grouping tests. Now, forensic scientists rely on DNA FINGERPRINTING.

forkhead transcription factors A family of transcription factors defined by a common DNA binding domain of about 100 amino acids called the forkhead domain, first described in a mutant of *Drosophila melanogaster*. Forkhead transcription factors have been found in a wide variety of species from yeast to humans, including FD1-5 (*D. melanogaster*), HNF-3 (mammalian), HTLF

(human), and HCM1 (*Saccharomyces cerevisiae*). Most forkhead factors are involved in embryonic development, but some factors have been shown to be involved in other functions, including regulation of circadian rhythm, cell-cycle control, and life span.

formamide One of the most commonly used chemicals for denaturing nucleic acids in hybridization techniques. Formamide has the chemical formula: $CONH_2$.

forming face The side of the Golgi stack where vesicles that have budded off from the rough endoplasmic reticulum fuse to the Golgi apparatus; the cis face of the Golgi apparatus.

forms I, II, and III, DNA The supercoiled, nicked circular, and linear forms, respectively, of circular episomal DNAs such as viral or plasmid DNAs. Forms II and III are not thought to be naturally occurring forms but are believed to be derived from native supercoiled DNA (form I) by nicking of one (form II) or both strands (form III) during the process of extraction.

formycin B A purine derivative that is used as an antiparasitic agent. Formycin B inhibits the ability of cells to use salvaged nucleotides from the extracellular medium for nucleic acid synthesis.

forward mutations Any mutation that causes a change from a normal functioning gene to an improperly functioning, or inactive, gene.

fos The viral oncogene (v-fos) carried by Finkel-Biskis-Jinkins (FBJ) murine osteosarcoma retrovirus. The fos homologue in human cells codes for a family of proteins consisting of four members: Fos, FosB, FosL1, and FosL2. These genes for leucine zipper transcription factors form dimers with the jun family of proteins to form the transcription factor complex AP1. As transcription factors, the fos proteins are implicated in cell proliferation, differentiation, and transformation. The human c-fos gene is located on chromosome 14q21-31.

fosfomycin (fosfonomycin) An antibiotic that acts by blocking an early step in the biochemical pathway by which the bacterial cell wall is synthesized. Fosfomycin, which is a structural analog of phosphoenol pyruvate (PEP), blocks the step at which PEP is required to create the pentapeptide that is used to construct the bacterial cell wall.

fos-related antigens (FRA) A group of nuclear phosphoproteins that are similar in structure to the product of the fos oncogene.

founder effect The presence of a chromosome, a portion of a chromosome, or even a particular allele in the members of a given population that can be traced back to a single individual.

four-strand crossing over Crossing over between two sister chromatids that involves breaking of both DNA strands on both chromosomes. This differs from the usual case that involves only one DNA strand from each chromatid. See CROSSING OVER.

F protein The Sendai virus–derived protein that is responsible for the ability of the virus to cause cell fusion. The F protein is used in the creation of fusogenic vesicles.

fps (fes oncogene) The oncogene carried by the Snyder-Theilen strain of feline sarcoma virus (FSV). The human homologue of fps (c-fps) encodes a cytoplasmic tyrosine kinase. The feline gene causes sarcoma tumors, but the association of the human fps gene with cancer has not yet been clearly established. The location of the human fps gene is chromosome 15q26.1.

fragile sites Sites on chromosomes that show a higher than normal probability of breakage and therefore more commonly are sites where chromosomal translocation is observed.

fragile X syndrome The second most frequent genetic cause of mental retardation after Down's syndrome. This disease belongs to a group of diseases that result from a repeating sequence of three bases (triplet) on a chromosome, caused by DNA polymerase slippage during DNA replication. Fragile X is caused by increasing the numbers of the triplet CGG on the X chromosome. The larger the number of repeats, the more severe the disease. Individuals with only a few repeats are carriers of the disease but in general do not have the symptoms.

frame-shift mutation A type of mutation in which nucleotide bases are inserted or deleted in the coding region of gene causing the triplet codons to be translated in the wrong reading frame.

Franklin, Rosalind (1920–1958) A physical chemist who carried out the high-resolution X-ray diffraction studies that led to the elucidation of the double helical nature of DNA for which James Watson and Francis Crick were awarded the Nobel Prize.

FRAP Fluorescence recovery after photobleaching; a technique whereby fluorescent molecules located in a specific cellular structure, for example, the nucleus or cell membrane, are bleached by a microscopic light beam. The bleached area is examined at various periods of time after the photobleaching to determine how fast the cellular structure regenerates the material in the bleached area. The FRAP technique is best known for its use in studies of cell membrane synthesis and fluidity.

free energy See GIBBS FREE ENERGY.

freeze-drying The removal of ice from a frozen cell sample to be examined by the freeze-etch technique by subjecting the sample to a vacuum as the temperature is slowly raised, thereby leaving the essential cell structural features behind. See LYOPHILIZATION.

freeze-etch A technique for examining cell structure by electron microscopy in which a frozen cell sample is cracked with a knife to reveal the cell contents. After freeze-drying, the sample is then shadowed and examined under the electron microscope.

freeze fracture A technique for examining the structure of the cell membrane by electron microscopy. The procedure is essentially the same as in freeze-etching, except that the sample is fractured along the plane of the cell membrane, which is then examined after freeze-drying and shadowing.

Frei test A clinical test to diagnose diseases caused by infectious microbes based on the appearance of a skin reaction when a killed preparation of the suspect microorganism is injected subcutaneously.

French pressure cell A device for lysing bacterial cells by subjecting them to hydrostatic pressure.

Freund's adjuvant An emulsion consisting of water, oil, and dead mycobacteria that, when mixed with an immunogen, enhances the immune response when the immunogen-Freund's adjuvant mixture is injected into an animal.

fructose An isomer of glucose found in citrus fruits. A phosphorylated form of fructose is an intermediate metabolite in the oxidation of glucose for energy production.

ftz An acronym for fushi tarazu.

fumarase The enzyme that catalyzes the conversion of fumarate to malate, an important step in the Krebs cycle phase of the metabolism of sugars.

fungi See MOLD.

fungicide An agent that selectively kills fungi.

Fura 2 A dye that fluoresces in the presence of calcium. Fura-2 fluorescence can be used to observe and quantitate the influx or efflux of calcium in the cytosol.

furanose A ring form of a sugar in which the ring is made up of four carbon atoms and one oxygen. The term designates a large group of sugars that form this type of ring when dissolved in water.

fushi tarazu gene A gene in the pair-rule locus of the fruit fly, *Drosophila melanogaster*. Mutants of the fushi tarazu (ftz) gene are missing every other segment. See PAIR-RULE MUTANTS and SEGMENTS, SEGMENTATION.

fushi tarazu mutation A mutation that causes a failure to produce the seven embryonic parasegments that appear at the BLASTODERM stage of development in *Drosophila melanogaster* (fruit fly). Experiments centered on this mutation helped identify the pair-rule class of genes.

fusidic acid An antibiotic that acts by blocking the translocation step in protein synthesis (translation) by blocking the release of the elongation factor (EF)-GDP complex.

fusion proteins Proteins that represent the product of the artificial splicing of two genes.

fusogenic vesicles Liposomes that contain, in the lipid bilayer, specialized fusion-inducing molecules (e.g., the F protein).

futile cycle In living systems, a combination of competing reactions in which the products of one set of reactions are reconverted to the original reactants. The term is generally applied to reactions of energy metabolism such as the interconversion:

$$\text{fructose-6-phosphate} \overset{\text{ATP->ADP}}{\underset{\text{ADP->ATP}}{\Longleftrightarrow}} \text{fructose-1, 6-bisphosphate}$$

G

G1 phase A segment of the cell cycle representing the time period during which there is an increase in cellular mass between the end of mitosis and the onset of DNA synthesis (S phase).

G2 phase A segment of the cell cycle representing the time period during which there is an increase in cellular mass between the end of DNA synthesis (S phase) and the onset of mitosis.

GABA Gamma *a*mino *b*utyric *a*cid; an inhibitory neurotransmitter derived from the amino acid glutamate. GABA acts to inhibit neural transmission by opening channels that admit chloride ions into the neuron. The GABA receptor is the target of pharmacologic agents, such as Valium and other diazepams, that act as depressants by potentiating the action of GABA.

GAG 1. Glycosaminoglycans. Long-branched chains of sugar molecules built from repeating dissacharide subunits containing amino groups. Glycosaminoglycans are present on the surfaces of eukaryotic cells where they are believed to play a role in cell-cell and cell-substate recognition.
2. Group-specific antigens. The proteins encoded for by the GAG gene of a retrovirus. The GAG proteins are the components of the virus capsid.

galactose An optical isomer of the sugar glucose. Galactose differs from glucose only at the fifth carbon and is converted into glucose through the action of an epimerase enzyme (UDP-glucose 4-epimerase) that acts on this carbon.

galactosemia A genetic disease caused by a deficiency of the epimerase enzyme that converts galactose into glucose. The disease is characterized by organ enlargement, mental retardation, and cataract formation resulting from the accumulation of D-galactose and D-galactose-1-phosphate in the bloodstream.

galactosidase Any one of the class of enzymes that catalyze the cleavage of the glycosidic linkage between galactose and another sugar. The galactosidases are divided into α and β galactosidases depending upon the type of glycosidic bond that is cleaved (i.e., α or β). See GLYCOSIDIC LINKAGE.

GALT *G*ut-*a*ssociated *l*ymphatic *t*issue; patches of lymphoid tissue in the small intestine; Peyer's patches.

gamete The mature product of the process of meiosis, for example, egg and sperm, in organisms that reproduce sexually.

gamma chain An immunoglobulin (Ig) chain that is found as a transmembrane protein on the surface of a B cell and is part of the B-cell antigen-receptor complex.

gamma interferon See INTERFERON.

gamma radiation A high-energy electromagnetic radiation that is produced during the process of nuclear decay in which one subatomic particle is converted into another; for example, the decay of a neutron into a proton and an electron also releases gamma radiation.

ganglion A collection of neurons that, in mammals, are centers of lower brain function outside the brain proper. In lower animals such as worms and invertebrates, which lack a brain, ganglia constitute the centers of all brain function.

gangliosides A class of cell membrane lipids found almost entirely in brain neurons. Ganglioside molecules form part of the brain-receptor complex for pituitary polypeptide hormones.

GAP GTPase-activating proteins. A group of proteins that inactivate the ras-GTP by inducing hydrolysis of the bound GTP to produce ras-GDP. The ras-GDP complex remains inactive until the GDP is exchanged for GTP. The inactivation of ras-GTP by GAPs is an important step in the signal transduction pathway mediated by ras.

gap genes A group of genes (hunchback, Kruppel, knirps) that play a key role in the development of segmentation in the embryo of the fruit fly, *Drosophila melanogaster.* Gap genes have been identified on the basis of mutations that result in the absence of segments in the midportion of the embryo.

gap junction A specialized channel that forms between two adjacent cells at their mutual point of contact and that connects the cells so that small molecules (about 2,000 kilodaltons or less in size) can pass between them. Gap junctions are believed to be a mechanism of intercellular communication involved in the control of cell growth and differentiation.

gap mutants Mutants of the fruit fly, *Drosophila melanogaster,* in which several adjacent segments fail to appear during the course of development. See SEG- MENTS, SEGMENTATION.

gasohol A mixture of 90 percent gasoline and 10 percent ethyl alcohol. Gasohol is purported to be a cleaner and more energy-efficient alternative to gasoline. It has been proposed that ethyl alchohol for this purpose could be cheaply obtained by bacterial fermentation.

GC box A sequence of nucleotides (GGGCGG) that is found in the promoters of mammalian cells and that appears to be a binding site for certain transcription factors.

GC content The fraction of the total nucleotides in a DNA molecule that are cytosine and guanine nucleotides, generally given as a percentage of the total.

gel A semisolid colloidal solution.

gel electrophoresis A technique for separation of substances, principally nucleic acids and proteins, in a mixture by using an electric field to induce them to migrate through a gel. Separation of the individual component substances in the original mixture is based on the size of the molecules (gel filtration) and/or the electric charge. Agarose gels are commonly used for separation of mixtures of nucleic acids, and polyacrylamide gels are generally used for separation of proteins and nucleic acids.

gel filtration A technique for separation of a substance from others in a mixture by passing the mixture usually through gel beads in a column. Separation is based on the size of the molecules and depends on the size of the spaces between the polymeric gel molecules (i.e., the pores). Substances whose molecules are smaller in size than the pores enter the gel, so their movement through the gel is slower than larger molecules that have less of a tendency to pass through the gel pores but, instead, pass around the gel beads. Each type of gel has a characteristic pore size that determines the exclusion size: the maximum molecular size that can enter the gel. Molecules larger than the exclusion size are completely excluded from the gel.

gel-retardation assay A technique for determining the presence of DNA protein complexes in a given DNA fragment by observing whether or not the rate of movement of the fragment in an

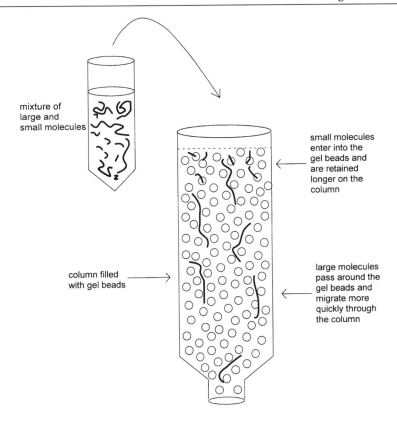

mixture of large and small molecules

small molecules enter into the gel beads and are retained longer on the column

column filled with gel beads

large molecules pass around the gel beads and migrate more quickly through the column

Gel filtration

electric field is slowed in the presence of a particular protein.

gelsolin A protein that, in the presence of calcium, liquefies gels formed from actin and filamin. The liquefaction is accomplished by cleavage of the actin filaments and subsequent capping of the cleaved filament ends.

gel transfer A term applied to the general process of transferring substances separated by gel electrophoresis from the gel to a membrane for analysis, for example, for Southern, northern, or western blot analysis. Blotting is one type of gel transfer.

GenBank A national database of nucleic acid and protein sequences contributed by various investigators around the world, currently maintained at the National Library of Medicine. The GenBank, which is the most comprehensive U.S. national database of this type, is divided into 13 sequence categories: primate, mammal, rodent, vertebrate, invertebrate, organelle, RNA, bacteria, plant, viral, bacteriophage, synthetic, and unannotated. See DATABASES.

gene A sequence of DNA nucleotides that carries the complete code required for the biosynthesis of a polypeptide.

gene bank A group of genes that are coordinately controlled.

gene cloning The science of creating recombinant DNAs that can be inserted into and copied by a host microorganism. The term and the power of the technique derives from the ability to rapidly grow and easily manipulate large populations of microorganisms carrying the recombinant DNA from a single cell (i.e., a clone).

gene(s) codominant Different versions (alleles) of a particular gene, both of which are active in the heterozygous state.

gene conversion A mechanism proposed to explain COINCIDENTAL EVOLU-

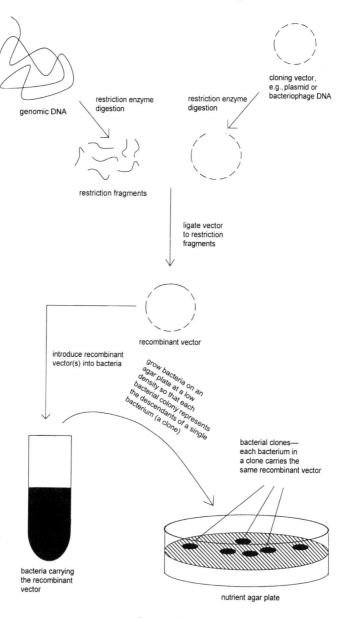

Gene cloning

TION of duplicated genes in which a DNA strand from one gene copy becomes paired with the complementary DNA strand from the other gene copy; any area of base mismatch (presumably representing a mutation that occurred in one of the gene copies) is then repaired by a mismatch repair system. See MISMATCH REPAIR.

gene(s) dominant A version of a particular gene (allele) whose expression obscures the expression of its recessive allelic form when both are present in the heterozygous state.

gene duplication An error in the normal process by which genes are copied and that results in a copy of a gene being placed in the same DNA strand as the gene from which it was copied.

gene expression Any gene activity. Gene expression may include gene transcription into mRNA, translation of mRNA into protein, or activation of a preexisting gene product (protein).

gene families A group of genes whose nucleotide base sequences show a high degree of sequence homology to one another. In evolution, gene families are believed to arise through gene duplication.

gene flow The tendency for gene frequency to appear in one population as the result of interbreeding with another population in which the gene is present.

gene frequency The relative percentage of occurrence of a particular gene relative to all versions (i.e., alleles) of that gene.

gene library See LIBRARY.

gene map locus The chromosomal location of a gene. Gene map locus is defined by chromosome number, p or q arm, band number, and sub-band number. Gene map locus is written as: Xp (or q) y.z where:
X is the chromosome number;
p or q designates the chromosome arm;
y is the band number; and
z is the sub-band number.

For example, 16p11.2 designates sub-band 2 of band number 11 on the p arm of chromosome 16.

gene(s) pseudogenes A variant form of a particular gene that has become permanently inactivated over time as the result of genetic drift.

generation time The average time between the appearance of parent and progeny generations in a population.

gene(s) recessive A version of a particular gene (allele) whose expression is obscured by the expression of its dominant allelic form when both are present in an organism (the heterozygous state).

gene splicing A term that is applied to the general area of recombinant DNA but that is also applied to the process of splicing of eukaryotic RNAs. See EXON, INTRON, and SPLICING.

gene therapy The treatment of human genetic diseases by the transfer of a wild type, or "good," gene into a person whose disease is the result of a faulty gene. Such therapies depend on creating the appropriate vector to bring the replacement gene into the appropriate tissue. Some blood diseases, such as thalassemia and hemophilia, are amenable to such treatment because wild-type genes missing in patients suffering from these diseases can be inserted into stem blood cells and introduced into the patients by bone marrow transplants. Gene therapy has been used to treat familial hypercholesteremia, a disease in which patients have a faulty gene for a protein that is responsible for clearing the blood of cholesterol. Because the process requires alteration of the genetic material of an individual and may be risky, such procedures require monitoring by local and federal oversight committees involving scientists, ethicists, and lawyers.

genetic code The sequence of three consecutive nucleotide bases in a strand of DNA or RNA that specifies an amino acid.

genetic disease A disease caused by an alteration or mutation in a gene that results in an aberrant form of the protein coded for by the gene. Because genetic diseases are based on alterations in genes themselves, genetic diseases are transmissable to offspring that receive the faulty gene. Hemophilia and sickle-cell anemia are examples of genetic diseases.

genetic drift The process of changes in structure of a gene in evolution as the result of a random substitutions, loss, or insertions of nucleotide bases in the DNA.

genetic engineering The manipulation of genes through the use of recombinant DNA techniques for the purpose of modifying the function of a gene or genes for a specific purpose.

genetic information Any information that is carried in terms of the sequence of nucleotide bases in DNA.

genetic map A map of the genome of an organism based on the relative distances between genetic markers. See LINKAGE MAP.

genetics The study of the process by which traits are transmitted from parent to offspring; the study of inheritance.

genome All the genetic information carried in the haploid number of chromosomes.

genomic DNA The DNA representing the entire genome. In laboratory terminology, the term *genomic DNA* is used to describe a pure preparation of total native DNA isolated from tissue or a cell culture.

genomic library A library created in a particular vector from genomic DNA such that the entire genome is included in the library.

genomics Computer analysis of gene sequences from different organisms to understand functional relationships between parts of genes and to assess evolutionary relationships between species. See ALIGNMENT and DATABASES.

genotype The set of all genes, including the different alleles, either expressed or not expressed, that is carried in the DNA of an organism.

gentamycin An aminoglycoside antibiotic isolated from *Actinomycetes* that is active against a number of Gram-positive cocci-type bacteria.

genus A subclassification of organisms between family and species. The standard classification categories in order of general to increasingly detailed biological characteristics is: kingdom, phylum, class, order, family, genus, species.

germacide Any agent, chemical or physical, that destroys microbes that cause disease.

germ cell A reproductive cell or any cell giving rise to a reproductive cell, such as an oocyte or spermatocyte.

germinal centers A region of the lymph node that contains a mass of rapidly dividing B cells. See B LYMPHOCYTES.

germination The growth of a plant from a spore or a seed.

germ line Embryologically, the cells that will, in the adult organism, give rise to germ cells.

germ-line therapy A gene therapy based on the introduction of new genetic material or on alteration of existing genetic material in cells that give rise to either sperm or egg. In germ-line therapy, the new or altered genetic material can be passed from parent to offspring.

ghost cells A cell, usually a bacterial or red blood cell, that lacks much or most of its internal contents as the result of lysis and resealing of the cell membrane.

Because ghost cells can fuse with other cells, ghosts have been used as a means of packaging and delivering drugs to other cell targets.

gibberllin A group of plant hormones that induce the plant growth and maturation in flowering plants.

Gibbs-Donnan effect The observation that charged molecules on one side of a semipermeable membrane often fail to ever become evenly distributed on both sides of the membrane. This effect is explained on the basis of the fact that other charged substances that cannot diffuse across the membrane produce an electric field that influences the migration of charged molecules.

Gibbs free energy The energy that is either released, or used by, a chemical reaction. The free energy absorbed or released during a reaction (ΔG) is the difference between the energy of the products and the energy of the reactants ($\Delta G = G_r - G_f$) and is given by:
$$\Delta G = \Delta H - t\Delta S$$
where:

ΔH is the energy released (or used) in chemical bond breakage (or formation) during the chemical reaction.

ΔS is a measure of the change in entropy (disorder) of the molecules involved in the reaction.

t is the temperature at which the reaction occurs.

For a bimolecular reaction, $A + B \rightarrow C + D$
$\Delta G = \Delta G^{o\prime} + RT \ln([C]^c[D]^d/[A]^a[B]^b)$
where $\Delta G^{o\prime}$ is the standard free energy for the reaction under standard biological conditions. See STANDARD TRANS-FORMED CONSTANTS.

Gilbert, Walter (b. 1932) Walter Gilbert became famous as a coinventor of the Maxam-Gilbert technique of DNA sequencing and for his research on the intron-exon structure of eukaryotic genes. He shared the Nobel Prize in chemistry in 1980 with Paul Berg.

Gleevec An anticancer pharmaceutical made by Novartis that inhibits the abl protein-tyrosine kinase that is activated in chronic myelogenous leukemia (CML) by formation of the Philadelphia chromosome. Gleevec is effective against CML and some gastrointestinal tumors (GISTs).

glial cell A cell type that occupies the spaces in between the neurons of the brain. Glial cells are divided into five major subclasses: Schwann cells, oligodendrocytes, microglia, astrocytes, and ependymal cells.

glial fibrillary acidic protein (GFAP) A protein that assembles into a cytoplasmic network of intermediate filaments found only in glial cells.

globin A group of proteins that form the subunits of the oxygen-carrying molecules hemaglobin and myoglobin. A mutation in the globin genes is responsible for the oxygen-transporting defect seen in sickle cell anemia.

Globotriaosylceramide A glycolipid molecule that accumulates in patients with Fabry's disease, which is caused by a deficiency of the enzyme α-galactosidase A.

globular actin (G-actin) The basic monomeric subunit that polymerizes to form the characteristic actin filaments in muscle. G-actin is a single polypeptide of 375 amino acids.

glucagon A small (29 amino acids) polypeptide hormone that stimulates the breakdown of glycogen into glucose primarily in the liver.

glucoamylase An enzyme, found largely in saliva but also in the juices of the lower digestive tract, that catalyzes the breakdown of complex sugars, for example, starch, by cleaving the bonds between two adjacent glucose subunits.

glucocorticoid One of the three classes of steroid hormones produced by the outer layer (the cortex) of the adrenal gland. The glucocorticoids cortisone, corticosterone, and cortisol regulate the metabolism of glucose. They also act as anti-inflammatory agents and are used to limit inflammation in certain chronic inflammatory conditions such as arthritis.

glucogenesis (gluconeogenesis) The process of creating glucose from its own metabolites. This pathway is active under conditions where the rate of metabolism of glucose is reduced.

glucose A six-carbon sugar that is the major source of energy for most of the animal kingdom. Energy is generated through the oxidation of glucose to yield carbon dioxide and water. See TRICARBOXYLIC ACID CYCLE.

glucose effect The blockage of the induction of the *lac* operon genes by the presence of glucose.

Glucose Galactose Malabsorption (GGM) A rare autosomal recessive disorder resulting from a defect in the SGLT1 gene that codes for a transporter for the sugars glucose and galactose. The condition is characterized by severe diarrhea and dehydration in neonates and can be fatal unless a sugar-free diet is maintained. In GGM most mutations are found to produce nonfunctional truncated SGLT1 proteins or in misplacement of the transporter so that glucose and galactose cannot be removed from the intestinal lumen. The sugars osmotically remove water from the body tissue into the intestinal space, which causes the diarrhea.

glucose isomerase The enzyme that catalyzes the conversion of glucose into the structural isomer fructose.

glucose oxidase An enzyme derived from certain molds that catalyzes the conversion of glucose into gluconic acid with the formation of hydrogen peroxide. Glucose oxidase is widely used for the determination of glucose in the urine in the diagnosis of diabetes because the hydro-

gen peroxide produced in the reaction can be used to oxidize certain aromatic compounds that form colored products.

glutamic acid An amino acid whose side chain is:
$$-CH_2-CH_2-COOH$$
The COOH group gives glutamic acid its acidic nature.

glutamine An amino acid whose side chain is:
$$-CH_2-CH_2-C{=}O$$
$$\backslash$$
$$NH_2$$
Glutamine plays an important role as an intermediate in the transfer of amino groups in the biosynthesis and degradation of a number of other amino acids.

glutamine-rich domains A region(s) found in certain types of eukaryotic transcription factors. Glutamine-rich domains are involved in interaction with other transcription factors with which they act in a synergistic manner to upregulate transcription; in human cells the glutamine-rich domains interact with acidic type transcriptional activators bound at a separate site. Sp1 and Oct1 are examples of transcription factors that contain glutamine-rich domains. A glutamine-rich domain in the transcription factor Sp1 makes contact with the dTAFII110 factor in the *Drosophila* TFIID complex to bring about transcriptional activation.

glutathione A molecule made up of three amino acids linked end-to-end (a tripeptide: glutamate-cysteine-glycine) that acts as a reducing agent. Glutathione plays a role in determining the proper folding of newly synthesized proteins through the cross-linking of cysteine residues.

GLUT transporters A group of transporters (GLUT1, GLUT2, GLUT3, GLUT4, and GLUT5) which function to transport glucose and other hexoses such as galactose and fructose into the cell. The GLUT transporters allow sugars to diffuse passively, down a concentration gradient, into cells. They are large integral membrane proteins in which a transport channel is formed from 12 membrane-spanning regions. The appearance of GLUT transporters on the cell surface is induced by insulin in liver cells.

glycerides (mono-, di-, tri-) A class of lipids in which one or more fatty acids is covalently attached to a glycerol molecule. Glycerides are divided into monoglycerides (one fatty acid molecule), diglycerides (two fatty acid molecules), and triglycerides (three fatty acids). Glycerides are important as storage vehicles of fat.

glycerol The simplest carbohydrate containing three carbon atoms with the structure:

$$H_2C-CH-CH_2$$
$$|\quad|\quad|$$
$$HO\ OH\ OH$$

Because of its ability to absorb water, glycerol is used commercially as a moisturizer.

glycine The simplest amino acid whose side chain consists only of a hydrogen atom.

glycocalyx The cell coat; an outer coating, rich in carbohydrates on the surface of most eukaryotic cells. The glycocalyx also contains some glycolipids and proteoglycans that may form part of the extracellular matrix (ECM).

glycogen A complex storage polysaccharide consisting of branching chains of glucose molecules. Glycogen is the primary source of glucose that is produced mostly from glycogen breakdown in the liver under conditions where the amount of free glucose is insufficient for the body's needs.

glycogen synthase kinase 3 (GSK3) A serine/threonine protein kinase that activates the enzyme glycogen synthase by phosphorylation. The β isoform of GSK3 (GSK-3β; one of the two isoforms) is also known as Factor A, which activates the phosphatase PP-1. GSK-3β is also a component of the Wnt signaling pathway, where it regulates levels of cyclin D by phosphorylation. Inhibitors of GSK3 are being investigated as therapeutic agents for the control of diabetes and to limit the degree of neurological damage in stroke patients.

glycolipid A sugar or polysaccharide covalently attached to a lipid. Glycolipids are important components of animal cell membranes. Cerebrosides and gangliosides that are derivatives of the lipid sphingosine, are important components of glycolipids in membrane receptors in the brain.

glycopeptide A polypeptide covalently linked to a sugar or polysaccharide. Glycopeptides are divided into two classes, depending on whether the sugar(s) are linked to the polypeptide by an oxygen atom (O-linked) or a nitrogen atom (N-linked). See PEPTIDOGLYCAN.

glycophorin A transmembrane glycoprotein (see TRANSMEMBRANE PROTEINS and GLYCOPROTEINS) found in ERYTHROCYTE membranes. The critical function that glycophorin serves is unknown, but it is believed that charged residues on the extracellular domain of the protein may help to prevent blood cells from sticking to one another. Glycophorin has been shown to be a site of attachment for the influenza virus and the malarial parasite, *Plasmodium falciparum*.

glycoprotein Proteins linked to sugars and/or polysaccharides that are prevalent on the outside surfaces of cell membranes. Glycoproteins are components of specialized receptor molecules and the extracellular matrix (ECM) in eukaryotic cells.

glycosaminoglycans See GAG.

glycoside A compound formed between a sugar and some other type of molecule, for example, a protein, lipid, or other organic molecule.

glycosidic linkage (bond) A covalent bond between the anomeric carbon of a sugar and another molecule, usually another sugar, protein, or lipid.

glycosylation The process of adding polysaccharides to polypeptides that are destined to become glycoproteins. Glycosylation takes place primarily in the interior of the endoplasmic reticulum (ER) during the synthesis of the polypeptide that is to be glycosylated.

glyoxylate cycle A pathway used by plants and bacteria for obtaining energy from acetate and other two carbon compounds that are metabolized into acetate or acetyl groups. The glyoxylate cycle is analogous to the Kreb's cycle with many of the same intermediate steps.

glyoxysome An organelle in plants where the glyoxylate cycle, a biochemical pathway used to convert fats to sugar, is located.

Golgi apparatus (Golgi body) In eukaryotic cells, a series of membrane-bound vesicles arranged in a stack in which the polysaccharides of glycosylated polypeptides are progressively altered or processed prior to their being sorted or transported to the cell surface.

gonadotropic hormones (gonado-tropins) A group of polypeptide hormones made by the pituitary gland that stimulate accessory cells surrounding the oocyte to release progesterone, which in turn causes the oocyte to mature. In animals, gonadotropins begin to appear at the age of sexual maturity.

gonadatropin The collective name for follicle-stimulating hormone (FSH) and lutinizing hormone (LH); small polypeptide hormones that are made in the anterior portion of the pituitary gland (ade-

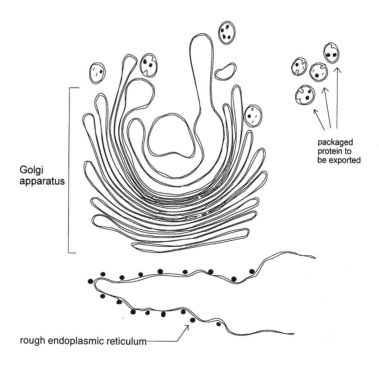

Golgi apparatus

packaged protein to be exported

rough endoplasmic reticulum

Golgi apparatus (Golgi body)

nohypophysis) and act to stimulate the reproductive organs in various ways.

gp120 A GLYCOPROTEIN found on the surface of the HIV virus. During viral infection, the attachment of gp120 to CD4, a receptor on the surface of the target LYMPHOCYTE, is a primary event. For this reason, antibodies created against gp120 epitopes are a major focus in the development of virus-neutralizing antibodies and anti-AIDS vaccines.

GPR14 G protein–coupled receptor 14; the membrane receptor for the neuropeptide hormone urotensin II, which regulates cardiovascular function and is hypotensive in mammals. GPR14 is predominantly expressed in the heart and pancreas and in low levels in portions of the brain. The gene map locus of human GPR14 gene is 17q25.3.

G protein(s) A class of cell membrane-bound proteins that bind GTP and/or GDP and act to alter certain metabolic pathways or gene expression when a specific ligand binds to a receptor on the outside of the cell.

graft v. host reaction A deleterious immune reaction in which lymphocytes present in grafted tissue attacks the tissues of the host.

gram A universally adopted measure of mass in the scientific world. A gram is defined as one thousandth of the mass of one liter of pure water at a temperature where its density is greatest, that is, just above the freezing point (0˚C).

gramicidins A class of polypeptide antibiotics isolated from *Bacillus brevis*. Gramicidins act as ionophores in which ions are carried across the bacterial cell wall in the interior of the circular molecule.

Gram stain A method for differentially staining bacteria developed by Christian Gram in 1884. Staining depends on the cell-wall properties of the bacteria. Bacterial cells are first stained with crystal violet and iodine and then decolorized with alcohol. Cells that retain the purple color of the crystal violet after the alcohol treatment have thick cell walls and are Gram-positive. Cells that lose the crystal violet after decolorization but then take up a pink counterstain, safranin, have thinner cell walls with an abundance of lipid in the cell envelope. These cells are called Gram-negative. The Gram-positive Gram-negative classification system is particularly useful because, not only does it help to identify bacteria, but in a wide variety of bacteria the Gram stain shows a correlation with sensitivity to antibiotics.

grana Stacks of thylakoid disks inside the chloroplast.

granulocyte macrophage-colony stimulating factor (GM-CSF) A recombinant protein produced in large scale and used as an adjunct to cancer chemotherapy since 1991. It stimulates the production of granulocytes to boost the immunity of patients taking chemotherapy.

granulocytes A class of leucocytes composed of neutrophils, eosinophils, and basophils. Granulocytes are active in allergic immune reactions such as allergic skin lesions and arthritic inflammation.

gratuitous inducer A molecule that, because it structurally resembles a certain inducer of transcription, can act to induce transcription in lieu of the authentic inducer, for example, IPTG in lieu of lactose as a gratuitous inducer of the *lac* operon.

gray matter That portion of the neural tissue of the spinal cord that is composed of the nerve cell bodies in contrast to the white matter, which is made up of the nerve-cell axons and dendrites. In cross-section, the gray matter is seen as a butterfly shaped structure that runs through the interior of the spinal cord.

GRB2 An intermediate in the RTK-ras SIGNAL TRANSDUCTION pathway. The

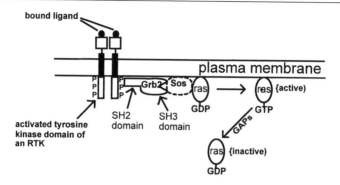

GRB2 is an adaptor protein that binds the TRK phosphotyrosines via an SH2 domain and the Sos protein via an SH3 domain. Sos then induces exchange of ras-bound GDP for GTP, thereby activating the ras-GTP complex that binds to the N-terminal end of cytosolic raf, a protein whose C-terminal end has serine/threonine kinase activity. GAP is a negative regulator of ras that acts to increase the rate of hydrolysis of ras-bound GTP to GDP.

SH2 domain GRB2 binds to the phosphorylated cytoplasmic DOMAIN of an RTK; the GRB2 SH3 domain simultaneously binds to a protein called Sos that, in turn, stimulates GDP bound to ras on the inner surface of the cell membrane to be exchanged for GTP.

Griffith, Frederick (1881–1941) A bacteriologist who demonstrated that heat-killed, pathogenic pneumococcus bacteria could transform live, nonpathogenic pneumococci into the pathogenic form when the two were mixed together. This experiment gave rise to the work of Avery, MacCleod, and McCarty that showed the transforming factor to be DNA.

griseofulvin An antifungal agent produced by *Penicillium griseofulvum*. Griseofulvin appears to act by inhibiting the movement of chromosomes during mitosis by interfering with the spindle apparatus.

group translocation A type of active transport in bacteria in which compounds that enter the cell by passive diffusion are immediately modified, for example, by phosphorylation such that they cannot passively diffuse back across the cell membrane. In this way, compounds

entering the cell are trapped in the cytosol where they accumulate.

growth curve A graph in which the number of individuals in some population of organisms, for example, cells in culture, animals in a herd, fish in a pond, and so on, is plotted as a function of time.

growth factors A group of small, secreted polypeptides that bind to receptors on certain specific target cells and stimulate cell division in those target cells. Growth factors have become the focus of intensive research because of their ability to influence the physiology of growth and also because many of them have been found to bear a close relationship to oncogenes.

growth hormone A growth factor produced by the anterior lobe of the pituitary gland that stimulates the growth of bone and muscle during childhood. Growth hormone was one of the first bioactive factors whose genes were cloned and expressed in transgenic animals, thereby demonstrating the feasibility of curing genetic disease by gene therapy.

growth media A synthetic solution of nutrients to support the growth of cells or

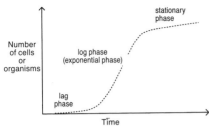

Growth phase

microorganisms in culture. See DEFINED MEDIUM and MINIMAL MEDIUM.

growth phases The different stages of growth of a culture of microorganisms (usually applied to cultures of bacteria) as reflected in the shape of the growth curve. There are three growth phases: (1) a period of slow growth just after the organisms are inoculated into fresh growth medium (lag phase), (2) a period during which the population doubles at a fixed interval of time (exponential phase or log phase), and (3) an indefinite period during which growth is slow or completely stopped as the culture becomes overcrowded and nutrients are displaced (stationary phase).

growth rate The change in the number of organisms in a population divided by the length of the time interval over which the change in the population number took place. For example, a culture of cells in which there were 2,500 organisms on one day and 7,000 organisms three days later would be said to have shown a growth rate of (7,000–2,500)/3, or 1,500 organisms/day.

Grunstein and Hogness method The technique of hybridization of a DNA probe to whole, lysed, bacterial colonies that have been transferred by blotting onto a nitrocellulose filter or other hybridization membrane; colony hybridization.

GST Glutathione S-transferase; a class of enzymes that aid in the detoxification of xenobiotics or toxic electrophiles

by acting on the agent after it has been bound to glutathione via the sulfur (S) atom attached to the cysteine residue in glutathione.

GT-AG rule The observation that intron sequences in DNA always begin with GT as the first two nucleotides and always end with AG as the last two nucleotides. The GT-AG rule plays a role in the mechanism of RNA splicing by which messenger RNA (mRNA) is created from heterogeneous nuclear RNA (hnRNA).

GTP-binding protein (G protein) A membrane-bound protein that acts as an intermediate between the binding of a stimulatory molecule, such as a hormone to a receptor on the outside of the cell, and the biochemical effect that ultimately takes place inside the cell. The mechanism by which G proteins transmit the signal (i.e., the binding of the stimulatory molecule to its receptor) from the outside of the cell to the inside is not entirely understood but is known to involve the exchange of a molecule of GDP for a molecule of GTP by the G protein as part of the process.

guanine One of the four purine bases normally found in DNA and RNA. See PURINE.

guanine nucleotide exhange factor (GEF) A protein factor required in the peptide-elongation step in protein synthesis. Following the hydrolysis of GTP bound to elongation factor Tu (EF-Tu), GEF is involved in the regeneration of the EF-Tu complex by mediating the exchange of bound GDP for GTP.

guanosine A ribonucleoside consisting of guanine and the sugar, ribose.

guanosine, 7-methyl A derivative of the normal nucleoside, guanosine, that is found in the cap of eukaryotic mRNAs. See CAPPED 5′ ENDS.

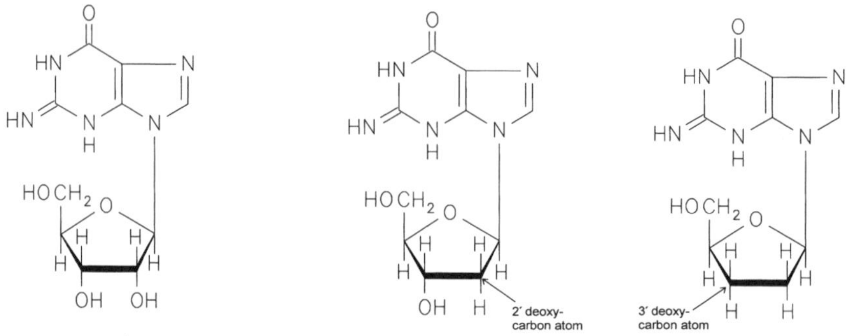

guanosine

deoxyguanosine

2' deoxy-carbon atom

3' deoxy-carbon atom

dideoxyguanosine

Guanosine nucleosides

guanosine monophosphate (GMP) The ribonucleotide containing guanine; a derivative of guanosine formed by phosphorylation of fifth carbon of the ribose.

guanosine triphosphate (GTP) Guanosine with three phosphate groups attached to the fifth carbon atom of the ribose sugar. GTP, together with ATP, CTP, and UTP, is a direct precursor of RNA and a high-energy compound that provides energy that drives other biochemical reactions.

H-2 histocompatibility The match of tissue proteins that, in the mouse, determines whether a tissue graft will or will not be rejected by the immune system of the host. H-2 compatibility is determined by a large gene complex that codes for cell surface glycoproteins. See MAJOR HISTOCOMPATIBILITY COMPLEX.

habituation The tendency of some neurons to require longer than normal refractory phases or stimulation by stronger-than-normal nerve impulses to trigger an action potential if action potentials have been triggered in that neuron in the recent past.

hairpin loop The folding back of a nucleic acid strand on itself. Hairpins are created by internal base pairing between purine and pyrimidine bases along two separate segments of the nucleic acid. See LOOPED DOMAINS.

half-register In repeated sequences in which the repeat unit can be divided into two halves, half-register refers to a misalignment of the two chromosomal copies such that the first half of a repeat unit is aligned with the second half of the other chromosomal copy. For example, if the two halves of the repeat units are designated X and Y, then the repetitive portion could be represented by
...XYXYXYXYXYXYXYX...
and in half register:
...XYXYXYXYXYXYXYX...
...XYXYXYXYXYXYXYY...

halophile A type of bacteria requiring sodium chloride (NaC1) as an essential nutrient.

hamartoma A mass comprised of the normal organ tissues in which it is found

but in which the tissue structure is poorly organized. These lesions are believed to represent developmental abnormalities rather than true neoplasms.

haploid The state of a cell having only one set of alleles as opposed to the diploid state in which a cell normally contains two copies of each allele.

haploid number Having one-half the number of chromosomes as are normally present in a diploid cell, thus being in the haploid state, for example, in gametes.

haploinsufficiency One copy of a gene is not sufficient to assure normal function.

haplotype A particular set of markers, for example, RFLPs or alleles, in a certain region of a chromosome. The term was originally applied only to clusters of alleles in the major histocompatibility complex (MHC) but is now applied to any specified genetic locus.

HA protein *Hem*agglutinin protein; a glycoprotein found on the surface of the influenza virus that binds to sialic acid residues on the cell membranes of cells that are infected by the virus; this binding initiates the process of infection. In a subsequent step, the HA protein mediates fusion between the viral membrane and the membrane of an endosome that encapsulates the virus. This observation has led researchers to utilize the HA protein as a tool to study the process of membrane fusion.

hapten A small molecule that can act as an immunogen only when combined with a larger molecule.

Hardy-Weinberg law A mathematical formulation describing how two different alleles are distributed among the individuals in a population. For two allelic forms of a certain gene, D and d, assuming that (1) mating between individuals is random, and, (2) if the frequency of a certain allele, D, in the population is p, and the frequency of the d allele form is q, then

1. The fraction of individuals homozygous for D (DD) will always be given by p^2;
2. The fraction of individuals heterozygous for D (Dd) will always be given by 2pq; and
3. The fraction of individuals homozygous for d (dd) will always be given by q^2.

harvesting Collection of cells, organisms, or growth medium from an experimental population, generally for purposes of analysis or extraction of biochemicals.

HAT medium A type of cell-culture growth medium containing hypoxanthine, aminopterin, and thymine used for negative selection (i.e., HAT selection) of certain kinds of mutant cells that cannot utilize hypoxanthine and/or thymine to make nucleic acids.

HAT selection The procedure for selecting cells in HAT medium. The procedure is based on the principle that only those cells that can utilize hypoxanthine and thymine supplied from outside the cell to make their nucleic acids will survive in the presence of the drug aminopterin or other folate antagonists, which prevents cells from synthesizing their own purine and pyrimidine nucleotides.

Haworth projection formula A type of representation of organic ring compounds that shows some of the characteristics of the three-dimensional structure, particularly the spatial arrangement of the substituent groups along the ring with respect to one another and the ring. Haworth projec-tion formulas are largely used to represent the furanose and pyranose rings of sugars to show the three-dimensional differences between structural, conformational, and optical isomers.

HB101 A substrain of the bacterium, *Escherichia coli,* that is widely used as a host in which to grow recombinant vectors because of its high efficiency of DNA uptake and transformation.

HCG *H*uman *c*horionic *g*onadotropin; an ovary-stimulating hormone produced in the placenta after the embryo implants into the wall of the uterus. HCG is referred to as the pregnancy hormone because antibodies to HCG are used to diagnose pregnancy.

HDAC *H*istone *deac*etylase(s); a group of 11 enzymes (isoforms) that catalyze acetylation/deacetylation of histones that are complexed with DNA in chromatin. Since aetylation of histones represses transcription, HDACs are considered to have a transcriptional repressor function. Because some of the genes whose expression is repressed by HDACs are tumor suppressors, inhibitors of HDACs are being tested as anticancer agents.

H DNA An unusual structure found in DNA segments with long stretches of polypyrimidines or polypurines in which there is also a palindromic repeat. In these regions DNA can fold back upon itself to form a hairpin in which three of the DNA strands are base paired with one another to form a triple helix while the fourth strand remains unpaired. The formation of H DNA structures is believed to play a role in the control of gene expression.

heat-shock genes A set of genes found throughout the animal kingdom that are suddenly and rapidly transcribed in a coordinated fashion when cells are subjected to certain kinds of stress, such as a sudden rise in temperature. Many of these heat-shock genes encode chaperons, proteins that aid in the folding or unfolding of proteins.

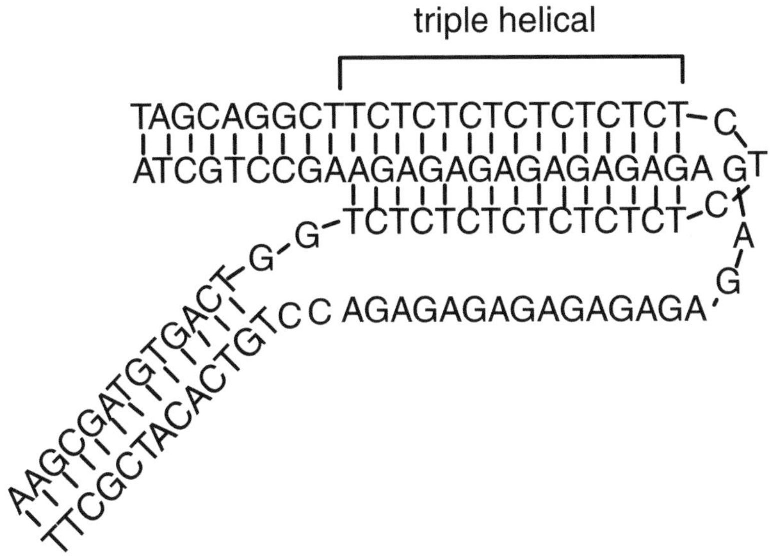

triple helical

TAGCAGGCTTCTCTCTCTCTCTCT−C
ATCGTCCGAAGAGAGAGAGAGAGA GT
−TCTCTCTCTCTCTCT−C

G−G−
ACT
GTGA
AAGCGGATG
TTCGCTACACTGTC C AGAGAGAGAGAGAGA

H DNA

heat-shock proteins (HSPs) A large group of proteins that are rapidly induced in response to various forms of stress such as exposure of cells to changes in temperature, sudden lack of availability of nutrients, oxygen deprivation, etc. The heat-shock proteins are also present under normal conditions, where they function as "molecular chaperones" that aid in correct protein folding and in shuttling proteins between intracellular compartments and to destinations outside the cell. HSPs are usually designated according to their molecular weight—for example, HSP70 or HSP40 (70 and 40 kilodaltons)—but may have unsystematic names such as GrpE and DnaJ. HSPs are believed to play a role in eliciting the immune response by presenting fragments of proteins (peptides) onto the cell surface to help the immune system recognize diseased cells.

heat-shock response element (HSE) A certain nucleotide sequence in the promoter of the heat-shock genes (CNNGAANNTCCNNG). The binding of a transcriptional enhancer protein to this sequence is the first event in the activation of the heat-shock genes.

heavy chain (immunoglobulin heavy chain) The longer of the two peptide chains that make up an antibody molecule, for example, IgG.

HeLa cell A line of tissue culture cells derived from a cervical cancer by Gey, Coffman, and Kubicek in February 1951. The cell-line designation is derived from the name of the tumor donor and was the first epitheliallike cell derived from human tissue to be placed into tissue culture.

helicase An enzyme(s) that catalyzes the unwinding of the DNA helix during DNA replication at a point just ahead of the replication fork.

helix-loop-helix A structural motif consisting of two helices separated by an oligopeptide loop found on many eukaryotic regulatory proteins. DNA binding is effected by a short stretch of basic amino acids just adjacent to the helix-loop-helix motif.

helix-turn-helix A structural motif consisting of two short (seven to nine amino acids) helices separated by a turn found on many prokaryotic DNA-binding regulatory proteins. One of the helices, called the recognition helix, interacts with DNA along the major groove. One example of a regulatory helix-turn-helix protein is the cro protein of lambda bacteriophage.

helper T cell A specialized type of T cell whose function is to stimulate other T lymphocytes (for example, cytotoxic T cells) that then go on to carry out various immune functions. Helper T cells are stimulated to divide after they are exposed to a foreign MHC antigen that is presented to it by specialized antigen processing cells; the stimulatory activity of T helper cells is mediated by interleukins.

helper virus A virus that provides a critical function to a defective virus when they both simultaneously infect the same cell. The oncogene-carrying retroviruses are examples of replication-defective viruses that can grow only when coinfected with the normal wild-type counterpart that does not carry an oncogene.

hemagglutinin Any substance that can cause red blood cells (RBCs) to agglutinate by binding to certain sites on the RBC membrane. Because the clumping (agglutination) of RBCs is easily seen even in the presence of small amounts of hemagglutinin, hemagglutination has been widely used as an assay for the presence of certain hemagglutinating viruses or other antigens.

hemagglutination-inhibition assay An assay for the presence of a hemagglutinating virus or other antigen by observing the loss of the ability of a test sample to agglutinate red blood cells after being treated with an antibody against the agglutinin whose presence is suspected.

hemagglutinin/neuraminidase protein (HN protein) A protein derived from the coat of the paramyxovirus, Sendai virus, that binds strongly to the cell membranes of many types of animal cells. For this reason, the HN protein is used to create liposomes that are intended to deliver agents (e.g., therapeutic agents) to various animal cells (fusogenic vesicles).

heme An organic, iron-containing, ring-shaped molecule that is the oxygen-binding group in hemaglobin.

hemicellulose The name given to a mixture of long polysaccharides with a celluloselike structure that, together with pectin, forms an amorphous matrix in which the cellulose fibrils of the plant cell wall are embedded.

hemidesmosome Literally meaning "half-desmosome," a hemidesmosome is specialized membrane-junctional complex in the epithelial cell membrane that structurally resembles a desmosome but, unlike a desmosome, is present at the site where an epithelial makes contact with the basal lamina.

hemizygous The cellular state of having one copy of a gene in a genome that is normally diploid for all genes. The term always refers to a particular gene or group of genes; for example, a cell is said to be hemizygous for gene x.

hemoglobin The large blood-borne molecule that carries oxygen and carbon dioxide between the lung and tissue. Hemoglobin consists of a heme group and four polypeptide chains; two alpha-globin chains and two beta-globin chains.

hemolymph The fluid found in the body cavity of insects that serves the same gas-exchange functions as blood.

hemolysins A group of bacterial toxins that cause hemolysis by attacking red blood cell membranes.

hemolysis The lysis of red blood cells.

hemolytic plaque assay An assay based on the localized hemolysis of red blood cells (RBCs) that appears as a plaque when the RBCs are spread out in a layer of agar. The hemolytic plaque assay is used to demonstrate the secretion of specific antibodies by antibody-producing cells that are mixed together with the RBCs.

hemophilia A genetic disease based on the inability of the afflicted individual to make a critical component (factor VIII) of the blood-clotting system. As a result, even minor cuts or bruises may result in dangerous, uncontrolled internal hemorrhage or bleeding.

hemopoiesis The generation of red blood cells by cell division of certain stem cells in the bone marrow.

Henderson-Hasselbalch equation A mathematical formulation that governs the pH of a given buffer solution. If pK is the negative logarithm of the equilibrium constant (K_d) for the ionization of the acid form of the compound that is used to buffer the solution for the reaction

$$HA \leftrightarrows H^+ + A^-$$

and

[HA] and [A$^-$] are the molar concentrations of the unionized and ionized forms of the buffer respectively,

then

$$pH = pK + log\ ([A^-]/[HA]).$$

heparin A sulfated glycosaminoglycan that is used medically to block blood clotting. The anticoagulant activity of heparin is based partly on the strong binding of the heparin molecule to antiprothrombin III, a blood protein that plays a critical role in the blood-clotting pathway.

hepatitis An inflammatory disease of the liver characterized by severe, chronic jaundice caused by accumulation of liver by-products and general malaise. Most cases of acute hepatitis are due to viral infections that currently fall into six main types: hepatitis A, B, C, D (but only with B-type virus present), E, and G. Other causes of hepatitis are alcohol abuse, various drugs, toxic agents, and autoimmune reaction.

hepatitis virus The viral agent, a small DNA virus, that causes the infectious form of hepatitis that infects a large fraction of the individuals in certain areas of the world. Hepatitis viruses fall into three subclasses, termed simply A, B, and C.

heptad repeat A tandemly repeated segment of seven amino acids in certain proteins. The heptad repeat is a prominent feature of intermediate filament proteins that are found in virtually all intermediate filament proteins throughout the animal kingdom.

herbicide A chemical agent that is toxic to plants.

Herceptin A monoclonal antibody directed against the HER2 receptor present on some tumors. Herceptin is active in inhibiting the growth of breast tumors that express HER2 by blocking access of growth factors to the receptor. Herceptin is only active against tumors that express high levels of HER2 (ca. 20 percent of all breast cancer patients) and whose growth is therefore dependent on growth factors.

hereditary disease See GENETIC DISEASE.

heredity The study of how physical traits are transmitted from parent to offspring: the study of inheritance. See MENDEL'S LAW.

heregulin A protein that stimulates the growth of breast cancer cells by binding to, and activating, the HER2 and HER4 receptors that are present on the surface of the cancer cells. In the more advanced stages of breast cancer, the HER receptors are often overproduced and may therefore be more responsive to the growth-promoting effects of heregulin. For this reason, heregulin is a target of therapeutic strategies employed in late-stage breast cancer.

Herpes A family of large DNA viruses that infect humans and produce both acute infections such as chicken pox and infections that result from persistent, latent infections, for example, shingles that results from the same virus, that is, *Herpes zoster*. The members of the herpes family of viruses are *Herpes simplex*, *Herpesvirus simiae*, *Varicella zoster*, cytomegalovirus, and Epstein-Barr virus.

Herpes simplex virus (HSV) A member of the herpes family of viruses that has been implicated as the causative agent in some cervical cancers.

Hershey-Chase experiment A classic experiment, conducted by Martha Hershey and Alfred Chase in 1952, that demonstrated that DNA was the hereditary material. In their experiment, bacteriophages containing ^{32}P-DNA and ^{35}S-protein were allowed to attach to host bacteria. When, after several minutes, the attached bacteriophages were stripped from the bacteria by strong mechanical agitation, it was found that the ^{32}P label and not the ^{35}S label had entered the host cells.

heterochromatin A very condensed form of chromatin, seen in the nucleus during interphase, that was found to be a transcriptionally inactive form.

heteroduplex Base pairing between nucleic acid strands from different sources, for example, RNA and DNA or DNA from two different species.

heteroduplex mapping A technique for the determination of the location of a particular sequence of nucleotide bases along a segment of a nucleic acid by creating a heteroduplex between the nucleic acid to be mapped and a reference nucleic acid strand.

heterofermentation (heterolactic fermentation) A type of fermentation characteristic of enteric bacteria (Enterbacteriaceae) in which only part of the fermentation product is lactic acid; the other part is formate and acetyl CoA.

heterogeneous nuclear RNA (hnRNA) The general name given to all the unprocessed RNAs in the nucleus. The name derives from the great heterogeneity in size and type of RNA present in the nucleus before the RNAs are processed and transported to the cytoplasm.

heterokaryon A multinucleated hybrid cell created either from the fusion between two or more cells or by cell division without cytokinesis.

heterolactic fermentation Heterofermentation.

heterologous gene expression The synthesis of foreign proteins in a host organism when that organism has been transformed with a vector carrying genes from a different organism. This can be a problem if DNA technology is to be used to produce proteins in bacterial hosts because some heterologous proteins are broken down by bacterial proteases, or are deposited into insoluble inclusion bodies, and/or do not fold properly. In addition, bacteria cannot add sugar residues to those proteins requiring glycosylation after synthesis.

heterotroph An organism, for example, an auxotrophic mutant, that is defective in the ability to synthesize essential complex organic molecules. A heterotroph therefore requires supplementation of the growth medium, diet, and so on, with either the essential compounds or certain precursors that it can use to make them.

heterozygote An individual who is heterozygous for a particular gene.

heterozygous The state of containing two different alleles of a particular gene. For example, a cell or an individual is said to be heterozygous for the trait that causes sickle cell anemia if both the normal and the abnormal copies of the globin gene is present.

hexose Any six-carbon sugar.

HGPRT *H*ypoxanthine-*g*uanine *p*hosphoribosyl *t*ransferase; an enzyme that catalyzes a major step in the formation of ATP and GTP from guanine. This pathway is the only means by which guanine or other purine analogs can enter into nucleic acids. Thus, as is true for thymidine kinase in the pyrimidine pathway, manipulation of this enzymatic step provides an important experimental tool for studying gene action by the incorporation of modified bases into DNA.

HGPRT marker A term that refers to the use of the gene that codes for the HGPRT enzyme as a selectable marker. Cells containing mutant HGPRT genes are resistant to purine derivatives that are toxic because they become incorporated into DNA via the HGPRT dependent pathway. See HGPRT.

high-frequency recombination strain Certain strains of donor-type bacteria in which the bacterial genomes are observed to undergo much higher rates of gene transfer and recombination than other bacteria in the same culture. The high frequency of recombination is based on the presence of an F factor that has integrated into the genome, which allows mating to occur between neighboring bacteria. See CONJUGATION, BACTERIAL.

highly repetitive DNA A fraction of the genomic DNA in eukarkyotic cells that consists of short sequences that are repeated thousands of times throughout the genome and that has been found to be equivalent to satellite DNA.

high-mobility group protein A heterogeneous group of proteins of unknown function that are part of the chromatin but are not histones.

Hill reactions The light-energy harvesting reactions in photosynthesis in which light energy is stored in the form of high-energy electrons. The Hill reactions were discovered by Robert Hill in 1939.

Hind III A restriction enzyme whose recognition sequence is

$$5'\ AAGCTT\ 3'$$
$$3'\ TTCGAA\ 5'$$

histamine A substance stored in the granules of mast cells that is released during allergic response. Histamine release causes smooth muscle contraction, secretory activity in mucous epithelium, and other symptoms of allergic reaction.

histidine An amino acid whose side chain is

$$-CH_2-C=CH$$
$$|\ \ \ |$$
$$HN\ NH$$
$$\backslash\ |$$
$$CH$$

histocompatability The state of similarity or dissimilarity between the proteins of a grafted tissue and proteins of the host on which the tissue is grafted. The degree of histocompatibility is the major factor in determining whether a host will accept or reject a tissue graft.

histones A group of proteins that are tightly associated with DNA to form structures known as nucleosomes. The histones fall into five subgroups: H2a, H2b, H3, H4, and H1. They appear to play a role in regulating the expression of genes.

HIV See HUMAN IMMUNODEFICIENCY VIRUS.

HLA antigens The proteins of the major histocompatibility locus in the human; an acronym for *h*uman *l*eukocyte-associated *a*ntigens.

hnRNA See HETEROGENEOUS NUCLEAR RNA.

Hodgkin's disease A type of cancer of the tissue of the lymphatic system, including the lymph nodes, spleen, tonsils, and thymus gland. The disease is characterized by fever, lymph node enlargement, and weight loss.

Hogness box A sequence of TATAAA that defines the part of the promoter region where RNA polymerase will bind in eukaryotic organisms. It is located

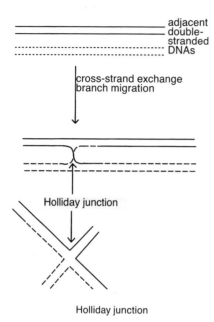

adjacent double-stranded DNAs

cross-strand exchange branch migration

Holliday junction

Holliday junction

about 30 bp upstream from the start of transcription. See TATA BOX.

Holliday junction The linkage of two homologous double-stranded DNAs by ligation of the broken end of one strand with the broken end of the corresponding strand on the homologous DNA. The Holliday junction is considered to be the essential intermediate in the process of recombination.

holoenzyme The complete and functional form of an enzyme; the polypeptide portion of enzyme plus any other factors necessary for normal enzymatic activity, for example, coenzymes and cofactors.

homeobox Those DNA sequences that are found in homeotic genes of the fruit fly, *Drosophila melanogaster,* that are also found in amphibians and mammals. The homeobox sequences are expressed in early development and appear to play a role in limb/appendage development similar to the role of the homeotic genes.

homeostasis The state of being in balance. As applied to biological systems, the

term is used to describe a group of biochemical reactions that act together as a system of checks and balances to prevent "overreaction" on the part of any reaction.

homeotic genes A cluster of genes that determines limb and appendage development in the fruit fly, *Drosophila melanogaster.* The homeotic genes have been defined in terms of mutations at certain genetic loci that may cause one type of appendage to develop in place of another, for example, an insect leg in place of an antenna. See SEGMENTS, SEGMENTATION.

homeotic selector genes The genes of the bithorax and the antennapedia complexes taken together.

homofermentation (homolactic fermentation) The type of fermentation characteristic of lactic bacteria in which sugar is oxidized by the fermentation pathway to a single product: lactate.

homogenate A crude slurry resulting from the disruption of cell/tissue structure by mechanical means, for example, by grinding.

homogeneously staining region (HSR) A region of a chromosome that contains multiple copies of a particular gene and that can be identified as a segment that stains homogeneously (as opposed to showing a number of bands as is normally true) with certain stains used to visualize chromosomes under the microscope. The amplification of the dihydrofolate reductase gene (DHFR) in cells exposed to the anticancer agent, methotrexate, is seen as a homogeneously staining region of giant chromosomes.

homolactic fermentation Homofermentation.

homologous chromosomes Two chromosomes, generally one from each parent, that are identical in terms of the genes they carry but which differ from each other in terms of the alleles of the genes they carry.

homologous recombination Recombination between two DNAs at certain sites where the two DNAs show sequence homology to one another.

homology In general meaning, similar but not necessarily identical to something. In molecular biology, the term *homology* is generally taken to mean sequence homology.

homopolymer In general meaning, any polymeric molecule containing a single type of monomer. In molecular biology, the term refers to a short nucleic acid segment that consists of a single type of nucleotide, for example, oligo dT.

homopolymer tailing The technique of attaching a nucleic acid homopolymer to the end(s) of a piece of DNA generally as an intermediate step in cloning. Homopolymer tailing has been widely exploited as a means of cloning cDNAs by attaching homopolymer tails to the ends of the cDNA that can then be easily annealed to complementary homopolymer tails attached to the ends of a vector molecule.

homoserine The precursor molecule from which, in plants and bacteria, the amino acids, methionine, threonine, and isoleucine, are made.

homozygous The state in which the genome contains two copies of the same allele of a particular gene. Organisms are said to be homozygous for a gene.

Hoogsteen base pairs An unusual type of base pairing that occurs in triple helical DNAs, a type of structure that is formed under special conditions. In the triple helix, one of the bases is hydrogen bonded to another base from each of the other two strands to form the unusual base pairs, such as: T=A-T or C=G-C.

hormone Any molecule that is made and secreted by a specific tissue and that causes or induces a specific action or behavior in another (target) tissue, for example, FSH, LH, or the steroid hormones, estrogen and testosterone.

hormone response elements (HREs) Nucleotide sequences to which nuclear steroid hormone receptors bind to regulate expression of a particular target gene. the HREs are usually located in the 5′ upstream regulatory region of a target gene and generally consist of two hexameric core sequences separated by a spacer sequence to form the complete binding site designated by the symbol RXR. The R sequences can be either inverted or direct repeats, and the spacer sequence (X) generally varies between one and five base pairs.

horse-radish peroxidase (HRP) An enzyme that causes the breakdown of peroxides by catalyzing the transfer of hydrogen atoms to an oxygen atom on the peroxide. The activity of the enzyme has been exploited as a technique for labeling proteins and nucleic acids colorimetrically instead of by radiolabeling. For this purpose, the molecule is attached to an HRP molecule that is then visualized using a substrate (S) that attains a color when it is oxidized by HRP:

$$H_2O_2 + Substrate\text{-}H_2 \xrightarrow{peroxidase} 2H_2O + Substrate_{ox}$$
$$\text{(uncolored)} \qquad \text{(colored)}$$

host cell Any cell that is infected by a virus is referred to as the host cell for that virus.

host range The group of all cell types that are susceptible to infection by a particular virus.

host restriction-modification Restriction enzyme systems that have been developed by bacteria that inactivate the DNA of infecting bacteriophages by cleaving them with restriction enzymes produced by the bacterial host, while at the same time protecting the host DNA from cleavage by the same restriction enzyme through another system that modifies the host DNA usually by methylation.

host-vector system A combination of host cell and virus vector to be used for cloning. For example, a particular strain of the bacterium *E. coli* and a particular strain of bacteriophage lambda (vector) that grows especially well in the host cell.

hotspot A genetic locus particularly prone to spontaneous alteration or mutation.

housekeeping genes A vernacular term to describe the genes necessary for basic cell functions required by and, therefore, expressed in all cells.

HOX genes Homologous genes in mammals to certain homeotic genes in *Drosophila*.

HPFH *H*ereditary *p*ersistence of *f*etal *h*emoglobin; a genetic state in which the normal adult α and β hemoglobin genes are absent, and so the fetal hemoglobin genes continue to be expressed past the time when they would normally be turned off.

HPLC High-performance liqid chromatography. A variation of liquid chromatography that uses small (three to 10 μm in diameter) silica beads to achieve high-resolution separations.

H-ras gene The human proto-oncogene homologue of the ras oncogene carried by the Harvey sarcoma virus.

HTLV A group of human retroviruses that cause human T cell leukemias; an acronym for *h*uman *T*-cell *l*eukemia *v*irus.

human growth hormone (somatotropin) A polypeptide hormone produced by the anterior pituitary that stimulates the liver to produce somatomedin-1, which in turn causes growth of muscle and bone.

human immunodeficiency virus (HIV) The retrovirus that causes acquired immunodefieiency syndrome (AIDS).

humoral antibody Secreted antibody produced by B cells circulating in the blood. Humoral antibodies mediate the immune response to soluble antigens as opposed to graft rejection.

huntingtin The protein encoded by the gene responsible for Huntington's disease, a genetic neurodegenerative disease. In the version of the gene that causes the disease, there are multiple repeats of a CAG triplet that leads to the insertion of an additional string of glutamine residues. Huntingtin is a cytoplasmic protein, but its function is unknown. The mutant huntingtin forms aggregates in nuclear inclusions in neurons. The gene for huntingtin is at gene map locus 4p16.3. See HUNTINGTON'S DISEASE.

Huntington's disease A genetic degenerative disease of the nervous system caused by a dominant gene. Symptoms that appear in midlife include involuntary movements of the face and limbs, mood swings, and forgetfulness. Disease symptoms progress until death within 20 years.

hyaluronic acid (HA) A type of glycosaminoglycan in which the repeating disaccharide consists of the sugars glucuronic acid and N-acetylglucosamine. The linear polysaccharide is a component of the extracellular matrix in vertebrates, where it forms the core of proteoglycan aggregates. It is also found in synovial

glucuronic acid (GlcA)   N-acetyl glucosamine (GlcNAc)

Hyaluronic acid

fluid that lubricates joints and in the vitreous humor of the eye.

hybrid An organism that is the offspring of parents of different genotypes.

hybrid-arrested translation An experimental technique that identifies a cDNA for a specific protein by making a hybrid between a cDNA and mRNA. If the cDNA does hybridize to the putative mRNA for the protein, then the mRNA containing the hybrid will not be able to be translated in an in vitro translation system, which will be indicated by the absence of that protein in a protein gel of the protein products.

hybrid cell A cell that was produced by cell fusion but that, after several cell divisions following fusion, now contains one nucleus with chromosomal material from the original parent cells.

hybrid dysgenesis The term that is applied to the inability of certain strains of the fruit fly, *D. melanogaster,* to interbreed because offspring resulting from matings between the strains are sterile.

hybridization The formation, in vitro, of a double-stranded nucleic acid segment by hydrogen bonding between two single strands. Experimental use of hybridization is the basis of DNA probe technology including Southern and northern blot analyses, primer annealing, and heteroduplex analysis.

hybridization probe Any labeled nucleic acid segment that is used in any of a variety of assays based on hybridization of the labeled nucleic acid to a target nucleic acid.

hybridization stringency A term used to describe the degree of mismatch tolerated by a specific set of hybridization conditions. Hybridization stringency is usually given in terms of the minimal percent base match that will be required for duplex formation between the hybridization probe and the target nucleic. Thus, the chemical and physical conditions under which a hybridization occurs can be adjusted so that the level of homology between the probe and the target is 85 percent, 90 percent, and so on. Levels of homology below about 70 percent are generally considered to be nonhomologous, and so hybridization conditions permitting duplex formation between nonhomologous nucleic acids are called nonstringent.

hybridoma An immortalized antibody-secreting cell created by fusing a myeloma cell to lymphocytes from the spleen of an animal that has been immunized to a particular antigen. Antibody-secreting hybridomas are the source of monoclonal antibodies.

hybrid vigor The state in which an offspring is genetically more robust and/or better equipped for survival than either parent as the result of heterozygosity, that is, receiving a beneficial combination of traits from its parents.

hydrocarbon Any organic molecule composed only of hydrogen and carbon. The most common hydrocarbons are those that are derived from a linear chain of carbon atoms.

hydrogen bonds Electrostatic attractions between positively charged hydrogen atoms and negatively charged atoms on other parts of a molecule or on other molecules. Hydrogen bonds are the major forces that stabilize the structures of many proteins and the DNA double helix.

hydrolase The general class of enzymes that catalyze reactions involving hydrolysis.

hydrolysate The product of the hydrolysis of a material, for example, a protein hydrolysate or a casein hydrolysate.

hydrolysis The breakage of any covalent bond that involves the insertion of a water molecule across the bond; for example:

```
-C=O    H₂O    -C=O    NH-
  \   ---------->  \   +  \
 NH-           OH       H
```

In the above example, the carbon-nitrogen bond is said to have undergone hydrolysis.

hydronium ion A water molecule to which a hydrogen ion is attached: $H^+ + H_2O \longrightarrow H_3O^+$. Hydronium ions are the form in which hydrogen ions are normally carried in aqueous solutions.

hydropathy plot A graph showing the degree of hydrophobicity of each amino acid in a polypeptide as a function of its location in the polypeptide. Hydropathy plots are often used to visualize the clustering of hydrophobic amino acids. If such clustering is observed, it may indicate that the polypeptide in question is a transmembrane protein, with the hydrophobic cluster representing a transmembrane domain.

hydrophilic The property of an atom, a molecule, or a molecular group that has an electrostatic attraction to water molecules. Hydrophilic groups tend to be soluble in water.

hydrophilic-signaling molecule A large class of highly water-soluble molecules that, because of their solubility, can diffuse easily across an aqueous medium between a cell from which they are secreted to a target cell where they trigger some specific event. Various growth factors and hormones are commonly encountered by hydrophilic-signaling molecules.

hydrophobic The property of an atom, a molecule, or a molecular group that has no or very little electrostatic attraction to water molecules. Hydrophobic groups tend to be insoluble in water.

hydroponics The science of growing plants in a synthetic, aqueous nutrient medium.

hydrops fetalis A type of thalassemia in which all four alpha-chains missing in the hemaglobin molecule are missing as a result of a defect in the DNA that codes for these proteins. Infants carrying the defect almost inevitably die at or before birth.

hydroxyapatite A form of calcium phosphate ($CaPO_4$) that is used for separation of single- and double-stranded nucleic acids by column chromatography in which the double-stranded form is preferentially retained on the hydroxyapatite matrix.

5-hydroxymethylcytosine An unusual derivative of cytosine that is used in place of cytosine in the DNA of the bacteriophage T4. This base protects the T4 DNA from nucleases, which the bacteriophage produces during its growth cycle.

hydroxyproline A derivative of the amino acid, proline, that is found in collagen and that helps to stabilize the molecule. Because the formation of hydroxyproline is dependent on vitamin C, a deficiency of the vitamin is manifest by a weakening of the collagen fibers, which results in the skin lesions characteristic of the disease scurvy.

hygromycin An aminocyclitol antibiotic produced by *Streptomyces hygroscopicus* with activity against both prokaryotes and eukaryotes. Hygromycin B acts by causing aminoacyl-tRNAs to be misread and also introduces errors into the ribosomal translocation process. The hygromycin B phosphotransferase (Hph) gene confers resistance to the antibiotic. By including the Hph gene in tranfection vectors, hygromycin can be used to select, from a large cell population, the subpopulation of cells containing a particular transfected gene.

hyperchromicity, hypochromicity The change in the optical density of a solution of a nucleic acid upon denaturation or renaturation. Denaturation of double-stranded DNA to the single-stranded state results in an increase in the optical density of the sample (hyperchromic shift) at 260 nm, and a reduction in optical density (hypochromic shift) accompanies renaturation of single strands to the double-stranded form.

hyperimmune A state in which an extreme immune response is provoked by an antigen present in quantities that are normally not effective in stimulating the immune system.

hypermutable phenotype Bacterial strains lacking the ability to remove uracil molecules that aberrantly arise in the cell DNA in place of cytosine. The persistence of uracil results in high rates of mutation in strains that carry this deficiency.

hyperproliferation A state in which cell division occurs at a greater-than-normal rate.

hypersensitive site, DNase I A region of the chromatin at which the DNA is accessible to the enzyme DNase I. Experimental evidence has shown that many of these sites are places where gene activity is regulated.

Hypoxanthine

hypervariable region A region of the immunoglobulin gene that shows a high degree of variability in sequence from one antibody to the next.

hypha A long, branching filament of connected cells in fungi. Hyphae may be either segmented, in which individual nuclei are separated from one another by a cell wall, or nonsegmented, in which many nuclei share a common cytoplasm (multinucleated).

hypoxanthine A purine intermediate in the degradative pathway of adenosine. Hypoxanthine can also serve as a precursor for nucleic acid synthesis by a series of reactions known as the salvage pathway.

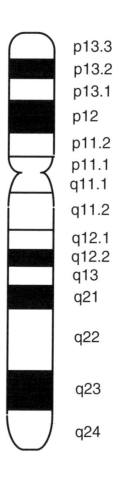

Ia antigen Proteins encoded by the I locus of the mouse H-2 histocompatibility complex.

ICE *Il*-1β *converting enzyme*; an alternative name for caspase 1 that is derived from the fact that it can convert pro IL-1β to its active form by proteolytic cleavage.

Ich-3 (caspase 5) A cysteine protease member of the Ced3/ICE family. Ich-3 is activated through cleavage by the serine protease, granzyme B, in cytotoxic T cells, which leads to apoptosis in these cells. Activating cleavage of Ich-3 generates the subunits p20 and p10, which share a high degree of homology to other caspases.

ideogram A type of chromosome map based on a pattern of chromosomal bands called G banding produced by staining with Geimsa. Each homologous chromosome pair will show a unique pattern of G bands that can be depicted by an internationally agreed upon schematic.

idiotope An antigenic peptide sequence located on the IgG antibody molecule near the antigen binding site; a specific idiotope is associated with a specific antigen binding site so that the same number of idiotopes exist as there are different antibodies.

idiotype The set of idiotopes on an IgG molecule.

i-gene The bacterial gene that codes for the *lac* operon repressor protein.

IgG See IMMUNOGLOBULIN.

p13.3
p13.2
p13.1
p12
p11.2
p11.1
q11.1
q11.2
q12.1
q12.2
q13
q21
q22
q23
q24

Chromosome 16

illegitimate recombination A rare event in which recombination occurs between two DNAs at an apparently randomly chosen site(s).

imaginal disk Disk-shaped structures symmetrically located on either side of the embryos of fly larvae that give rise to certain adult structures, for example, legs, eyes, and wings.

immortalized cells Cells that continue to divide indefinitely in tissue culture. Immortalization is a defining property of transformed cells; cells expressing the properties of cancer cells (i.e., the transformed phenotype).

immune response The proliferation of specific antibodies or cells of the immune system, such as macrophages, T and B lymphocytes, and so on, in response to a foreign antigen.

immune system The collection of all the cells and tissues (thymus, spleen, lymphocytes) that are involved in providing an immune response.

immunization The injection of an animal with an immunogen sometimes in combination with an ADJUVENT to induce the production of antibodies that will specifically bind to the immunogen injected.

immunoadsorbant A solid matrix to which antibodies are attached and that is used to purify the antigens from a biological preparation by allowing the antigens to bind to the matrix-bound antibodies.

immunoaffinity chromatography A technique for purifying antigens by passing a biological preparation or extract over a column containing an immunoadsorbent.

immunoassay Any technique for determination of the quantity of an antigen based upon the binding of the antigen to its specific antibody. See COMPLEMENT-FIXATION TEST, HEMAGGLUTINATION-INHIBITION ASSAY, and RADIOIMMUNOASSAY.

immunoblotting A technique for determination of the presence and properties of an antigen by reaction of labeled antibodies to the antigen after the antigen has been separated according to size and/or charge by gel electrophoresis and then transferred to a membrane.

immunodeficiency The state of impairment of the immune system resulting in the inability or lowered ability of the immune system to mount an immune response to a cell or particular antigen.

immunodiffusion A technique for determining the presence of an antigen by allowing an antigen and an appropriate antibody to diffuse into a gel where an immune precipitate forms at the point where antigen and antibody meet.

immunoelectrophoresis A variation of the immunodiffusion technique in which the antigen is subjected to electrophoresis in a gel that is then used for assay by immunodiffusion.

immunofluorescence A technique for visualizing structures in a cell or a tissue through the use of antibodies attached to a fluorescing label that bind to antigens in the target structures.

immunogen Any substance capable of provoking an immune response.

immunogenicity The property of being an immunogen, that is, the property of being capable of provoking an immune response.

immunoglobulin Any of the globular serum proteins secreted by cells of the immune system for the purpose of dealing with foreign antigens. Immunoglobulins are divided into five classes: IgM, IgG, IgA, IgD, and IgE.

immunoglobulin gene switching Developmental changes in the class of immunoglobulin (such as from IgM to IgG) produced by a single lymphocyte as the result of the expression of different genes. Gene switching is accomplished by recombination and changes in the way

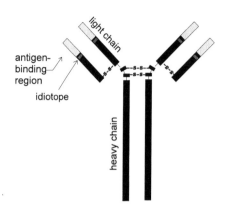

antigen-binding region

idiotope

light chain

heavy chain

Immunoglobulin

RNA is spliced so that the products of different portions of the immunoglobulin genes are fused to one another in different arrangements.

immunolabeling The technique of labeling molecules and/or biologic structures through the use of antibodies bound to other molecules that serve as labels for the antibody-antigen complex.

immunology The study of the immune system.

impermeable junction A term used to describe any cell-cell junctional complex that connects cells together so that even small molecules cannot diffuse between the connected cells. Impermeable junctions are generally used as synonymous with tight junctions.

inclusion bodies Clumps of material that accumulate in the nucleus or cytoplasm of virus-infected cells. Inclusion bodies consist of aggregates of viral structural components such as virion proteins.

indirect end labeling A technique for demonstrating the unique nature of a DNA fragment by hybridizing the fragment to a probe representing an end piece of the sequence that it is proposed to rep-

resent. For example, if it is supposed that the nucleosome-binding DNA sequence is a unique sequence, then a probe may be constructed from the end portion of this sequence, and any DNA that is isolated from a nucleosome should hydridize to the probe. See NUCLEOSOME PHASING.

inducer Any agent, chemical or physical, that brings about the expression of a gene or gene cluster.

inducible enzyme An enzyme whose expression requires the presence of a specific inducer.

induction, gene Transcription of a gene(s) brought about in the presence of a specific agent that is referred to as the inducer.

induction, phage The production of lytic bacteriophage in bacteria that carries a lysogenic prophage, brought about by treatment with some chemical of physical agent.

informative meiosis A mating that generates a crossing over between two genetic markers so that linkage between the two markers can be determined. Informative meioses have been used in the genetic analyses of genetic disease such as cystic fibrosis to establish linkage between RFLPs known to be associated with the disease.

infrared spectroscopy An analytical technique for determination of the chemical structure of an unknown by observing how much light is absorbed by a sample of the unknown at different light wavelengths in the infrared spectrum (>1,000 nm). The absorption of light in this region of the spectrum reflects, and is analyzed in terms of, the presence of certain types of chemical bonds, for example, C=O, C–OH, and C–C, that produce characteristic absorption patterns.

inhibitory postsynaptic potential A membrane potential across the postsynaptic membrane in a neuron that inhibits the generation of an action potential in that neuron. Inhibitory potentials are those in which the membrane is more polarized (hyperpolarized) than in the resting state.

initiation codon An AUG codon in messenger RNA that codes for the first amino acid (methionine) in a polypeptide. All polypeptides in both prokaryotic and eukaryotic cells begin with a methionine coded for by an initiating AUG codon during the process of translation.

initiation factor (IF) Certain proteins that catalyze the formation of the initiation complex between the mRNA and the ribosome in the process of translation.

INK4 A gene locus that codes for a group of tumor suppressors designated INK4A (CDKN2A), INK4B (CDKN2B), INK4C (CDKN2C), and INK4D (CDKN2D). The INK4 proteins act to inhibit the progression of cells into S phase by blocking the activity of the cyclinD-cdk4/cdk6 complex. Two of the INK4 tumor suppressors, 4A (p16^{INK4A}) and ARF (p19^{ARF}), are produced by alternative splicing of a single transcript that acts upon different secondary tumor suppressors, Rb and p53, respectively. p16^{INK4A} acts directly to prevent phosphorylation of Rb by the cdk4/cdk6 kinase complex, while p19^{ARF} is an antogonist of mdm2, which in turn leads to enhancement of p53 activity.

inosine The purine base in inosine monophosphate (IMP), the nucleotide from which the normal purine nucleotides, adenosine monophosphate (AMP) and guanosine monophosphate (GMP), are synthesized biologically.

inositol A five-carbon sugar that is a major constituent of the phospholipids found in cell membranes. The release of inositol in the cell membrane resulting from the action of certain growth factors and other effectors of cell growth and differentiation is now believed to be an important step in the pathway by which cells respond to extracellular signals.

insert In molecular biology, the term *insert* refers to a piece of DNA ligated into a specific site in a vector for molecular cloning. The resulting recombinant molecule composed of vector and insert is designed to allow replication of the recombinant in an appropriate host.

insertional inactivation The loss of gene activity as the result of insertion of a segment of DNA into a region critical to the expression of the gene. Insertional inactivation of genes coding for resistance to an antibiotic by insertion of a cloned DNA is often used as a means by which bacteria containing the recombinant plasmid are selected.

insertion mutation A mutation caused by the insertion of a nucleotide or oligonucleotide sequence into the coding region of a gene. Insertion of oligonucleotide sequences containing any number of nucleotides not evenly divisable by three will result in a frame shift.

in situ In the natural setting or environment; generally, the intact tissue as opposed to a biochemical extract or preparation.

Inosine

131

in situ hybridization Nucleic acid hybridization carried out on sections of intact tissue or chromosomes.

insulin A polypeptide hormone secreted by a part of the pancreas known as the islets of Langerhans, which controls the entry of glucose into cells. A deficiency of insulin production is the underlying cause of diabetes.

int 1 gene An oncogene activated by the nearby integration of the mouse mammary-tumor virus (MMTV) that produces mammary tumors in mice. The int 1 gene homologue in *Drosophila melanogaster* has been shown to play a crucial role in wing development, suggesting that the mammalian int 1 gene may play a regulatory role in development.

intasome A complex between bacterio-phage DNA and two proteins (Int and IHF) that is required for bacteriophage DNA to integrate into the bacterial host DNA when a bacteriophage enters into lysogeny.

integral membrane protein (intrinsic protein) A protein that is integrated into the cell membrane; that is, it penetrates into the membrane lipid bilayer. Most eukaryotic receptors (e.g., RTKs) are integral membrane proteins.

integrant A cell in which a transfected gene has become stably integrated into the genome of the recipient. See TRANSFECTION.

integrase The enzyme that catalyzes the site-specific recombination of lambda bacteriophage DNA with the bacterial host DNA that results in integration of the bacteriophage DNA.

integration In molecular genetics, the insertion of a foreign DNA into the genome of a recipient cell. The term is most often applied to the integration of viral DNA into the genome of an infected host, for example, integration of the prophage in a bacterium infected by a bacteriophage.

integrin-linked kinase (ILK) An enzyme associated with the cytoplasmic domains of integrins and also attached to the actin cytoskeleton at regions of the cell membrane called focal adhesions (FAs). Integrin-linked kinases act to transduce

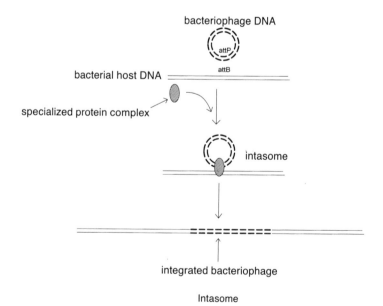

Intasome

signals from ligand-activated integrins by phosphorylation of other signal transduction components; for example, Akt and GSK-3.

integrins A very large family of transmembrane proteins that act as adhesion molecules in cell-cell and cell-extracellular matrix interactions. The integrins are dimeric glycoproteins comprised of one α (16 identified) and one β subunit (eight identified) to form at least 22 different integrins. The integrins can bind to a variety of extracellular matrix components as ligands, particularly fibronectin and laminin. Integrins also function as receptors in signal transduction, which, upon ligand binding, can stimulate signaling through a number of signaling pathways, involving activation of integrin-associated kinases such as FAK (focal adhesion kinase) and MAPK kinase.

interbands Lightly staining regions in polytene chromosomes.

intercalate In biochemistry, to fit a molecule in between biomolecules that are part of an array. The term is commonly applied to certain dyes that stain nucleic acids by inserting themselves in between the purine and pyrimidine bases arrayed along the nucleic acid backbone. See ETHIDIUM BROMIDE.

interferon(s) A group of small glycoproteins produced by virus-infected cells that act to inhibit viral infection. The interferons are heterogeneous both with respect to structure and mode of action. The genes for various interferons have been cloned and have been tested as therapeutic agents for various diseases, including Kaposi's sarcoma in HIV-infected individuals. Gamma interferon has been found to induce MHC class II antigens in B cells, macrophages, and endothelial cells.

interferon regulatory factors (IRFs) A small group of transcription factors that are activated by interferons. Interferons mediate the various functions of interferons, including cytokine signaling, antiviral defense, regulation of cell growth, and immune activation by inducing the transcription of genes.

interleukins (lymphokines) A subgroup of soluble proteins called cytokines. Interleukins act as signal molecules to control immunologic and inflammatory responses. At present there are at least 18 known interleukins, most of which have only recently been discovered. Most cytokines, with the exception of IL-1, transmit intracellular signals via the Jak-STAT signaling pathway. Interleukins are mostly secreted by white blood cells (leukocytes) and stimulate a variety of responses in other leukocytes, including stimulating proliferation, inducing or inhibiting the release of other cytokines, and activating themselves. The cellular targets of the known interleukins (IL-x) and their actions are summarized below:

- IL-1—Macrophages stimulate secretion of IL-2 from T cells.
- IL-2—Helper T cells cause activated T cells and B cells to divide. Induces antibody synthesis
- IL-3—T cells induce proliferation of other leukocytes; induce hematopoietic stem cells to differentiate into leukocytes.
- IL-4—Helper T cells stimulate growth of T cells and B cells. A factor in the production of IgE antibodies
- IL-5—Helper T cells stimulate growth of B cells and eosinophils. Induce proliferation of B cells that produce IgA antibodies
- IL-6—T cells and macrophages induce B-cell differentiation together with alpha-interferon.
- IL-7—Stromal cells induce differentiation of lymphoid stem cells into progenitor T cells and B cells.
- IL-8—T cells and neutrophils help to recruit these cells to the site of an inflammation.
- IL-9—Helper T cells stimulate growth.

- IL-10—Produced by T cells, B cells, monocytes; represses the production of gamma interferon, TNF-alpha, IL-1, and IL-6
- IL-11—Plasmacytoma cells stimulate growth.
- IL-12—Produced by macrophages and B cells; stimulates T cells and natural killer cells to proliferate
- IL-13—Produced by T cells; induces B-cell differentiation and inhibits production of inflammatory cytokines
- IL-14—Produced by T cells; enhances memory B-cell proliferation
- IL-15—Stimulates T-cell proliferation
- IL-16—An adhesion molecule and activator for T cells. Plays a role in asthma and autoimmune diseases
- IL-17—T cells activate neutrophils.
- IL-18—Stimulates the release of interteron gamma and Th1 cytokines

internal guide sequence A nucleotide sequence in group I introns that plays a key role in precisely localizing the 3′ splice site during the process of RNA splicing. The mechanism of splice-site localization involves base pairing between the internal guide sequence and sequences at the 5′ splice site.

intermediary metabolism That part of biochemistry that deals with how energy is derived from nutritive biomolecules and how that energy is used in the metabolism of other biomolecules.

intermediate filaments A type of filament that makes up one kind of cytoskeleton in mammalian cells. Intermediate filaments are distinguished by virtue of their size (approximately 8–10 nm in diameter), which places them in a range intermediate between the actin and microtubule type cytoskeletal filaments. Intermediate filaments are divided into six classes: keratins, desmins, vimentins, glial filaments, neurofilaments, and nuclear lamins.

interphase The period between mitoses. The interphase is divided into the G1, S, and G2 phases of the cell cycle.

intracellular Within the cell.

intramuscular Located in or directly administered to muscle tissue.

intraperitoneal Located in or directly administered to the cavity between the internal organs of the abdomen and the abdomen wall. Intraperitoneal inoculation of transformed cells into mice is widely used to promote the growth of transformed cells or to derive large quantities of substances they secrete, for example, monoclonal antibodies.

intravenous In a venous blood vessel (vein); for example, the route of an intravenous injection.

intrinsic factor A glycoprotein secreted by the parietal cells of the lining of the intestines (the gastric mucosa) that plays a critical role in the absorption of vitamin B_{12} (cobalamin) in the intestine. The inability to produce or utilize the intrinsic factor leads to the condition known as pernicious anemia caused by vitamin B_{12} deficiency.

intron The nucleotide sequences in between the exons of a gene. Introns in the genomic DNA are copied during transcription but are removed by the process of splicing.

inulin A long polysaccharide composed largely of repeating fructose subunits. Because it is a large molecule and largely inert, inulin is used experimentally to control osmotic flow across membranes and as a diagnostic aid for kidney function.

inversion The alteration of cellular DNA sequences in which the orientation of a DNA segment is reversed; placed into an inverted orientation. Inversions are frequently caused by the movement of transposons, especially in cells carrying two copies of a transposons in opposite orientation to one another.

invertase The enzyme β-D-galactosidase that catalyzes the cleavage of lactose into the monosaccharides, glucose, and galactose. The enzyme derives its name from the fact that action of the enzyme causes the resulting sugars to undergo conversion to the opposite optical isomer (i.e., from D form to L form). A lack of this enzyme is responsible for lactose intolerance.

inverted terminal repeats Segments of DNA at the ends of an insertion element, such as a transposon, that are inversions of one another, for example, AACGCTTCG and GCTTCGCAA. Inverted terminal repeats are essential for the transposability of the insertion sequence.

in vitro Biological material outside of the normal setting, for example, in cell or tissue culture or in cell or tissue extracts.

in vitro fertilization The process of carrying out fertilization with egg and sperm outside the body in a petri dish or a similar vessel. Fertilization is observed under the microscope, and the fertilized egg is then implanted in the uterus for normal fetal development and birth. In vitro fertilization is a procedure widely used to acheive pregnancy in certain types of infertility.

in vitro mutagenesis Mutagenesis of cells in vitro, that is, by exposure of cells in tissue culture to mutagenic agents.

in vitro packaging The formation of the viral coat around a viral nucleic acid using biological preparations or extracts in an artificial environment, for example, a mixture containing biological extracts, salts, and necessary biological molecules.

in vitro protein synthesis The synthesis of a polypeptide using mRNA that codes for that polypeptide in an artificial mixture containing ribosomes and all the molecular components necessary to reproduce the normal process of polypeptide biosynthesis as it occurs in the intact cell.

in vitro transcription/translation The synthesis of both mRNA and its encoded protein in an artificial mixture containing the appropriate DNA, ribosomes, and all the molecular components necessary to reproduce the normal processes of transcription and polypeptide biosynthesis as they occur in the intact cell.

in vivo In the intact cell or tissue.

iododeoxyuridine (IUdR) A synthetic nucleoside that is an inducer of Epstein-Barr virus (EBV) gene expression in EBV-infected cells that otherwise produce no viral proteins.

ion-exchange chromatography A technique for separating substances in a mixture, based on passing the mixture through a column containing a matrix that binds the substances in the mixture according to their electric charge.

ion-exchange resin A material used to separate substances in a mixture by ion exchange chromatography. Ion-exchange resins are generally in the form of beads that are composed of an inert polymeric substance, such as cellulose or sepharose, that is covalently attached to molecules that are electrically charged.

ionizing radiation Any electromagnetic radiation that can knock electrons from molecules, thereby producing ions. Alpha, beta, gamma radiation, and X-rays are all considered to be ionizing radiation.

ionophore Any of a number of relatively small organic molecules that act to allow the osmotic passage of ions and other molecules across cell membranes that would otherwise be impermeable to them. Ionophores have been widely used experimentally to study the function of ion gradients across membranes, for example, the Na^+–K^+ gradients in neurons.

ion-selective electrode An instrument for measuring the concentration of ions of one specific atom in a solution by measuring the current produced when a probe containing the oxidized or reduced form of the atoms to be tested is immersed in the solution.

IPTG *Isopropyl-β-D-thiogalactopyranoside*; a synthetic analog of naturally occurring galactosides, for example, lactose. IPTG is widely used in place of lactose as a potent inducer of the *lac* operon and, unlike lactose, is not acted on by the enzyme, alpha-galactosidase, that it induces.

islets of Langerhans Small clusters of cells scattered throughout the pancreas that produce the hormones insulin and glucagon. Because loss of ability of the islet cells to produce sufficient quantities of insulin is a cause of one form of diabetes (insulin-dependent diabetes), introduction of insulin genes targeted to the islet cells is a strategy being developed to treat this disease through gene therapy.

isoaccepting tRNAs Different tRNAs that carry the same amino acid. See ADAPTOR MOLECULE.

isoantigen (alloantigen) An antigen that is produced by only some members of a species but not others and that is capable of eliciting an immune response in the individuals of the species that lack the antigen. Blood group antigens are examples of alloantigens.

isoelectric focusing A variation of polyacrylamide gel electrophoresis in which proteins in a mixture are separated on the basis of their individual isoelectric points.

isoelectric point The pH at which the net charge (the sum of the charges on all the individual subgroups) on a molecule is exactly zero.

isoleucine An amino acid whose side chain is

$$-CH_2-CH-CH_2-CH_3$$
$$\backslash$$
$$CH_3$$

isomerase Any of a class of enzymes that catalyzes the rearrangement of atoms in a molecule.

isoprene An organic molecule that appears in polymeric form in a number of important molecules that act as intermediates in electron transfers in various metabolic reactions in intermediary metabolism, for example, ubiquinone (coenzyme Q) and chlorophyll. The structure of isoprene is

$$CH_2=CH-C=CH_2$$
$$\backslash$$
$$CH_3$$

isoschizomer Any one of a group of different restriction enzymes that recognizes the same nucleotide sequence.

isotonic point The point at which the concentration of all solutes in a solution results in an osmotic pressure across a membrane that is exactly the same as the osmotic pressure of a reference solution. This term or synonyms for it are frequently used to describe solutions that can be introduced into a biological system without causing osmotic lysis of the cells; for example, solutions that are injected into the bloodstream without causing hemolysis are called isotonic solutions.

isotope One of any alternative forms of an element that differ from one another in terms of the number of neutrons in the nuclei of their atoms.

J

jagged 1 (JAG1) A gene that codes for a protein called jagged 1. The jagged 1 protein binds to notch proteins that are receptors on the surfaces of certain cells. The formation of the jagged 1–notch complex sets in motion a series of signaling reactions that controls the development of various cell types in an embryo, including the heart, liver, eyes, ears, spinal column, and blood cells. Certain mutations in JAG1 can cause Alagille syndrome, which is characterized by missing or narrowed bile ducts in the liver, heart defects, and characteristic facial features. Other mutations in JAG1 result in various other abnormalities, including a heart defect called Tetralogy of Fallot, deafness, and a liver condition called extrahepatic biliary atresia (EHBA). The JAG1 gene is located on chromosome 20 at gene map locus p12.1-11.23.

JAKs *J*anus *k*inases; tyrosine kinases that are one of the two main components of the JAK/STAT signaling pathway. JAKs are activated by certain receptors for cytokines, lymphokines, and growth factors. Ligand binding causes dimerization of the receptors, which then act to phosphorylate, and activate, JAKs. The activated JAKs in turn phosphorylate the cytoplasmic ends of the receptor, which then serves as a docking site for STATS. Mutations in the genes for JAKs are associated with several leukemias, including polycythemia vera, thrombocythemia, and myeloid metaplasia.

JC virus A human virus member of the Papova group. JC virus, which is closely related to the monkey virus, SV40, was first isolated from the brain tissue of a patient with progressive multifocal leukoencephalopathy (PML), which the virus is believed to cause.

jnk *J*un *N*-terminal *k*inase; the protein is a member of the MAP kinase family. MAP kinases act as an integration point for multiple biochemical signals and are involved in a wide variety of cellular processes, such as proliferation, differentiation, transcription regulation, and development. This kinase is activated by various cell stimuli and targets specific transcription factors, and thus mediates immediate-early gene expression in response to cell stimuli. The activation of this kinase by tumor-necrosis factor alpha (TNF-alpha) is found to be required for TNF-alpha-induced apoptosis. This kinase is also involved in ultraviolet radiation–induced apoptosis, which is thought to be related to cytochrome c-mediated cell-death pathway. Studies of the mouse counterpart of this gene suggest that this kinase plays a key role in T-cell proliferation, apoptosis, and differentiation.

joining gene (j gene) A DNA segment in the immunoglobulin gene cluster that joins the constant and variable immunoglobulin gene regions during B cell maturation. During the maturation process, antibody diversity is generated by joining a constant region with a large number of different variable regions.

jun An oncogene transduced by a chicken retrovirus that causes fibrosarcoma tumors. The jun proto-oncogene

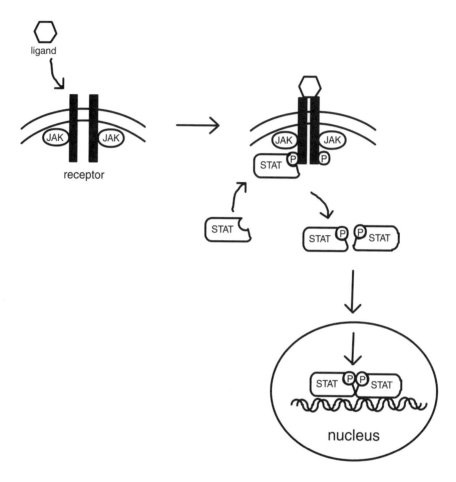

The JAK-STAT pathway

has been found to share identity with the transcription regulation factor, AP-1, and apparently exerts its oncogenic effects by inducing aberrant gene transcription.

junin virus A member of the tacaribe subgroup of the arenaviruses. Junin virus, which causes hemorrhagic fever, is carried by bats and rodents.

K

Kallmann syndrome protein The protein implicated in causing Kallman syndrome, a congenital condition characterized by anosmia, small genitalia, and sterile gonads. Kallman syndrome results from the failure of gonadotropin-releasing hormone-secreting neurons to migrate into the brain from the olfactory placode during development. The defect is due to mutations in a gene called KAL-1, which codes for a protein called anosmin-1. Sequence analysis predicts anosmin-1 to be a cell-adhesion protein that may mediate axon-axon adhesion. KAL-1 is at gene map locus Xp22.3.

kanamycin A broad spectrum antibiotic active against both Gram-positive and Gram-negative bacteria and a number of types of mycoplasma. Kanamycin is an aminoglycoside derived from the soil bacterium *Streptomyces kanamyceticus*.

kappa light chains One of the two types of light chains in IgG antibody molecules. The lambda-type light chain is distinguished from the kappa-type on the basis of antiserum to light chain proteins; immunoreactivity of the light chains show that they are of either the kappa or lambda type, but not both. Kappa and lambda light chains are secreted by myeloma tumors and appear in the urine of myeloma patients where they were originally called Bence-Jones proteins. This discovery helped to elucidate the structure of the IgG molecule. See IMMUNOGLOBULIN.

karyogamy The fusion of two nuclei; for example the fusion of pronuclei that occurs during fertilization of the egg.

karyokenesis The division of the nucleus during cell division.

karyoplast The cell fraction containing the nucleus surrounded by a small ring of cytoplasm in cells enucleated by treatment with cytochalasin.

karyotype The characterization of the chromosomes of a cell type normally including chromosome morphology, chromosome number, chromosome banding patterns, and any abnormalities of these characteristics.

keratin A type of intermediate filament found almost exclusively in the epithelial cells of mammals and the feathers of birds. The mammalian keratin proteins are a large and diverse family of proteins of which more than 30 different polypeptides are known, of which only a small number is required for filament formation in any one epithelial cell subtype.

keratinocyte Any mammalian epithelial cell.

keratinocyte growth factor (KGF) A cytokine that stimulates the growth of epithelial cells (keratinocytes). Human KGF-1 (also called FGF-7) is produced by recombinant DNA techniques (FGF-7) as a polypeptide chain of 164 amino acids and is considered to be one of a class of synthetic drugs called biological response modifiers. KGF is being tested as treatment for mouth sores (oral mucositis) caused by radiation or chemotherapy.

ketone body Any of either acetone, acetoacetate, or beta-hydroxybutyrate, produced as the result of the accumulation of acetyl CoA as might be caused by blockage of normal glucose metabolism via the Krebs cycle, for example, as occurs in diabetes.

ketose Any sugar containing a keto (as opposed to an aldo) group:

$$
\begin{array}{l}
R_1 \\
\ \ \backslash \\
\ \ \ C{=}O{\leftarrow}\text{---------}\ \text{keto group} \\
\ \ / \\
R_2
\end{array}
$$

Khorana, Har Gobind (b. 1922) A biochemist who carried out experiments using synthetic oligoribonucleotides to program synthesis of peptides. The amino acid composition of the resulting peptides allowed assignment of amino acids to specific codons and thus the deciphering of the genetic code. For this work he received the Nobel Prize in medicine in 1968.

kilobase (kb) A measure of the length of a nucleic-acid-strand equivalent to 1,000 nucleotides.

kilodalton (kD) A measure of the size of large biomolecules, but generally applied to proteins, that is equivalent to the molecular weight of the molecule divided by 1,000. For example, a protein of 125,000 molecular weight correspond to 125 kD.

kinase A class of enzymes that catalyze the transfer of a phosphate group from one substrate to another. Phosphorylation is a means of regulating the activities of a number of other enzymes, and so kinases may control a wide variety of biochemical pathways through a single phosphorylation.

kinesin A protein involved in the movement of small vesicles along a microtubule in the axons of nerves and perhaps in other cell types. Kinesin is an ATPase that uses the energy released by ATP hydrolysis to induce movement.

kinetochore A dense structure in the centromeric region of a chromosome to which the spindle fibers are attached.

kinetochore fibers The microtubules extending between the kinetochore of the chromosome(s) and polar bodies that pull the chromosomes to the poles of a dividing cell during mitosis.

kininogen A factor in the intrinsic blood-coagulation (clotting) pathway. Kinogen is one of two factors reqired for activation of factor XII (Hageman factor).

kinins (cytokinins) Plant hormones that, in combination with auxins, stimulate cell division and differentiation in a variety of plant tissues. Chemically, the cytokinins are purines with terpenoid side chains.

kirromycin An antibiotic that acts by inhibiting protein synthesis on the bacterial ribosome. Kiromycin forms a complex with a tRNA that prevents the elongation of the growing polypeptide chain.

Kirsten sarcoma virus (Ki-MuSV) A retrovirus that infects rats and that produces sarcomas and erythroleukemia in the infected host. Ki-MuSV carries the Ki-ras oncogene.

KISS-1 (KISS-1 metastasis-suppressor) A gene that has been shown to suppress chemotaxis, invasion, and metastasis of cancer cells in human melanoma and breast carcinoma cells. The KISS-1 gene codes for a peptide that binds to a G-protein-coupled receptor named hOT7T175. Lymph node metastasis is the most important predictor of prognosis in esophageal squamous-cell carcinoma (ESCC). Recently, KISS-1 was cloned as a human metastasis suppressor gene, and an orphan G-protein-coupled receptor (hOT7T175) was identified as the endogenous receptor of the KISS-1 product. However, the clinical importance of KISS-1 and hOT7T175 gene expression in ESCC remains unclear.

Kleibsiella An important nitrogen-fixing soil bacterium. See NITROGEN FIXATION.

Klenow fragment, enzyme A sub-fragment of DNA polymerase I produced by proteolytic cleavage of the 103 kD enzyme by subtilisin. The Klenow fragment is the larger (68 kD) of the two subfragments produced by subtilisin treatment. This fragment retains the normal DNA polymerase and $3' \rightarrow 5'$ exonuclease activities but lacks the $5' \rightarrow 3'$ exonuclease of the intact enzyme.

Klett unit A unit of light absorbtion used in measuring bacterial cell number in terms of the turbidity of bacterial liquid cultures. Klett units, measured at wavelengths between 490 and 550 nm, are approximately proportional to cell number during logarithmic growth.

Klinefelter's syndrome A chromosomal aberration involving the sex chromosomes in which cells contain two X chromosomes and one Y chromosome. Afflicted individuals have the physical appearance of males but are infertile and have underdeveloped testicles and other physical abnormalities.

Kornberg, Arthur (b. 1918) The discoverer of DNA polymerase I, which he isolated from *E. coli* bacteria in work dating from 1956. The discovery of the first enzyme known to be responsible for synthesis of DNA won him the Nobel Prize in physiology and medicine in 1959. In 1967 Kornberg stunned the scientific community by creating a biologically active virus ΦX174 from isolated components, the first time a virus had been produced in the laboratory.

Kornberg enzyme DNA polymerase I, the bacterial enzyme discovered by Arthur Kornberg. Originally believed to be responsible for the bulk of DNA synthesis. This was later disproved.

K-ras The oncogene carried by the Kirsten sarcoma virus. K-ras is a member of the ras oncogene family that contains Ha-ras and N-ras. The family is defined by base sequence homology of the members to one another.

Krebs cycle See TRICARBOXYLIC ACID CYCLE.

KRPs *k*inesin-*r*elated *p*roteins; a class of proteins related to kinesin and serving the same basic function, that is, providing the motor that moves cytosolic structures along microtubules. However, kinesin-related proteins differ from kinesins in that KRPs are involved in spindle assembly and chromosome segregation during mitosis, while kinesins are involved in the movement of transport vesicles.

Kruppel gene One of the homeobox gap genes identified in *Drosophila melanogaster*. In mutants of the Kruppel gene, abdominal segments are deleted from the larva.

KSS1, FUS3 Yeast MAP kinases that function in the same way in a pheromone-responsive signaling pathway that activates functions required for mating. Fus3 is activated by phosphorylation carried out by Ste7p. Either KSS1 or Fus3 can activate Ste12, a transcription factor, through phosphorylation.

lac **operon (lactose operon)** The operon that contains the three genes coding for proteins that are involved in the metabolism of the sugar, lactose, and other beta-galactosides: beta-galactosidase, galactoside permease, and galactoside acetylase.

lac **repressor protein** A protein (produced by the i gene) that blocks transcription of the genes in the *lac* operon by binding to the operator region of the *lac* promoter.

lactam antibiotics A class of antibiotics whose molecular structure is derived from the lactam ring, usually penicillin and its synthetic derivatives, for example, oxacillin, nafcillin, and benzyl penicillin (penicillin G).

lactamase An enzyme made by penicillin-resistant bacteria that cleaves the bond between NH and C=O in the lactam ring.

$$
\begin{array}{ccc}
\text{C=O} & \text{lactamase} & \text{COOH} \\
/\ \backslash & \text{------------->} & / \\
(CH_2)_n\text{- NH} & & (CH_2)_n\text{-}NH_2
\end{array}
$$

lactam ring Any molecule with the general structure

$$
\begin{array}{c}
\text{C=O} \\
/\ \backslash \\
(CH_2)_n\text{- NH}
\end{array}
$$

lactate dehyrogenase The enzyme responsible for catalyzing the conversion of pyruvate, in the presence of NADH (the reduced form of NAD), to lactic acid.

lactic acid The product formed from pyruvate by lactate dehydrogenase when sugars are oxidized under anaerobic conditions such as occur in muscle tissue after prolonged exercise or in bacteria that thrive in low-oxygen environments. See FERMENTATION.

lactic acid bacteria anaerobic bacteria that generate lactic acid during the process of sugar oxidation. The production of acid by lactic acid bacteria is responsible for the souring of milk and the sour taste of sauerkraut.

lactoperoxidase labeling A technique for labeling proteins on the outside of cell membranes with radioactive isotopes of iodine (e.g., ^{125}I). Lactoperoxidase catalyzes the transfer of iodine from iodoacetamide to the tyrosine residues of the protein to be labeled.

lacZ gene The gene that codes for the enzyme beta-galactosidase in the *lac* operon of bacteria. The lacZ gene is incorporated into many cloning vectors as a means of determining whether recombinant vec-

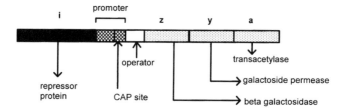

Lac operon

tors have been stably introduced into a recipient cell or clone of cells. In a population of transfected cells, expression of the lacZ gene in a transfectant(s) can be determined from the appearance of blue color in the cells when they are exposed to the synthetic galactoside X-gal (5-bromo-4-chloro-3-indoyl-β-D-Galactopyranoside), which is hydrolyzed by the enzyme to produce a visible blue precipitate.

Lafora disease Lafora disease is a form of epilepsy caused by an autosomal recessive mutation(s) in the EPM2A gene carried on chromosome 6q24. The EPM2A gene is believed to code for a protein tyrosine phosphatase. Lafora disease is a stimulus-sensitive myoclonic epilepsy that falls into two subtypes: Unverricht (earlier onset, more severe) and Lundborg (later onset, less severe). The disease is associated with inclusion bodies in the neurons of the brain (Lafora bodies); in particular in the cerebral and cerebellar cortex and in the brain stem. Lafora bodies are mostly glucose, 80–93 percent in $\alpha(1\text{->}4)$ and $\alpha(1\text{->}6)$, but also contain protein (ca. 6 percent).

lagging strand During DNA synthesis, the DNA strand whose synthesis begins at the replication fork. Because the replication fork is continually moving, DNA synthesis on the lagging strand must be continually reinitiated resulting in a series of contiguous but not covalently joined fragments. See OKAZAKI FRAGMENTS.

lag phase The period of slow growth between the time when a microorganism is inoculated into a nutrient broth and the time when those microorganisms enter into logarithmic growth.

lambda exonuclease An enzyme that catalyzes the cleavage of single nucleotides with 5′ phosphate groups from the 5′ ends of double-stranded DNA.

lambda light chain One the two types of light chains in IgG antibody molecules; a Bence-Jones protein. See KAPPA LIGHT CHAIN.

lamella A thin membrane or plate dividing certain biological compartments, for example, the region in between the cell walls of opposing cells of certain plants (i.e., the middle lamella).

lamellipoda A cytoplasm-containing protrusion or villus extending out of the leading edge of an animal cell during its movement along some substrate and oriented along the axis of movement.

laminar flow A uniform, eddy-free flow of air or liquid. Laboratory work requiring particular care to avoid chemical or microbial contamination is carried out in specialized, laminar flow hoods that maintain a continuous stream of filtered air.

laminin A protein component of the basement membrane that forms underneath epithelial cells where the cells adhere to the basement membrane or other substrate.

lampbrush chromosomes Enlarged chromosomes seen in amphibian oocytes during meiotic prophase. Lampbrush chromosomes are characterized by large protruding loops of transcriptionally active DNA.

lariat An intermediate stage in the splicing out of introns during the formation of mRNA in the nucleus. In a lariat, the intron of an mRNA precursor is cut at one end; the cut end then forms a covalent bond to a nucleotide in the interior of the intron to form the lariat structure. (See figure on next page.)

laser Light amplification by stimulated emission of radiation; laser light is created by causing a group of atoms to emit photons in synchrony. Lasers are used in varous types of molecular biological analyses, for example, flow cytometry.

Lassa fever virus An RNA-containing virus member of the Arenavirus family. First discovered in Nigeria, the virus is known to cause an acute infection characterized by fever, malaise, throat lesions, and pneumonia.

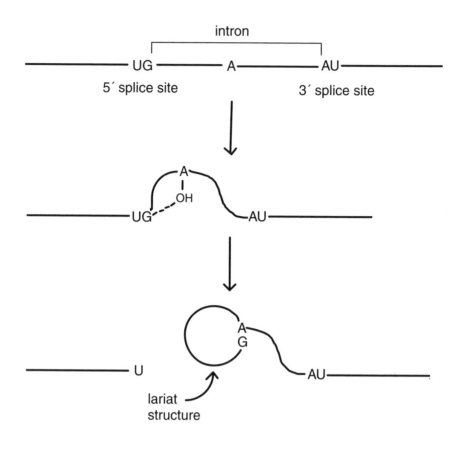

Lariat

late genes In viral infection, a set of genes that are always expressed late in the life cycle of the virus. Generally, the late genes code for proteins required for packaging of the viral DNA that is replicated early in the viral life cycle.

lateral meristem Meristem tissue lining the plant stem. Because the meristem is mitotically active, the lateral meristem is responsible for growth in diameter of the plant stem.

LAV *L*ympho *a*denopathy *v*irus; another name for HIV, the virus that causes AIDS. The designation LAV was used by the French discoverers of the virus at the Pasteur Institute.

LD-50 The dose of a test drug that is fatal to 50 percent of test animals to which it is administered.

LDL receptor A transmembrane protein in liver cells that binds specifically to low-density lipoproteins (LDLs), after which the LDLs are taken into the cells by endocytosis and degraded. Uptake of LDLs by this mechanism is one of the major pathways for the metabolism of LDL-associated cholesterol.

leader peptidase An integral membrane protein that catalyzes the cleavage of the leader sequence during the insertion of preproteins into their target membranes.

leader sequence In RNA transcripts for mRNAs, an RNA segment upstream from the part of the mRNA that encodes the protein. This term is also used to refer to a short segment at the beginning of a newly synthesized peptide that serves as a signal that the peptide is to be transported outside the cell, that is, secreted or deposited on the outer surface of the cell membrane. See SIGNAL SEQUENCE.

leading strand During DNA replication, the strand that is synthesized continuously in the 5'–3' direction.

leaky mutant A mutant microorganism in which the normal properties continue to be expressed at a low level or in which the mutation is only partly expressed.

Leber's Hereditary Optic Neuropathy (LHON) A genetic disease carried in the mitochodrial genome characterized by blindness resulting from degeneration of the optic nerve. Because the genetic defects are carried in the mitochondria, the pattern of inheritance is always maternal. The vast majority of LHON cases carry point mutations in the region of the mitochondrial genome that codes for the ND1 subunit of complex I of the electron chain (NADH-ubiqinone oxidoreductase), and, for this reason, the disease is believed to result from intracellular ATP deficiency.

lecithin The common name for the membrane phospholipid, phosphatidyl choline. Lecithin is believed by some to possess detergent properties capable of dissolving cholesterol present in arterial plaques.

lectins Plant-derived proteins that bind to specific polysaccharides. Lectins are used to label the cell surfaces of, or to agglutinate, cell types that bear the particular polysaccharides.

Lederberg, Joshua (b. 1925) A bacterial geneticist who is credited with discovering the phenomenon of transduction by bacteriophage. He also introduced the technique of replica plating to study genetic linkage and recombination in bacteria. His work led to the elucidation of the process by which various genes, especially those responsible for conferring antibiotic resistance, are transferred between bacteria. He was awarded the Nobel Prize in physiology and medicine in 1958.

legume A member of the pea family of plants. The roots of leguminous plants maintain a symbiotic relationship with nitrogen fixing bacteria so that the legumes are a rich source of nitrogen stored in the form of nitrates.

Lehninger, Albert (1917–1986) A biochemist famous for his work on the process of oxidative phosphorylation. Lehninger is best known for identifying the mitochondrion as the site at which the Kreb's cycle, fatty acid oxidation, and oxidative phosphorylation all take place. His classic work on the subject was published in 1948 together with the noted biochemist Eugene Kennedy, who was then his graduate student. The duo are also known for helping to perfect a technique for the isolation of mitochondria, which made their discoveries possible. Lehninger made a number of other contributions to the study of bioenergetics, including the elucidation of important differences in metabolism between normal and cancer cells.

lentivirus Literally, "slow virus." A type of retrovirus that produces a chronic, generally subclinical infection, for example, a visna virus that infects the brain cells of sheep. However, infection may invoke an immune response that can result in demyelination of the nerve cells. It is believed that some demyelinating diseases in humans may follow the same paradigm.

leptin A polypeptide protein hormone produced by adipose tissue that acts as a signal to the brain in the regulation of appetite and metabolism. Leptin binds to receptors in the hypothalamus, where it inhibits the actions of neuropeptide Y

(NPY) and agouti-related peptide (AgRP) and by enhancing the actions of alpha-melanocortin stimulating hormone (α-MSH). Binding of leptin to its receptor alters gene expression by the JAK/STAT signaling pathway, which is coupled to the receptor. As a natural regulator of appetite, leptin and/or components of leptin action are potential targets of new antiobesity drugs.

Lesh-Nyhan syndrome A genetic disease characterized by mental retardation and loss of coordination. The disease, which becomes manifest by the age of two, is due to the lack of the enzyme hypoxanthine-guanine phophoribosyltransferase (HGPRT) and was identified by Michael Lesch and William Nyhan in 1964.

lethal locus A genetic locus where mutations tend to prove lethal to organisms carrying the mutation(s).

lethal mutation Any mutation whose presence leads to the death of the organism in which the mutation is present.

leucine The amino acid that contains as a side chain:

$$-CH_2-CH \begin{array}{c} CH_3 \\ / \\ \backslash \\ CH_3 \end{array}$$

leucine zipper A structural motif on DNA binding regulatory proteins. Leucine-zipper proteins are helices in which hydrophobic amino acids are arrayed along one side of the helix. In the active dimeric form, leucine residues at approximately every seventh position are hydrophobically bonded to the leucine residues in a second helix so that the two helices are wound around each other in a coiled-coil arrangement.

leukemia A cancer of the blood characterized by the uncontrolled proliferation of white blood cells (leukocytes).

leukocyte A white blood cell. The white blood cells are largely composed of the cells of the immune system.

leukotrienes One of a class of hormonelike biochemicals called eicosanoids. Leukotrienes cause a number of wide-ranging effects on tissues distal from where they are produced, including the contraction of smooth muscle that lines the airways; overproduction of leukotriences leads to asthma attacks.

levorotatory isomer One of the two main classes of optical isomers. When polarized light is passed through a solution of a levorotatory isomer, the plane of polarized light is rotated in a counterclockwise direction from the point of view of the observer.

library A large set of DNA sequences from some specified source, for example, cDNAs or fragments of chromosomal DNA derived from a certain tissue or cell type. This term is also applied to peptides as in libraries created in automated peptide synthesizers that contain a large number of different peptides.

ligand Any molecule that is bound by a specific receptor for that molecule.

ligand-gated channels Channels in cell membranes that permit the passage of ions through the membrane when, and only when, a specific ligand is bound to its membrane receptor ("ligand gating"). Ligand-gated channels are the means by which nerve impulses are propagated when a neurotransmitter produced by one neuron binds to a receptor on the membrane of another neuron.

ligase A type of enzyme that catalyzes the formation of a covalent bond between the free ends of two nucleic acids.

ligase, DNA Catalyzes the linkage between a 5′ terminal phosphate group at the end of one DNA and a 3′ terminal hydroxyl group at the end of another.

ligation The chemical linking of the free ends of two nucleic acids to form one larger strand out of two smaller ones.

light chain Either of the two shorter peptides that make up an immunoglobulin molecule.

light-dependent reactions Those chemical reactions in photosynthesis that require light. The so-called light reaction(s) involves the capturing of light energy by pigments, including chlorophyll, in the form of high-energy electrons. The activated electrons are derived from water that is split during the process to liberate free oxygen.

light-harvesting complex (LHC) A complex of pigments associated with the photochemical reaction centers in chloroplasts that serve to collect the photons falling on an area of the thylakoid disk. Then the photon energy is transferred to a chlorophyll molecule in the form of an excited electron whose energy is then used to store the energy as ATP or NADPH (the reduced form of NADP). Light-harvesting complexes contain a variety of pigments that act as "antenna molecules" for collecting light. These pigments include carotenoids, phycobillins, phycoerythrins, and chlorophylls.

light-independent reactions Those chemical reactons in photosynthesis that are carried out in the absence of light. The so-called dark reactions are responsible for the trapping of carbon dioxide and water to create sugars.

lignin A phenolic polymer that forms a matrix in which the cellulose fibers of the plant cell wall are embedded. Lignin forms the "cement" that holds the fibers in place and also lends tensile strength to the cell wall.

limit digest The product of a degradative enzymatic reaction in which the substrate has been digested to the maximal extent possible. Limitation of the extent of digestion may be imposed by physical constraints, for example, clumping of the substrate material or failure of the substrate to enter into solution completely.

lincomycin (lincocin) An antibiotic produced by *Streptomyces lincolnensis*. The antimicrobial action of lincocin is due to its binding to the large subunit of the ribosome that prevents synthesis of peptides by blocking elongation of partially synthesized peptide chains.

LINEs Long *in*terspersed *e*lements; a repeated sequence of intermediate redundancy (~50,000 copies/genome in human cells) that acts as a mobile element using a reverse transcription mechanism similar to retroposons. LINEs contain genes encoding proteins similar to the reverse transcriptases of retroviruses and may contain other open reading frames as well. LINEs also differ from retroposons in that the coding regions are flanked by short direct repeats rather than long terminal repeats.

Lineweaver-Burk plot A plot of the reciprocal of substrate concentration (1/S) versus the reciprocal of the reaction velocity (1/V_o) for an enzyme catalyzed reaction; also called a double reciprocal plot. This type of plot is useful for graphical determination of K_m and V_{max} for a given enzymatic reaction.

linkage The degree to which any two genetic markers are associated with each other as determined by the frequency with which the two markers appear together in the same individual during genetic transmission (e.g., in offspring or in microorganisms in which the genetic markers have been transferred by transduction or conjugation). Genetic linkage is related to, but is not the same as, the physical distance between the two markers.

linkage disequilibrium A term to indicate the tendency of some genes, or genetic loci, to remain linked to one another; in other words to show patterns of inheritance that are statistically different from what would be expected if the genes recombined at random. If there is no statistical deviation in the distribution of genes from what would be expected from random recombination, the linkage disequilibrium would then be zero.

linkage map A genetic map based upon genetic linkage as opposed to actual physical distances.

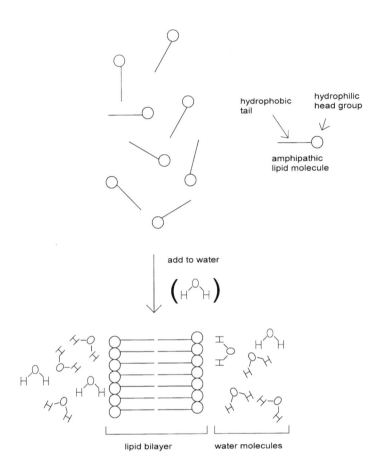

Lipid bilayer

linked genes Genes that are located within close enough physical proximity to one another that they appear together in virtually all organisms to which one or the other is transmitted.

linker A synthetic molecule that serves as a molecular bridge between two other molecules, for example, a synthetic oligonucleotide that joins together two DNA fragments.

linker-scanner mutations The replacement of a segment of DNA with a synthetic oligonucleotide (a linker) that contains mutations whose effect on the activity of a promoter is to be tested. The promoter sequence that has been altered by insertion of the linker is tested for activity in vitro, for example, by CAT assay.

linking number The number of complete helical turns in a circular DNA molecule. The linking number is related to the degree of supercoiling of the DNA.

linking-number paradox In experimental determinations of the number of winds of DNA around each nucleosome, the linking-number paradox refers to the discrepancy between the values obtained by different experimental methods. For example, digestion of nucleosomes by endonucleases gives about 1.8 DNA coils per nucleosome, but measurements based on supercoiling of DNA after the nucleosomes are removed give a value of about 1 coil per nucleosome. The discrepancy is of theoretical significance for models of the helical structure of the DNA molecule.

lipase Any of a variety of enzymes that catalyzes the breakage of an ester linkage in a lipid, thereby participating in the breakdown of that lipid.

lipid Any of a variety of oily, highly insoluble biomolecules associated with cell membranes and fatty tissues. Lipids are divided into six major classes: fatty acids, triglycerides, phosphatides, spingosines, waxes, and cholesterol derivatives.

lipid bilayer A thin film of regular thickness that, under certain conditions, is spontaneously formed by amphipathic lipids when they are placed in water. The plasma membranes of animal cells is formed, in large part, from naturally occurring bilayers of phospholipids.

lipopolysaccharide Lipids that are bound to polysaccharides. Lipopolysaccharides are found attached to the outsides of many cell membranes, including bacterial cell membranes.

lipoprotein Complexes of lipids and proteins. The most important and abundant examples of lipoproteins are the lipoproteins that function to transport fats in the blood and that are classified mainly in terms of their densities: high-density lipoproteins (HDLs), intermediate-density lipoproteins (IDLs), low-density lipoproteins (LDLs), and very-low-density lipoproteins (VLDLs).

liposome A synthetic structure composed of a lipid bilayer that completely encloses an interior cavity in which may be carried various substances of interest. Because the lipid bilayer of the liposome can spontaneously fuse with the lipids of

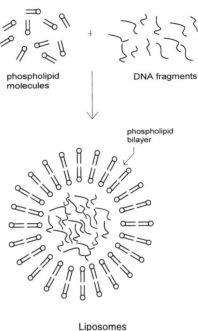

phospholipid molecules + DNA fragments

phospholipid bilayer

Liposomes

the cell plasma membrane, liposomes are being studied as vehicles specifically to introduce drugs or other bioactive chemicals directly into target cells.

lipotropic agents Lipid solvents or chemicals with hydrophobic properties that permit them to form nonpolar bonds with lipid molecules. Certain lipotropic agents have application as virus inactivating drugs.

locus The position occupied by a gene or a genetic marker on a chromosome.

LOD score A numerical value used to indicate genetic linkage between two markers. The LOD score is defined as
$$LOD = log_{10}(P_{linked}/P_{unlinked})$$
where

P_{linked} is the probability that the frequency with which two markers segrate from one another in the offspring of a mating could have occurred if the markers were linked.

$P_{unlinked}$ is the probability that the frequency with which two markers segrate from one another in the offspring of a mating could have occurred if the markers were unlinked from one another.

By convention, a LOD score of 3 is considered the threshold value for declaring that two markers are linked to one another.

logarithmic [growth] phase The term used to describe the growth of a culture of microorganisms under conditions where, on the average, one organism gives rise to two daughters at a consistent uniform rate. Logarithmic phase of growth follows lag phase during which some organisms may not reproduce or give rise to only one daughter during reproduction.

long QT syndrome (LQTS) An hereditary disorder, usually seen in children, that affects the heart's electrical rhythm such that the interval between the Q and T parts of the contraction wave (contraction-relaxation of the ventricles) is abnormally long. This leads to a rapid heart rhythm (arrhythmia) called *Torsade des pointes*, which can cause fainting in a matter of seconds. The biological bases of LQTS are mutations in genes that code for ion channels that cause the channels

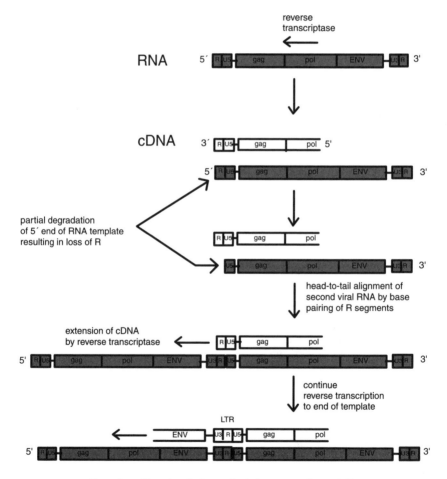

Formation of long terminal repeats during reverse transcription

to malfunction. At least five genes are known to be involved in LQTS: 1. the KCNQ1 gene on chromosome 11 (encodes a potassium channel), 2. the HERG gene on chromosome 7 (encodes a potassium channel), 3. the SCN5A gene on chromosome 3 (encodes a sodium channel), 4. the LQT5 gene (also called MinK, or KCNE1) on chromosome 21 (encodes part of the potassium channel together with the KCNQ1 gene product; mutations of this gene—like those of KCNQ1—produce the LQT1 form of LQTS), and 5. the MiRP1 gene on chromosome 21 (encodes part of the potassium channel together with the HERG gene product).

long terminal repeat (LTR) Specialized sequences located at the 5′ and 3′ termini of the genome of retroviruses. LTRs mediate the integration of the retrovirus into the host genome and regulate the transcription of the retrovirus genes. Because many LTRs contain strong enhancer elements, they are often used in synthetic constructs for the purpose of expressing foreign or engineered genes in recipient cells.

loop, looped domains A single-stranded region in either RNA or single-stranded DNA that forms a hairpin structure. The sequence between inverted repeating sequences forms looped domains because of base pairing between the inverted repeated.

low-density lipoprotein (LDL) A class of lipoprotein particles that carry cholesterol esters to cells that have specialized LDL receptors. Receptor-bound LDLs are taken up into the cell where the cholesterols are metabolized.

L-phase variants Bacterial variants that lack a cell wall. L-phase variants are produced under conditions in which cell-wall synthesis is inhibited, for example, the presence of penicillin. The L-phase variants pass through filters that retain normal bacteria.

luciferase An enzyme isolated from fireflies that is responsible for their characteristic light flashes. Luciferase catalyzes the decarboxylation of the substrate, luciferyl andenylate, to generate light. This reaction has been exploited as a nonradioactive means of labeling molecules for analytical purposes.

$$\text{luciferin} \xrightarrow[+AMP]{ATP} \text{luciferyl adenylate} \xrightarrow[oxygen]{luciferase} \text{oxyluciferin} +CO_2 + AMP + LIGHT$$

lucifer yellow A fluorescent dye used as an intracellular tracer to visualize living cells. Lucifer yellow and other tracer dyes have much application in neurobiology for discriminating individual nerve cells in a cluster.

luminometer A device for measuring emitted light. Luminometers have particular application as an analytical tool for determining the cellular content of ATP and other energy-containing nucleotide phosphates by utilizing biochemical reactions that emit light, for example, the ATP-luciferase reaction.

Luria, Salvador (1912–1991) A geneticist whose work with bacterial and bacteriophage mutants led to the elucidation of gene structure. He carried out a famous experiment in 1943 that demonstrated the inheritance of antibiotic resistance in bacteria and showed that the pattern of inheritance followed Darwinian rather than Larmarckian principles. Luria won the Nobel Prize in medicine in 1969.

luteinizing hormone (LH) A glycoprotein hormone released from the anterior pituitary. LH stimulates oocyte maturation and ovulation and progesterone secretion in the ovary.

luxury genes A vernacular term to describe the genes that are not essential for basic cell functions, but rather are made by only a certain cell type and that perform a function necessary to the functioning of the organism as a whole, for example, hemaglobin genes. See DIFFERENTIATION ANTIGEN.

lymphocyte The subclass of white blood cells responsible for carrying out the immune response. Lymphocytes are further subdivided into T and B lymphocytes and are found mostly in the thymus, lymph nodes, the spleen, and the appendix.

lymphokines (interleukins) Hormones secreted by certain antigen-processing cells of the immune system that cause T cells specific for the antigen to proliferate.

Lyon effect The silencing of expression of the genes on one of the two X chromosomes in the cells in a female animal. Silencing of the particular X chromosome occurs at random and is maintained in all the progeny cells.

lyophilization The removal of water from a frozen biological specimen by placing the specimen in a vacuum; often referred to as freeze-drying.

lysate The resultant mixture of cell debris and soluble cytoplasmic substances that results from mass cellular lysis of a cell culture or tissue.

lyse (lysis) A breaking open of cells by damage to the cell membrane by any mechanical, biological, or chemical agent.

lysine An amino acid with an amino butyl side chain: $-(CH_2)_4-NH_2$. The amino group makes lysine a basic amino acid.

lysogen A bacterial strain harboring a lysogenic virus (bacteriophage). See PRO-PHAGE.

lysogenic The property of certain bacteriophage mutants to integrate into and remain dormant in the bacterial host DNA. Although lysogenic bacteriophages do not immediately cause lysis, they may be induced to do so by various chemical agents or by ultraviolet light.

lysogeny The state of being lysogenic.

lysosome A cytoplasmic organelle in eukaryotic cells in which digestion of particulate material brought into the cell via endocytosis or phagocytosis occurs. Lysosomes are characterized by a highly acidic internal environment and the presence of enzymes (acid hydrolases) that carry out the digestive process.

lysozyme An enzyme that derives its name from its ability to cause certain bacteria to lyse. Lysozyme acts by cleaving polysaccharides in the bacterial cell wall. Lysozyme is used as a reagent in preparing DNA from bacteria and in the creation of protoplasts.

lytic cycle The events in the growth cycle of a lytic virus. The term is usually applied to bacteriophages.

lytic virus Any virus that as part of its life cycle, causes lysis of the host cells that it infects.

M

Machado-Joseph disease protein A neurological genetic disorder, also known as spinocerebellar ataxia-3, that is transmitted to offspring in an autosomal dominant. The disease is caused by degeneration of specific groups of neurons and is characterized by spinocerebellar ataxia, limited eye movement and rigidity. The gene involved in Machado-Joseph disease (MJD) is MJD1 (ataxin 3), and the associated disease-causing mutation is an expansion of CAG repeats in the coding region from the normal number of 13–36 to 68–79. While the function of the disease gene is unknown, it appears that pathogenesis is associated with misfolding and aggregation of the protein in the nuclei of affected neurons. MJD1 is located at gene map locus 14q24.3-q32.2.

macrolides A group of antibiotics with a large aliphatic ring structure with many hydroxyl and keto groups. Erythromycin is the best-known member of this group of antibiotics.

macromolecule A large molecule made up of many individual units such as amino acids in proteins, nucleotides in nucleic acids, and unit sugars in large carbohydrates.

macrophage A large phagocytic white blood cell. Macrophages travel in the blood but are capable of leaving the blood to enter tissue. They function as a defense by ingesting invading bacteria and other foreign cells as well as by removing particulate debris.

macroporous gels A chromatography matrix composed usually of cellulose or agarose gels that are useful for ion exchange or affinity chromatography or size fractionation of large proteins or substances such as viruses.

magic spot nucleotides Nucleotides with two or more phosphate groups on both the 3′ and 5′ carbon atoms that accumulate in bacterial cells during the stringent response.

main band The band that corresponds to the bulk of DNA, as opposed to satellite DNA, when a preparation of mammalian genomic DNA is subjected to density gradient centrifugation analysis.

major facilitator superfamily (MFS) One of the two largest families of membrane transporters. The MFS is found in both prokaryotes and eukaryotes and functions to transport a wide variety of ions, sugars, amino acids, drugs, and other solutes. Members of the MFS are grouped together on the basis of sequence homology as a result of sequence analyses of public databases completed in 1997. Phylogenetically, 17 distinct subfamilies can be delineated within the MFS, each of which usually transports one type of compound. All 17 subfamilies contain 12 or 14 transmembrane alpha-helices and show a highly conserved motif between transmembrane linking peptides 2 and 3. The MFS is believed to have evolved from a common ancestral gene through a process of gene duplication.

major histocompatability complex (MHC) A cluster of genes present in the genomes of most higher vertebrates that code for cell surface proteins that are recognized as the main transplantation antigens when cells are transplanted to a foreign environment. The MHC proteins are therefore the antigens mainly responsible for provoking graft rejection in animals receiving foreign tissue grafts.

malaria A chronic disease characterized by periodic acute attacks of chills and fever. The disease is the result of infection by the sporozooite parasite, plasmodium, that lives in red blood cells and is transmitted to humans via the *Anopheles* mosquito.

maltase The enzyme that catalyzes the breakdown of the malt sugar, maltose

$$\text{maltose} \xrightarrow{\text{maltase}} 2 \text{ glucose}$$

maltose-binding protein A protein produced by the bacterium *Escherichia coli* that is used to transport the disaccharide sugar maltose across the bacterial plasma membrane.

mammalian cell culture The maintenance of mammalian cells outside the body using synthetic media to meet the nutritional requirements normally supplied by the blood.

mammalian expression systems The term for expression vectors that are specifically designed to express cloned genes in mammalian cells. See EXPRESSION SYSTEM.

mannose A sugar that is an optical isomer of the main energy-producing sugar, glucose. Because mannose differs from glucose at only one of the six carbons, mannose is called an epimer of glucose and can be converted directly into glucose by enzymes.

Manton-Gaulin homogenizer An apparatus used for large-scale breakage of cells to release their internal contents, using the mechanism of liquid shear. See CELL DISRUPTION.

map distance A means of defining distance between two markers on a segment of chromosomal DNA. In bacterial systems, map distance is measured in terms of map units, defined as the recombination frequency between two genetic markers and expressed as a percentage. In eukaryotic chromosomes, map units are given in centimorgans.

MAP kinase A class of enzymes that phosphorylate MAPs. Phosphorylation (and dephosphorylation) of MAPs regulates the polymerization of microtubules and therefore functions as a means to control the entry of cells into mitosis. MAP kinases are activated by signaling pathways, such as those mediated by receptor tyrosine kinases (RTKs). For this reason MAP kinase in the context of signaling pathways is an acronym for *m*itogen-*a*ctivated *p*rotein kinase.

MAPs See MICROTUBULE-ASSOCIATED PROTEINS.

MAR (SAR) Matrix *a*ttachment *re*gions (*S*caffold *a*ttachment *r*egions); specific DNA sequences at which attachment to the nuclear scaffold network occurs.

Marburg virus A virus discovered as a contaminant of tissues from African green monkeys in Marburg and Frankfort, Germany, in 1967. The classification of Marburg virus is unclear. The disease is characterized by fever, rash, gastrointestinal upset, and central nervous system involvement and is potentially fatal.

Marfan syndrome An inherited disorder of connective tissue that affects the skeleton, lungs, eyes, heart, and blood vessels. The disease is characterized by unusually long limbs. Marfan syndrome is carried as an autosomal dominant linked to mutations in the FBN1 gene on chromosome 15q21.1, which encodes a glycoprotein called fibrillin that is a major building block of microfibrils. Microfibrils are structural components of the supporting matrices of the tissue of the eye, aorta, lung airways, and the dura of spinal column. Mutations in FBN1 produce abnormal fibrillin-1 monomers that prevent microfibril formation. This is an example of a dominant-negative effect because the mutant fibrillin-1 disrupts microfibril formation even with normal fibrillin is being produced by the other allele.

marker Any genetic element that produces a variation in expression of a trait (e.g., hair color) and that resides at a particular locus.

Marshall, Barry J. (1902–1992) One of the codiscoverers of the ulcer-causing bacterium *Helicobacter pylori*. This discovery earned him the Nobel Prize in physiology or medicine in 2005, which he shared with Robin Warren. Marshall became well known for having tested the ulcer-causing properties of the bacterium on himself.

mass spectrometry An analytical technique for determining the molecular structure of an unknown compound by observing the paths that fragments of the molecule take when they are forced to migrate in a magnetic field.

mast cell A connective tissue cell located near capillaries and most abundant in the lung, the skin, and the gastrointestinal tract. Mast cells possess receptors for IgE and release histamine when bound to IgE. The release of histamine is responsible for the runny nose, itchiness, and other respiratory symptoms of allergy.

master regulatory genes A cluster of genes that governs the development of the major structural features during embryogenesis, for example, the bithorax complex in *Drosophila melanogaster* that is responsible for development of the abdominal and thoracic segments.

maternal effect Characteristics controlled by the mother and expressed in the offspring. Usually, maternal effects are caused by mRNA or a transcription factor produced by the mother and passed into the egg.

maternal inheritance Inheritance of genes of extrachromosomal factors such as the mitochondria that are transmitted through the egg cytoplasm.

mating type One of two alternative states (α or a mating types) that the haploid (budding) form of yeast can assume for the purpose of mating. During mating, diploid cells are formed by fusion of haploid cells of opposite mating types. The mating process converts yeast cells from haploid asexual cells to diploid sexually reproducing cells.

mating-type locus (MAT) A locus containing master regulatory genes in yeast that determine the male and female mating types.

maturing face [trans face (Golgi)] The outermost membrane in a Golgi stack. The maturing face of the Golgi stack is the place where proteins that have been processed in the Golgi exit for various cellular destinations. See GOLGI APPARATUS.

Maxam-Gilbert sequencing A technique for determining the sequence of a nucleic acid by chemical treatments that cleave the nucleic acid strand at only one of the four nucleotide bases (i.e., adenine, cytosine, guanine, or thymine), depending on the chemicals used. The fragments produced by chemical cleavage are then separated by electrophoresis, and the sequence is determined from the size of the different fragments.

MBP vector Maltose binding protein vector; an expression vector designed to facilitate the purification of the proteins that are expressed via the vector. In MBP vectors, the gene to be expressed is fused to the gene coding for the maltose binding protein. The fusion protein expressed by the vector can be purified by running a cell extract over an amylose column.

McClintock, Barbara (1902–1992) A geneticist whose work on the cytogenetics of maize led her to postulate the idea that genes could be transposable, both within a chromosome and between chromosomes. Her work on transposable elements won her the Nobel Prize in medicine in 1983.

MDM2 Murine double minute 2; MDM2 is a nuclear phosphoprotein with an apparent molecular mass of 90 kD that forms a complex with the p53 tumor-suppressor protein. Human MDM2 was identified as a homologous product of the murine double minute 2 gene. The MDM2 gene enhances the tumorigenic potential of cells when it is overexpressed and encodes a putative transcription factor. Forming a tight complex with the p53 gene, the MDM2 oncogene can inhibit p53-mediated transactivation, and MDM2 also binds to p53 protein. Inactivation of tumor-suppressor genes leads to deregulated cell proliferation and is a key factor in human tumorigenesis. p53 can be subjected to negative regulation by the product of a single cellular proto-oncogene. The interference of binding to p53 prevents the interaction of MDM2 and its regulation of the transcriptional activity of p53 in vivo. Direct association of p53 with the cellular protein MDM2 results in ubiquitination and subsequent degradation of p53. MDM2 p53 complexes were preferentially found in S/G2M phases of the cell cycle. The MDM2 gene is alternatively spliced, producing five additional splice-variant transcripts from the full-length MDM2 gene. The alternatively spliced transcripts tend to be expressed in tumorigenic tissue, whereas the full-length MDM2 transcript is expressed in normal tissue.

mDNA The DNA that represents genes that are expressed in a variety of tissues. The mDNA is presumed to represent genes that are required for processes required in all cell types.

medium The nutrient broth used to grow cultures of cells, bacteria, or microorganisms.

meiosis The process of cell division that takes place in the reproductive tissue and that produces gametes (sperm and egg in animals). The meiotic process leaves each daughter cell that becomes a gamete with half the number of chromosomes as are found in other cell types in the body.

MEK An intermediate in ras signal transduction pathways. In these pathways, phosphorylation of MEK by raf activates MEK. Activated MEK is a phosphorylase that phosphorylates a MAP kinase, the next intermediate in the pathway. See SIGNAL TRANSDUCTION.

melanin The brown, reddish, or black pigment that, in mammals, gives skin and hair their characteristic color(s). Melanin is derived from the amino acid, tyrosine, and is also found in parts of the brain and the eye, where its function is unknown.

melanocyte A specialized cell type found beneath the epidermal layer of skin that produces melanin for the purpose of skin pigmentation. Melanin made in the melanocyte is passed on to the upper layers of skin through dendritic cell processes.

melanoma A highly malignant skin cancer originating in the melanocytes.

melting of DNA The breakage, by heating, of the hydrogen bonds that hold the double-stranded helical structure of DNA together. On melting, DNA changes from double stranded to single stranded.

melting temperature Defined as the temperature at which 50 percent of the double-stranded DNA is turned into single-stranded DNA.

membrane In general a flexible sheet or layered material that separates two chemically different environments. Biologically, membranes are made up primarily of lipids and proteins and are the structures that define the compartments of the cell, the organelle, and the nucleus. Synthetic membranes are used biochemically to separate chemically different liquids and gases from one another for experimental purposes.

membrane filtration A method of clarifying microbial cell extracts. The procedure may not be optimal because microbial preparations can be gelatinous and block the filter. To compensate for this problem, membranes with an asymmetric pore structure have been used successfully in large-scale operations to isolate certain enzymes from lysed microorganisms.

membrane potential The electric potential (voltage) created by the difference in ion concentration on different sides of a membrane. Membrane potentials drive certain kinds of transport systems through the membrane and are responsible for nerve impulses. See PROTON GRADIENT.

membrane ruffling A wavelike movement observed at the leading edge of a cell membrane during movement; the location of the ruffled portion of the membrane indicates the direction of cell movement.

memory, immunologic The ability of the immune system to respond to antigens to which it has previously been exposed; the maintenance of immunity to an antigen over long periods of time.

memory cells Small populations of B cells and T cells of the immune system that produce antibodies or have receptors for an antigen that appears after an initial exposure of the organism to that antigen.

Mendel, Gregor (1822–1884) The founder of the field of genetics. His experiments on crossbreeding of peas led to the first formulation of the principles of heredity based on the idea of *independent assortment* of genetic units (chromosomes) that are still in use today.

Mendelian genetics Genetics based on the concepts originally proposed by Gregor Mendel, that is, that genetic traits are contained on completely independent, randomly reassorting genetic elements (i.e., chromosomes). Recombination is not taken into account in classical Mendelian genetics.

Mendel's law The principle that genetic elements controlling individual traits can reassort themselves independently of one another during the reproductive process. The postulated genetic elements were later found to be based on physically discernible structures: chromosomes.

mercaptoethanol A widely used reducing agent, particularly useful in biochemical procedures where breakage of disulfide bonds is desirable.

$$-S-S- \text{-------------}> -SH\ HS-$$

6-mercaptopurine A purine analog whose DNA-damaging effects particularly target the production of T lymphocytes. For this reason, 6-mercaptopurine is given as an immunosuppressant drug to prevent graft rejection.

meristem The mitotically active tissue in higher plants that, through cell division, forms new plant tissues. Meristematic cells are found in the root (apical meristem) and along the outside of the stem (lateral meristem).

meristem culture A technique used to produce pathogen-free plants. The meristem is a dome of actively dividing cells that is resistant to contamination by microbes. When the lab growth of meristems is combined with micropropagation techniques, large numbers of disease-free plants can be cultured.

merozygote The stage in bacterial conjugation in which the recipient bacterium, prior to division, contains two bacterial chromosomes.

Meselson, Matthew (b. 1930) A biochemist who, in collaboration with Franklin Stahl, carried out the critical experiments that demonstrated the semiconservation nature of DNA replication in 1957.

mesophile Bacteria that grow in the narrow temperature range of the mammalian body; from about 37°C to 44°C.

messenger RNA (mRNA) A ribonucleic acid strand that carries the genetic code for a protein. The mRNA is copied from DNA in the nucleus, is processed, and is then transported to the cytoplasm where its code is read on ribosomes and translated into a polypeptide.

metabolic disease A disease stemming from a defect in an essential metabolic pathway.

metabolic pathway A series of successive biochemical steps in the metabolism of a nutrient molecule. In a metabolic pathway, the input molecule is progressively altered until a specific final metabolite is produced.

metabolism The process of altering a nutrient molecule via a metabolic pathway for the purpose of energy production or the creation of important biomolecules, for example, amino acids, hormones, and nucleic acids.

metabolite One of the intermediate molecules generated during the metabolism of a nutrient.

metabolomics A recently formed subdivision of bioinformatics devoted to high throughput analyses of metabolites according to their physical and chemical properties using various techniques, including gas chromatography, high-pressure liquid chromatography, capillary electrophoresis, and mass spectrometry.

metalloenzyme An enzyme requiring a metal atom(s) for normal activity.

metallothionein Any of a class of metal binding proteins that play a role in preventing toxicity due to metal accumulation in cells. Because metallothionein synthesis is induced by the presence of metal ions, the metallothionein promoter has been widely used to control the expression of genetically engineered genes.

metamerism A concept that usually applies to arthropod development but which refers to unique features on serially repeating segments; for example, the segments of the embryo of *Drosophila melanogaster,* which are initially identical in appearance but later develop appendages associated with the head, thorax, and abdomen. Metamerism is the result of the action of various segmentation genes such as the homeobox genes.

metamorphosis The maturational process in amphibians and insects as exemplified by the transition from tadpole to adult frog.

metaphase The phase in mitosis in which the chromosome pairs are aligned along the axis of the cell just prior to telophase.

metastasis The spread of cancer cells from a primary lesion to other parts of the body.

methane The simplest hydrocarbon made up of one carbon and four hydrogen atoms (CH_4). Methane is a gas that is generated from carbon dioxide by certain bacteria during the oxidation of fatty acids.

methanogenic bacteria Bacteria that generate methane gas during metabolism.

methanol The alcohol of methane (CH_3-OH), also known as wood alcohol.

methanophile (methanotroph) Any of a class of bacteria that derives its energy from the metabolism of methane.

methicillin A synthetic derivative of penicillin created by the addition of a dimethoxyphenyl group to the side chain of penicillin. Because methicillin is not susceptible to the action of penicillinase, methicillin can be used in cases of infection by penicillin-resistant bacteria.

methionine One of the two sulfur-containing amino acids whose side chain is: $-CH_2-CH_2-S-CH_3$. Methionine is an essential amino acid that is important as a methyl group donor.

methionine-enkephalin (met-enkephalin) A short peptide with the structure: Tyr-Gly-Gly-Phe-Met. Met-enkephalin is one of a class of pain inhibiting neuropeptides known as endorphins that act by binding to the opiod receptor in the brain.

methotrexate An antibiotic and chemotherapeutic agent that blocks the enzyme dihydrofolate reductase (DHFR), which is necessary for purine biosynthesis. Because certain cancer cells have a high requirement for this enzyme, methotrexate can specifically target metabolism of these cancer cells to inhibit their growth. Resistance to methotrexate arises when the DHFR mutates to a form that no longer responds to methotrexate. However, it has been shown that when methotrexate is added to certain cell lines, resistance arises due to the amplification of the gene for DHFR, thus producing more copies of the protein. This phenomenon has been exploited by biotechnologists, and methotrexate is used to amplify genes or parts of chromosomes cloned near the gene for DHFR.

methyl-accepting chemotaxis proteins (MCPs) A class of bacterial transmembrane proteins involved in chemotaxis. The portion of the MCPs that extend into the bacterial cytosol becomes methylated when the portion of the protein that extends outside the cell binds an attractant substance. However, it becomes demethylated if a repellent substance is bound.

methylation of nucleic acids The addition of methyl groups to nitrogen atoms on the bases in nucleic acids. Methylation of nucleic acids is known to serve at least three functions:
(1) Methylation of the bases in DNA is believed to be a mechanism for controlling gene expression (methylated DNA is not expressed).
(2) Methylation of the 5' terminal guanine in mRNA is required for the mRNA to be functional.
(3) Methylation of restriction enzyme sites in the DNA of bacterial cells that make restriction enzymes as a defense against invading bacteriophage; methylation of the sites on the bacterial DNA prevents cleavage of the DNA by its own restriction enzymes.

methylmalonic academia An autosomal recessive genetic disorder of the metabolism of any of four amino acids (methionine, threonine, isoleucine, and valine) in which the blood and body tissues become acidic. The acute form is characterized by drowsiness, coma, and sometimes seizures. Over the long term, mental retardation may be a consequence. The metabolic defect is the inability to convert methylmalonyl-CoA to succinyl-CoA, which leads to the accumulation of methylmalonic acid. The enzyme that catalyzes this reaction is methylmalonyl-CoA mutase, encoded by a gene located at gene map locus 6p21.

methyl tetrahydrofolate A form of the B vitamin folic acid that acts as a coenzyme in methyl group transferring reactions in the synthesis of purines.

methyl transferases A set of enzymes that catalyzes the transfer of methyl groups from S-adenosylmethionine (SAM) to another substrate, particularly nucleic acids or nucleic acid precursors.

methylcellulose An inert polymeric substance used to increase the density of culture medium to maintain growing microorganisms in suspension.

5-methylcytosine (5MeC) A modified form of cytosine to which a methyl group has been added by a methylase. These modified residues are found at specific sites along the DNA and provide hotspots for transition type mutations. They are readily spontaneously deaminated, resulting in the conversion of 5MeC to thymine, leaving a mispaired G-T base pair. On subsequent DNA replication, one newly synthesized strand will contain the A-T mutation.

MHC See MAJOR HISTOCOMPATIBILITY COMPLEX.

micelle A more or less spherical structure that amphipathic lipids spontaneously assume when mixed with water. In a micelle, the polar portion of the lipid is oriented outward in contact with the water molecules, but the hydrophobic portions of the lipids are in the interior of the spheroid.

Michaelis-Menten constant (K_M) A reaction rate constant pertaining to enzyme-catalyzed reactions:

$$E + S \overset{k_1}{\underset{k_{-1}}{\longleftrightarrow}} ES \overset{k_2}{\longrightarrow} E + P$$

where
E = free enzyme
S = substrate
ES = enzyme-substrate complex
P = product of the reaction

k_1, k_{-1}, k_2 = reaction rate constants for the individual reactions

K_M is defined by $K_M = (k_2 + k_{-1})/k_1$.

Michaelis-Menten equation Defines the relationship between K_M, the reaction rate velocity (v), the maximum reaction rate attainable at a given concentration of enzyme, and the concentration of substrate [S]

$$v = V_{max}[S]/([S] + K_M)$$

microaerophile A microorganism that is neither aerobic or anaerobic but can grow under conditions of very limited oxygen.

microarray, cDNA A tool for analysis of patterns of gene expression in which the levels of gene products of a large number of genes are assayed simultaneously. Microarrays consist of cDNAs (cDNA microarrays), proteins, or antibodies to proteins (antibody microarrays) that are bonded to a solid substrate (such as glass, plastic, or nylon membranes) in a regular array of spots. The arrays are reacted with labeled probes representing the proteins or RNAs present in a population of cells or tissues and scanned to quantitate the relative signal intensity corresponding to probe bound to each spot. Computer analyses of signal intensities can generate gene expression profiles that can be used to characterize cellular responses to drugs, genes expressed in disease states, or genes involved in normal regulatory processes.

microbe A microorganism. Often used as synonymous with *germ*.

microcapsule A very thin version of the capsule that covers the bacterial cell wall; a gellike, largely polysaccharide matrix that protects the bacterium from phagocytosis.

micrococcal nuclease An endonuclease isolated from *Staphylococcus aureus* that cleaves DNA strands by breaking the deoxyribose-phosphate backbone of DNA at the 5' carbon atom. Micrococcal nuclease is sometimes used in place of DNase I for mapping protein binding sites on DNA. See FOOTPRINT.

microfibrils A bundle of fine cellulose fibers that make up the plant cell wall.

microfilament An actin-containing filament that makes up one type of cytoskeleton in mammalian cells. Microfilaments are believed to be the basis for movement of cells in culture.

microglobulin A short peptide that is noncovalently bound to the class I major histocompatibility complex glycoprotein.

microgram 0.000001, or 10^{-6} grams.

microheterogeneity Slight variation in the nucleotide sequences of a repeated unit of DNA. For example, the spacer regions in the histone genes are copies of one another, but there is some slight variation in their restriction fragment profiles that is referred to as microheterogeneity.

microinjection The injection of materials directly into individual cells using a small glass micropipet.

micromanipulator An instrument for guiding extremely small instruments, for example, microinjection pipets and microelectrodes into individual target cells under a microscope.

micron (μm) 0.000001, or 10^{-6} meters.

microporous gels A chromatography matrix made up of cross-linked dextrans or polyacrylamide used to fractionate proteins.

micropropagation A technique used in plant breeding to produce many genetic clones of the same plant. Small pieces of a plant are taken from a shoot tip, leaf, lateral bud, stem, or root tissue and are grown in culture medium. Subculturing of the buds or shoot is repeated many times until many plants are produced, all having the genetic characteristics of the original plant.

microsatellite markers A type of marker, used in DNA fingerprinting, that is composed of short, repeat DNA

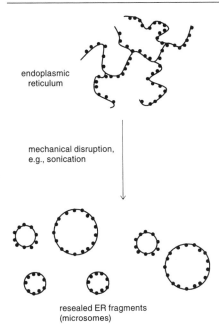

endoplasmic
reticulum

mechanical disruption,
e.g., sonication

resealed ER fragments
(microsomes)

Microsomes

sequences interspersed throughout the genome. See DNA FINGERPRINTING.

microsomes An experimental preparation derived by fragmentation of the endoplasmic reticulum (ER); the small, roughly spherical bodies consisting of bits of ER membrane that form spontaneously when animal cells are broken. Micosomes are categorized as either rough or smooth microsomes depending on whether they are derived from rough or smooth ER. The experimental significance of microsomes stems from the fact that, whereas the intact ER itself is difficult to isolate, microsomes are not and thus provide a convenient means of studying ER function.

microspikes Very thin (0.1 μm diameter × 5–10 μm long), actin-containing projections that protrude out of the membrane of cultured animal cells.

microtiter agglutination test A microscale test for the presence of an antibody or an antigen that is based on the presence of a precipitate formed when an antigen and its corresponding antibody bind to one another. The test is carried out in a series of small wells in a plastic tray (microtiter tray) in which a fixed amount of antigen (or antibody) is added to wells containing dilutions of a sample of an unknown amount of the corresponding antibody (or antigen). Semi-quantitative results are obtained as the highest dilution at which a precipitate can no longer be detected by eye.

microtome An instrument for creating extremely thin slices (sections) of biological specimens for microscopic examination.

microtubule Small tubules composed of a protein called tubulin that are attached to the centromeres of chromosomes and are responsible for the segregating movement of chromosomes during mitosis.

microtubule-associated proteins (MAPs) A large class of proteins that is believed to stabilize microtubules by binding to their tubulin subunits. Because MAPs can bind to several tubulin molecules at once, MAPs also accelerate the rate at which microtubules are polymerized from the tubulin subunits. MAPs are also believed serve as a binding material that glues microtubules to other proteins and cellular structures such as the chromosome centromere.

microvillus A fingerlike projection that is actually an actin filament–filled outpocketing of the cell membrane. Because microvilli are especially abundant on absorptive cells such as intestinal epithelial cells, microvilli are thought to function as a mechanism for increasing the absorptive surface area of the cell membrane.

migration-inhibitory factor (MIF) A factor(s) produced by certain T cells after stimulation by an antigen that inhibits the chemotactic response in macrophages.

mil The oncogene of the chicken sarcoma virus. Mil is believed to function as a serine kinase.

milk agent Mouse mammary tumor virus (MMTV), a retrovirus that causes cancers in mice. It was first identified as the agent that causes cancers in suckling mice nursed by mothers carrying the virus and was among the first tumor causing viruses to be described.

milligram   0.001, or 10^{-3} grams.

Milstein, Cesar (1927–2002)   The researcher who, together with Georges Kohler, developed the technique of creating hybridomas for the production of monoclonal antibodies by fusion of mouse spleen lymphocytes with myeloma cells. This work won him the Nobel Prize in medicine in 1984.

minicells   A daughter cell that is produced by cell division of a certain type of bacterial mutant that lacks a chromosome. Because minicells lack the bacterial chromosome, minicells that are found to contain DNA have been used to study aberrations of DNA replication and the properties of nonchromosomal DNAs such as plasmids.

minichromosome   The nucleosome-bound form of polyoma or SV40 DNA that is found in the nuclei of the virus infected cells.

minimal medium   A bacterial medium containing inorganic salts, inorganic nitrogen, and a simple sugar; the minimal requirements necessary to support bacterial growth. Minimal medium was classically used for the detection of mutants that were unable to synthesize an essential biochemical, for example, an amino acid or nucleoside. See DEFINED MEDIUM.

minisatellite DNA   Tandem repeats of a short core sequence. The number of repeats varies greatly between individuals, and the lengths of the minisatellites have been used as a characteristic in DNA profiling.

minisatellite variant repeat mapping   A technique used in DNA profiling in which not only the length of the minisatellite is analyzed but also the sequence variation, which may be only one base difference between individuals that is examined.

mismatch repair   A type of excision repair process that targets any region of DNA in which damage or mutation has resulted in a region where nucleotide bases on complementary DNA strands are improperly paired with one another. In the bacterium *Escherichia coli*, mismatch repair involves the actions of the genes, mutH, mutL, mutS, and mutU. See EXCISION REPAIR.

missense mutation   A point mutation that results in the replacement of one amino acid with another in the protein that is coded for by the gene in which the mutation occurs.

mitochondrion   A cytoplasmic organelle that is responsible for the bulk of energy production in eukaryotic cells. The mitochondrion is the site at which the electron transport process and the Krebs cycle portions of sugar metabolism take place.

mitogen   Any agent, such as a growth factor, that stimulates a cell to divide.

mitomycin C   One of a class of anti-tumor antibiotics (the mitomycins) that is isolated from the soil bacterium *Streptomyces caispitosus*. Mitomycin C exerts its antibiotic effects as an inhibitor of DNA synthesis.

mitosis   The orderly parceling out of replicated chromosomes to daughter cells. See M PHASE.

mitotic apparatus   A term used to describe the mitotic spindle apparatus that consists of microtubule bundles attached at one end to the centromere of the chromosome and at the other to the centriole located at one of the two cell poles.

mitotic index   The percentage of cells that are in mitosis at any given moment.

mitotic recombination   Crossing over of chromosomal segments between homologous chromosomes in a somatic cell. Such recombination is a normal event in meiosis but occurs rarely in somatic cells;

the recombined chromosomes are not passed on to progeny.

mitotic shake-off A method for obtaining a cell-cycle synchronized population of cells in tissue culture. The method depends on the fact that cells engaged in mitosis are not well attached to the bottom of the tissue culture vessel; therefore, the subpopulation of cells that are easily detached by light shaking are, for the most part, those in mitosis. See SYN-CHRONOUS CULTURE.

mitotic spindle The microtubule part of the mitotic apparatus.

MLH1, MSH2, and MSH6 Genes involved in hereditary nonpolyposis colon cancer (HNPCC), a form of colon cancer with an early age of onset that is transmitted as an autosomal dominant. There is a high frequency of association with mutations in MLH1, MSH2, and MSH6, three genes that code for proteins involved in excision repair—about 35

percent of cases have mutations in MSH2 and about 60 percent of cases have mutations in MLH1.

MN blood group A group of red blood cell surface glycoproteins (oligosaccharide derivatives of the protein glycophorin) that form a blood group family, distinguishable on the basis of naturally occurring antibodies, which is distinct from the ABO blood group.

mobile genetic element(s) Insertional elements (IS).

modafinil A drug used to treat narcolepsy in which its activity is based on its ability to act as an agonist of receptors for a class of neuropeptides known as orexins. Orexins stimulate wakefulness and mood by binding to receptors in a group of specialized neurons in the lateral hypothalamus. Modafinil is also currently being used to treat some of the symptoms of Alzheimer's disease and depression.

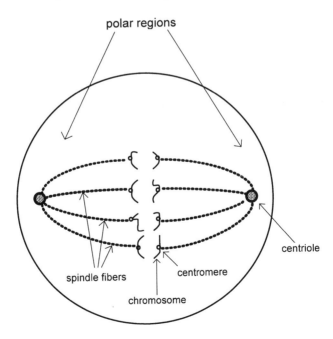

Mitotic spindle

molar (M) The concentration of a solution given as moles of the solute per liter of solution.

mold The filamentous, multicellular subfamily of the fungi.

mole A measure of the amount of a particular molecule such that 1 mole = 6.023 × 10²³ molecules (Avagadro's number). One mole of a substance has a weight, in grams, that is equal to its molecular weight.

molecular evolution The field of study devoted to establishing evolutionary relationships between species by analysis of the relatedness (homology) of the sequences of the nucleic acids or proteins of different organisms.

molecular genetics The study of the molecular basis of genetics; the structure and function of the DNA sequences in genes and control of gene transcription and expression.

molecular symmetries In a macromolecule such as a protein consisting of identical multimers, a term applied to different arrangements of the subunits in such a way that the resulting structure can be divided into identical halves along an axis. Multimeric proteins can display a number of different types of symmetry, including rotational, helical, cyclic, dihedral, and icosohedral.

molecular weight A measure of the mass of a molecule based upon a system where the mass of the hydrogen atom is taken as 1 and all other atoms are then assigned a molecular weight relative to hydrogen. See DALTON.

molecule Any group of covalently bonded atoms.

molt In arthropods, a shedding of the outer covering (the exoskeleton) during maturation to accommodate growth of the body.

monocistronic RNA A bacterial mRNA that codes for a single protein.

monoclonal antibody An antibody produced by the daughter cells derived from a single antibody-producing cell.

monocotyledon A subclass of the higher plants known as angiosperms, characterized by the presence of a single, as opposed to a double, seed leaf.

monocyte A large leukocyte with phagocytic properties. It is distinguished from other phagocytes by its size and the presence of small cytoplasmic granules.

monomer The single subunit of a polymeric molecule.

monosaccharide A single, or simple, sugar. A term generally referring to a subunit of a polyasaccharide.

Morgan, T. H. (1866–1945) A geneticist whose classic studies on the genetics of the fruit fly, *Drosophila melanogaster*, confirmed the Mendelian laws of transmission of traits from parent to offspring and gave rise to the concept of the gene. He was awarded the Nobel Prize in medicine in 1933.

morphogenesis The developmental process by which an appendage, limb, or organ comes to assume a specific form and structure.

morphogens Certain factors of maternal origin that are present in an egg that help to determine the location where limbs and other structures of the mature organism will appear in the developing embryo.

morphology The study of morphogenesis.

mosaicism The property of a tissue being a mixture of clonal populations of cells. By the Lyon effect, only one X chromosome of the pair will be expressed in any one cell lineage; therefore, the presence of cells with different X chromosomes activated in a cell population indicates that the cell population is a mixture of cell clones. Conversely, the presence of cells in which the same

X chromosome is always active indicates that the cell population is of clonal origin. This type of analysis has been used to demonstrate that most tumors probably derive from a single cell.

mos oncogene An oncogene that is found in a retrovirus that causes sarcoma tumors in mice. The name is an acronym derived from: *mo*loney *s*arcoma *v*irus.

MPF Maturation-*p*romoting *f*actor; a factor isolated from the cytoplasm of progesterone-stimulated *Xenopus laevis* oocytes that was shown to stimulate meiotic cell division in unstimulated oocytes when microinjected into the cytoplasm. The maturation-stimulating factor in MPF was later shown to consist of a cyclin B-cdk complex.

M phase The period of the cell cycle covering mitosis. M phase is divided into five subphases: prophase, prometaphase, metaphase, anaphase, and telophase.

M-phase promoting factor A protein factor isolated from eggs of the frog, *Xenopus laevis,* that forces cells at any stage in the cell cycle into mitosis (M phase).

msr, msd Loci that code for an unusual RNA-DNA hybrid molecule in which RNA transcribed from msr is covalently linked to msd DNA. This type of structure was first discovered in myxobacteria but has also been discovered in the soil bacterium *Stigmatella aurantiaca.*

mtDNA Mitochondrial DNA.

MTF-1 Metal-regulatory *t*ranscription *f*actor 1; a zinc finger transcription factor that is activated by various heavy metals such as zinc, cadmium, and copper but also by stressors such as oxidative stress and hypoxia. MTF-1 activates a number of genes, including metallothionein genes and other target genes.

MTOC *M*icro*t*ubule *o*rganizing *c*enter; an amorphous mass to which microtubules are found to be attached in a "hub-and-spoke"–like structure in the cytosol of interphase cells. The MTOC, which is also the centrosome in most cells, serves as a nucleation center for the polymerization of microtubules.

MuD phage A variant of the bacteriophage, Mu that has been engineered as a vector to be used for the determination of promoter activity using the beta-galactosidase gene as a reporter gene.

multidrug-resistant gene Genes that confer resistance to the lethal effects of certain drugs, particularly chemotherapeutic agents, for example, methotrexate. Multidrug-resistant genes arise through massive amplification of a single copy gene and are present in homogenously staining regions or double minute chromosomes.

multilocus probes (MLP) A technique used in DNA profiling in which the DNA from an individual, blotted to a membrane (see SOUTHERN BLOT), is mixed with a probe under conditions that will allow it to bind not only to the target but also to similar sequences as well.

multinucleate Having more than one nucleus within the same cytoplasm. See HETEROKARYON and SYNCITIUM.

mung-bean nuclease An enzyme that catalyzes the breakdown of single-stranded DNA into single nucleotides and short oligonucleotides that have phosphate groups on their 5′ ends.

murine Of or pertaining to the genus *Mus,* which includes mice, rats, and other rodents.

murine leukemia virus (MuLV) A retrovirus that causes leukemia in mice and that carries the abl oncogene.

murine sarcoma virus Also known as the Moloney sarcoma virus, the retrovirus that carries the mos oncogene.

mutagen Any chemical, physical, or biological agent that causes permanent, heritable alterations in the sequence of bases in DNA.

mutagenesis The process of introducing mutations by exposing cells to a mutagen(s).

mutagenesis, site-directed A technique for introducing specific changes in base sequence in a cloned segment of DNA, usually by replacing a portion with a synthetic oligonucleotide.

mutagenesis in vitro Mutagenesis of cells in culture that may differ from mutagenesis using the same mutagen in the intact organism in terms of the type and number of mutations.

mutations, somatic Mutations that occur in the cells of the body as opposed to the germline, or reproductive cells. Somatic mutations are not passed on to an organism's offspring, whereas germline mutations are.

mutator loci Genes whose function is associated with the fidelity of DNA synthesis; for example,
mutD = subunit of DNA polymerase III
mutU = DNA helicase
mutY = endonuclease that cleaves mismatches between A and G

mutator phenotype A strong tendency to undergo mutation. The mutator phenotype is expressed by any organism that carries a mutation in a gene involved in maintaining the fidelity of DNA synthesis.

muteins Modified therapeutic proteins with improved or novel biological activities that have been produced by mutagenesis in the lab.

myasthenia gravis An autoimmune disease of the nervous system characterized by a progressive paralysis of the motor nerves. The disease is caused by the formation of antibodies to one's own neurotransmitter receptors that act at the neuromuscular junction.

myb The oncogene carried by the avian myeloblastosis virus that causes myeloblastic leukemia in chickens. The human proto-oncogene (c-myb) codes for a transcription factor that is believed to be required for hematopoietic cell proliferation. The human myb gene is located at 6q22-23.

myc The oncogene carried by the avian leukosis retrovirus that causes a type of leukemia in birds. The human myc proto-oncogene encodes a nuclear transcription factor. Activation of the myc gene by translocation to regions of chromosomes 2, 14, or 22 causes a type of leukemia known as Burkitt's lymphoma. The myc gene is located on chromosome 8q24.12-q24.13.

mycelium A mass of hyphae.

mycobacteria A group of rod-shaped acidophilic bacteria that includes the bacteria that cause leprosy and tuberculosis.

mycology The study of molds (fungi).

mycoplasma The smallest independently growing organisms known. Mycoplasmas were isolated and characterized on the basis of their role in causing a type of pneumonia. Mycoplasmas are also known as PPLO (pleuropneumonia-like organisms).

mycotoxin Any toxic substance produced by a fungus or mold. Among the mycotoxins that have been used in molecular biological research are muscarine, phalloidin, ergotamine, and various antibiotics.

myelin A class of lipids that comprise the outer sheath that surrounds the axon of neurons. In this location, it serves as an electrical insulator without which nerve impulses could not be propagated.

myelin basic protein (MBP) A small (20 kDa) protein that is the most abundant component of the myelin sheath of the neurons of the central nervous system. MBP is believed to play an important role(s) in assembly and stabilization of the myelin sheath. There are at least six isoforms of MBP: C1, C2, C3, C4, C5, and C6. Tests that measure the level of MBP in the cerebrospinal fluid (CSF) are used as indications of disease states involving breakdown of the myelin sheath.

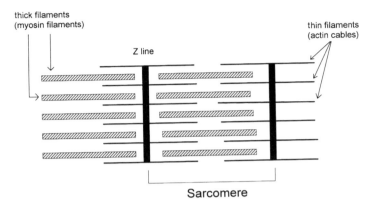

Sarcomere

Actin and myosin filaments in muscle

myeloid cell The collective term for all classes of blood cells, not including T and B lymphocytes; that is, all bone marrow–derived blood cells.

myeloma A tumor of the antibody secreting lymphocyte cells. Also known as plasmacytoma.

myeloma proteins An antibody molecule of the Ig class secreted by a myeloma tumor. Each myeloma antibody represents the specificity of a single antibody cell (monoclonal antibody).

myoblast Embryonic cells that fuse with one another to form mature muscle cells.

myoclonic epilepsy and ragged-red fiber disease (MERRF) A maternally inherited disorder of muscle functioning caused by a mutation in the gene that encodes lysyl tRNA, which in turn leads to malfunctioning of a number of other important proteins in mitochondria (mitochondrial encephalomyopathy) that impair the functioning of the electron transport chain. The disorder is characterized by myoclonic seizures, ataxia, speech difficulty (dysarthria), hearing loss, and dementia. Biopsies of muscle show ragged-red fibers.

myoD One of a family of proteins, referred to as myogenic proteins, that induces cells to differentiate into muscle cells. myoD exhibits the helix-loop-helix motif characteristic of certain transcription factors and is believed to act, at least in part, by inducing the transcription of another mygenic gene, myogenin.

myoglobin The protein that carries and exchanges oxygen for CO_2 in muscle tissue. Like hemaglobin, myoglobin carries oxygen on a heme group attached to a single polypeptide (globin), but, unlike hemaglobin, is present only as a monomer.

myosin One of the two contractile proteins of muscle. Myosin bundles interdigitate with actin bundles, and muscle contraction is the result of the two proteins sliding over one another.

myxobacteria Bacteria that normally live in the soil as individual cells but that, under conditions where nutrients become limiting, form multicellular aggregates similar to primitive multicellular organisms.

myxovirus A class of viruses that was identified and characterized on the basis of their role in causing influenza. Myxoviruses are divided into two families: orthomyxoviruses and paramyxoviruses.

NAD, NAD(P) Nicotamide adenine dinucleotide (phosphate). A cofactor for enzymes involved in oxidoreductions and electron transfer in numerous biochemical reactions, but particularly those involved in the oxidative metabolism of sugars for energy production. NAD is a combination of two nucleotides, one of which is derived from the B vitamin niacin.

NAD (nicotamide adenine dinucleotide)

nalidixic acid A synthetic antibiotic that mimics the action of the natural antibiotic novobiocin.

NANA N-acetylneuraminic acid. The molecule attached to a glycoprotein that forms the red-blood-cell membrane receptor for influenza virus that accounts for the ability of the virus to cause hemagglutination. Neuraminidase cleaves off the NANA residue, thereby inactivating the receptor.

nanogram 0.000000001, or 10^{-9} grams.

nascent protein A polypeptide in the process of being synthesized on ribosomes.

National Center for Biotechnology Information (NCBI) Created as a national resource for molecular biology information, this center creates public databases and conducts research in computational biology.

National Human Genome Research Institute (NHGRI) An institute of the National Institutes of Health created to head up the Human Genome Project. It supports not only large-scale sequencing and analysis of the human genome but also comparative genome projects, genome informatics, and programs that anticipate and address the ethical, legal, and social issues that arise as the result of human genome research.

native conformation The three-dimensional shape that a biomolecule naturally assumes in its normal biological environment.

natural killer cells A type of thymocyte (T lymphocyte) found in the spleen that is responsible for a particular immune response to various tumor cells.

N-CAM Neural cell adhesion molecule; a protein present on the surface of neurons that causes them to aggregate. N-CAM is expressed at specific times during development suggesting that it plays a role in the development of neural structures such as ganglia.

nebulin A large filamentous actin-binding protein in the muscle sarcomere that is believed to maintain the actin filaments at a uniform distance from the myosin containing thick filaments and that also may help to determine the length of the actin filaments.

negative complementation The suppression of a normal gene by a mutant of that gene.

negative regulation (negative regulators) A term applied to regulation of gene expression based on repression rather than stimulation of gene expression. In negative regulation, the target gene is normally expressed to its maximal extent, and under appropriate environmental conditions, expression is controlled by lowering its level of expression. See LAC OPERON.

negative selection The selection of cells or organisms within a population that expresses a desired trait by elimination of members of the population that do not express the trait. Negative selection is commonly used to select microorganisms that express a gene such as resistance to an antibiotic that allows them to survive in the presence of that antibiotic while other microorganisms that do not have, or express, the gene are eliminated. See HAT SELECTION.

neomycin A synthetic antibiotic derived from streptomycin. Neomycin and other related antibiotics act on the ribosome to inhibit bacterial protein synthesis.

neoplasia Any abnormal growth of an adult tissue.

Nernst equation An equation that relates the free energy change for a given reaction

$$R_1+R_2+R_3+\ldots+Rn \longrightarrow$$
$$P_1+P_2+P_3+\ldots+P_n$$

(where R_i represents a reactant and P_i represents a product) to the concentrations of the reactants and products:

$$\Delta G = \Delta G° + 2.303RT \times log([R_1][R_2]\ldots$$
$$[R_n])/([P_1][P_2]\ldots[P_n])$$
and
R = universal gas constant
T = temperature in degrees Kelvin
$\Delta G°$ = a constant for a given reaction

nerve growth factor (NGF) A factor, originally isolated from a tumor transplanted into a chicken embryo, that causes selective outgrowth of sensory and sympathetic neurons. NGF is essential for proper growth and survival of these types of neurons during development.

nested deletion A deleted region of a nucleic acid that occurs within a region covered by a second, larger deletion of the same nucleic acid. The smaller deletion is said to be nested within the larger deletion.

neu oncogene An oncogene isolated from rat cells by transfection into 3T3 cells. neu is not carried by a retrovirus and is apparently activated by a point mutation of the proto-oncogene form of neu. The neu sequence is homologous to the erb-B oncogene.

neuraminidase A glycoprotein present as a spike on the outside of the influenza virus envelope. Neuraminidase acts to break down an inhibitor of the influenza virus hemagglutinin (HA) protein.

neuroblastoma Cancerous growth of the neuroblast cells, or the cells of the developing embryo that go on to form the nervous system.

neurofilament proteins Three proteins that are the subunits of the neurofilaments, a type of intermediate filament found in neurons. The function of neurofilaments is unknown.

neuron The brain cell type that carries the nerve impulses involved in higher thought and movement.

neuropeptide Any of a class of short polypeptides that function as neurotransmitters. Thyrotropin-releasing hormone (TRH) and luteinizing hormone–releasing hormone (LHRH), which act to trigger the release of hormones from the pituitary gland are examples of neuropeptides.

neuropeptide Y (NPY) A small (36 amino acid) peptide neurotransmitter that potentiates the effects of noradrenergic neurons in causing vasoconstriction. NPY binds to six classes of receptors, Y1,

Y2, Y3, Y4, Y5, and Y6. NPY binding to the Y3 receptor inhibits catecholamine synthesis while the Y2 subtype causes inhibition of catecholamine release. NPY release stimulates appetite in a region of the hypothalamus called the arcurate nucleus. The gene for NPY is located at gene map locus 7p15.1.

Neurospora A genus of mold that has been used as a tool in genetic experiments because it is normally haploid and because it forms a structure (the ascus) from which single-celled spores can be readily isolated. *Neurospora crassa* was the organism originally used to demonstrate that genes coded for individual proteins.

neurotransmitter A type of chemical, usually a small organic molecule, released from the terminal button of an axon of one neuron that induces a nerve impulse in an adjacent neuron when it binds to a specific receptor in the dendritic membrane of the second neuron. Different neurotransmitters are associated with different neural functions, e.g., dopamine (movement and mood), acetylcholine (movement), gamma aminobutyric acid (GABA; arousal relaxation), and serotonin (mood).

neutral substitution A change in a nucleotide base in the coding region of a gene that does not produce a change in the activity of the protein.

neutrophil One of the three subclasses of white blood cells known as granulocytes, also known as polymorphonuclear leucocytes (PMNs). Neutrophils contain large multilobed nuclei and phagocytose small invading organisms such as bacteria.

nick A gap in the sugar-phosphate backbone of a nucleic acid.

nick translation A technique for labeling double-stranded DNA fragments to be used as hybridization probes. DNA polymerase is used in the presence of labeled deoxyribonucleotides to fill in nicks produced by treatment with low levels of the endonuclease DNase I.

nicotinic receptors One of the two types of cholinergic receptors in brain

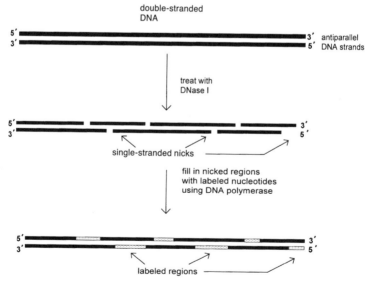

Nick translation

cells; named because of its sensitivity to nicotine.

ninhydrin A chemical that forms a purple pigment after reacting with amino groups. This reaction is used in the detection and quantitation of proteins.

ninjurin Nerve *injury-induced protein*; a protein that is upregulated after nerve injury in neurons of the dorsal root ganglion in Schwann cells. It has properties similar to other biomolecules that serve as adhesion molecules and is believed to play a role in nerve regeneration. The gene that encodes ninjurin is NINJ1, located at gene map locus 9q22. A second gene that codes for the protein ninjurin-2 is a homologue of ninjurin that contains the same transmembrane region as ninjurin but does not contain the same adhesion motifs. Ninjurin-2 shows the same nerve injury-induced upregulation as ninjurin.

Nirenberg, Marshall (b. 1927) A biochemist who carried out the first experiments using synthetic oligoribonucleotides that ultimately led to the deciphering of the genetic code by others using variations of the technique pioneered by Nirenberg (see KHORANA, HAR GOBIND). He was awarded the Nobel Prize in physiology and medicine in 1968.

nitrifying bacteria Bacteria that carry out the process by which nitrogen gas in the atmosphere becomes converted into ammonia. This is the only means by which nitrogen can become incorporated into biologically useful molecules such as proteins and nucleic acids. See NITROGEN FIXATION.

nitro-blue tetrazolium (NBT) A chemical that forms a blue precipitate when certain substrates are acted upon by alkaline phosphatase. For this reason, NBT is used as a colorimetric indicator in techniques that utilize alkaline phosphatase labels. See ALKALINE PHOSPHATASE.

nitrocellulose filter A thin flexible membrane made of nitrocellulose, a material that noncovalently binds tightly to a variety of biological materials, including proteins and nucleic acids. Because of this property, nitrocellulose is widely used as a binding material for carrying out colony and plaque hybridizations and Southern, northern, and western blots.

nitrogen cycle The global process of nitrogen recycling. The nitrogen cycle involves the breakdown of organic materials with the liberation of ammonia and other inorganic nitrogen-containing compounds. Ammonia is converted into nitrates and nitrites and then into nitrogen gas by soil bacteria. Atmospheric nitrogen is then converted back into ammonia by nitrogen-fixing bacteria associated with the roots of certain plants, thereby completing the cycle.

nitrogen fixation The process by which nitrifying bacteria convert nitrogen gas in the atmosphere into ammonia. Nitrogen fixation utilizes two systems: (1) the reductase that generates electrons and (2) the nitrogenase that uses the electrons generated from the reductase to reduce nitrogen (as N_2) to ammonia (NH_3).

NMDA receptor A specialized type of glutamate receptor on the postsynaptic membrane of neurons that is known to mediate long-term potentiation, a process involved in memory. NMDA is derived from N-methy-D-aspartate, a synthetic glutamate analog used to study this receptor.

NMR See NUCLEAR MAGNETIC RESONANCE.

nonautonomous controlling elements Defective transposons that are unable to transpose but that can do so when a normal transposon is also present.

noncompetitive inhibition Inhibition of enzymatic activity by a substance that acts at site on the enzyme different from the active site.

nondisjunction An error of chromosome segregation during cell division (either meiosis or mitosis) in which the daughter chromosomes (sister chromatids) fail to move to opposite poles.

nonessential amino acid An amino acid that can be synthesized from other amino acids or other precursors. For this reason, the nonessential amino acids, in contrast to the essential amino acids, can be omitted from the diet without causing death.

nonheme iron Iron atoms that are found in iron-sulfur proteins as opposed to heme groups. The iron-sulfur proteins are electron carriers in the process of electron transport.

nonidet P40 (NP40) A nonionic detergent (octylphenyl-ethylene oxide) that selectively breaks open the plasma membrane in animal cells but not the nuclear membrane. For this reason, NP40 is often used for rapid isolation of nuclei and cytoplasmic fractions of cells.

nonpermissive cell A host cell type that can become infected by a particular virus but does not support its replication. For example, monkey cells are permissive for SV40 virus but mouse cells are nonpermissive for SV40.

nonpolar group A small group of atoms held together by linkages where electrons are more or less equally distributed among the constituent atoms so that the linkages have no polar character. Such groups are important in biological systems because they are not soluble in water.

nonreciprocal recombinant chromosomes The result of RECOMBINATION between misaligned chromosomes such that a gene on one chromosome receives a duplication of part of the gene, while the copy of that gene on the other chromosome suffers a deletion of the same material.

nonsense mutation A mutation that causes a normal codon to change into one of the three termination codons (TAA, TAG, or TGA). This type of mutation causes premature termination of synthesis of the protein in which it occurs.

nonsense suppressor A mutant tRNA that permits the placement of an amino acid in a polypeptide when one of the nonsense codons is encountered during translation of an mRNA. Such tRNAs can reverse the effects of nonsense mutations (suppressor effects).

nontranscribed spacer See SPACER DNA.

noradrenaline (norepinephrine) A chemical that is both a neurotransmitter and a hormone released by the adrenal glands. As a neurotransmitter, noradrenaline acts on certain neurons of the sympathetic nervous system. As a hormone, noradrenaline acts on cells of the liver and muscle to stimulate both sugar mobilization and storage.

Noradrenaline (norepinephrine)

northern blot An analytical technique in which RNA is run on an agarose gel, blotted onto a membrane where it is hybridized to a specific probe. This technique is particularly useful as a means of detecting the expression of a particular mRNA. See HYBRIDIZATION and BLOT.

NOS Nitric oxide synthase; a group of enzymes that catalyze the formation of nitric oxide (NO) from L-arginine. NO acts as a neurotransmitter in the brain and as a signaling molecule in endothelial cells that induces vasodilation, platelet aggregation, and cardiovascular homeostasis. NOS is designated bNOS (brain) or nNOS (neuronal), and there are two forms produced by endothelial cells: eNOS, which is constitutive, and an inducible form designated iNOS. The activity of NOS in both the brain and endothelium requires binding of the calcium/calmodulin complex and is regulated by calcium levels.

notch protein A transmembrane signal transduction protein that functions in development of the *Drosophila* nervous system. Binding of critical ligands to the extracellular domain (receptor region) of notch is believed to result in selection of the cell carrying the bound ligand for development as a sensory organ cell.

novobiocin An antibiotic produced by *Streptomyces niveus*. Novobiocin acts by preventing ATP binding to the enzyme DNA gyrase, thereby stopping DNA synthesis in the infectious bacteria.

NSAIDs Nonsteroidal *a*nti-*i*nflammatory *d*rugs; a class of anti-inflammatory drugs that block the formation of the pro-inflammatory prostaglandin hormones by inhibiting a critical step in the synthesis of prostaglandins that is mediated by activity of enzymes known as cyclooxygenases (COXs). The popular NSAIDs aspirin, acetaminophen, ibuprofen, celecoxib (Celebrex), rofecoxeb (Vioxx), and naproxen are all COX inhibitors. As suggested by their name, these agents provide an alternative to steroid anti-inflammatory drugs that often have debilitating side effects.

NSF A tetrameric cytosolic protein that mediates the fusion of a transport vesicle and a vesicle of the Golgi.

NTG An acronym for *n*eomycin, *t*hymidine kinase, *g*lucocerebroside.

NTG vector A mammalian expression vector that combines the enhancer/promoter activities of the long terminal repeats (LTRs) from the Moloney murine leukemia virus with the bacterial-derived neomycin resistance gene (neo^r) as a selectable marker. See APH.

NtrB, NtrC proteins Regulatory proteins in the control of expression of genes involved in nitrogen metabolism under conditions of nitrogen deficiency in bacteria. When nitrogen-containing compounds in the bacterial medium drop to low levels, the NtrB protein becomes activated; the activated form of NtrB is responsible for phosphorylation of NtrC protein that stimulates the gene transcription. See NITROGEN FIXATION.

nuclear lamina A thin matrix composed of dense filaments just beneath the envelope that surrounds the cell nucleus.

nuclear lamins The intermediate filaments that comprise the nuclear lamina.

nuclear localization signal (NLS) A sequence of amino acids in some proteins that serves as a signal for their import into the nucleus. After synthesis of a nuclear protein in the cytoplasm, it is actively transported through nuclear pore complexes through a process that involves the Ran GTPase. Most of the NLSs are rich in basic amino acids (e.g., lysine or arginine).

nuclear magnetic resonance (NMR) A type of spectroscopy based on the magnetic properties of atoms. In this technique, a compound is placed in a magnetic field, and the energy (i.e., the electric current required to create a certain magnetic field) required to change the magnetic orientation of individual atoms, usually only the hydrogen atoms, is measured. Because the electric environment of different atoms differs from one another depending upon the other atoms to which it is attached, each compound gives a different "fingerprint" (spectrum) of field strengths at which individual atoms reorient in the magnetic field. In biological preparations, the absorption characteristics of a molecule is influenced by its three-dimensional structure and the biochemical environment that can be revealed by the NMR spectrum of a biochemical sample.

nuclear membrane (nuclear envelope) A double-walled membrane that forms the enclosure surrounding the nuclear compartment. The outer membrane is joined to and may actually be thought of as part of the endoplasmic reticulum.

nuclear pore complex An octagonal array of large protein granules that surround pores that perforate the double-layered nuclear membrane at various points. The nuclear pore complex is a specialized channel through which nucleic acids and other materials shuttle between the nucleus and the cytoplasm.

nuclear scaffold A fibrous network that extends from the inside of the nuclear membrane and is distributed all throughout the nucleus and that is attached to the cellular DNA at specific sites. The nuclear scaffold is seen by electron microscopy in isolated nuclei that are carefully treated with nucleases and either salt or detergents to remove histones, some nonhistone proteins, and the free (i.e., unattached) DNA strands from the chromatin. The function of the nuclear scaffold is unknown but is believed to play a role similar to the chromatin itself in control of gene expression.

nuclear transplantation A technique for removing the nucleus from one cell and placing it into the foreign cytoplasm of a second cell (i.e., an enucleated cell).

nuclease Any of a class of enzymes that catalyze the breakdown of a nucleic acid(s) by cleavage of the phosphodiester bonds of the sugar-phosphate backbone.

nucleic acid A molecule of either DNA or RNA.

nucleohistone (histone) Any of the five different proteins that make up a nucleosome, designated: H2A, H2B, H3, H4, and H1.

nucleoid body The analog of the nucleus in bacteria. The nucleoid body, which contains the bacterial genomic DNA, is not enclosed in a membrane, and the DNA is not complexed with chromatin but is distinguishable as a large centrally located mass that appears less dense than the surrounding cytoplasm.

nucleolus A large structure within the nucleus of eukaryotic cells consisting of numerous loops of chromatin-bound DNA that contains clusters of tandemly repeated ribosomal RNA genes. The nucleolus is therefore the structure responsible for the production of rRNA and is continually engaged in high levels of synthesis of rRNA.

nucleophilic group Any cluster of covalently linked atoms that tends to donate electrons in a chemical reaction. Nucleophilic groups often initiate important biochemical reactions by attacking electron-deficient carbon atoms that are attached to oxygen atoms.

nucleoplasmin An acidic nonhistone protein that binds to histones H2A and H2B during histone assembly with free DNA to form nucleosomes.

nucleoprotein particles Complexes of protein and RNA, primarily in the nucleus, that play a role in the processing of RNA.

nucleor organizer region A cluster of rRNA genes on a DNA loop in the nucleolus.

nucleoside A ribose or deoxyribose molecule that is attached to any purine or pyrimidine base via the first carbon atom of the sugar. The common nucleosides that are found in DNA and RNA are: (deoxy-) adenosine, (deoxy-) cytidine, (deoxy-) guanosine, thymidine (deoxyform only), and uridine.

nucleoside antibiotic An antibiotic that is a nucleoside containing an analog of a purine or pyrimidine base. These antibiotics act by inhibiting the normal mechanisms of DNA and RNA synthesis in rapidly growing microorganisms. The nucleoside antibiotics (e.g., cytosine arabinoside), unlike other types of antibiotics, are active against viral infections.

nucleosome An octameric structure that is complexed around a strand of DNA. Nucleosomes are evenly spaced along the DNA strand, forming a linker (nonhistone-complexed) region of 142 base pairs. Nucleosomes are composed of two each of the different "H" histones

Nucleosomes

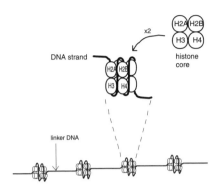

the third carbon of the ribose or deoxyribose sugar.

nucleus The central, membrane-enclosed structure that contains the cell DNA in the form of chromatin in eukaryotic cells.

null DNA The DNA that represents genes that are only expressed in single cell or tissue type. Null DNA is presumed to represent genes for specialized proteins that are unique to a specific cell type, for example, hemoglobin genes in red blood cells.

null mutation The result of a mutational event that results in the complete elimination of a gene.

nurse cell Accessory cells in the ovary that surround an oocyte and supply a variety of macromolecules and nutrients to the oocyte via cytoplasmic bridges. Ribosomes, mRNAs, and proteins are passed to insect oocytes in this way.

nystatin An antibiotic with a polyene structure (i.e., containing many carbon-carbon double bonds) that is active against fungal infections. Nystatin is active against *Candida* infections when applied topically.

and are believed to play a role in regulating the expression of the genes with which they are complexed.

nucleosome phasing A model of nucleosome structure in which a certain DNA sequence is always located at a certain position on the nucleosome. If this can be shown to be true, it implies that some mechanism exists for aligning nucleosomes with certain sequences on the DNA in the nucleosome complex.

nucleotide A nucleoside with a phosphate group attached either to the fifth or

O-antigen A branched polysaccharide attached to a specific lipid (lipid A) on the outer surface of the cell envelope of the pathogenic bacterium *Salmonella typhimurium* and other Gram-negative bacteria.

obligate anaerobe Bacteria that have an absolute requirement for oxygen and are not fermenting (e.g., tuberculosis).

ochre codon The nonsense codon TAA that is a signal for termination of polypeptide synthesis.

ochre mutation Any mutation that produces an ochre codon in place of a codon for an amino acid.

ochre suppressor See OCHRE CODON, OCHRE MUTATION, SUPPRESSOR GENE, SUPPRESOR MUTATION, SUPPRESSOR tRNA.

Okazaki fragment A short (1,000–2,000 bp) DNA fragment produced during DNA replication of the lagging (5′ terminating) strand of the template DNA. Okazaki fragments are initiated by a short RNA primer that is hybridized to the lagging strand and then destroyed after synthesis of the Okazaki fragment is complete. See LAGGING STRAND.

oligonucleotide A short strand of either DNA or RNA with a length in the range of two to about 30 bases. The term generally refers to synthetic polynucleotides.

oligopeptide A polypeptide of anywhere between approximately two and 10 amino acids.

oligosaccharide A chain of sugars containing anywhere between approximately two and 10 monosaccharide subunits.

oligotrophic An environment in which nutrients are in low abundance.

onc function The property acquired by a proto-oncogene of inducing a cancer or promoting tumorigenesis when it becomes an oncogene.

oncogene The activated form of a proto-oncogene. Mechanisms by which proto-oncogenes become activated include transduction by a retrovirus, mutation, and chromosomal translocation whereby the proto-oncogene is placed into a new genetic environment.

oncogenic Pertaining to any agent—chemical, physical, or biological—that causes cells to undergo changes characteristic of cancer cells.

oncogenic virus Any of a broad range of viruses that cause cells to undergo changes characteristic of cancer cells or to cause tumors in animals. See ONCOGENE and RNA TUMOR VIRUS.

oncostatin M (OSM) A cytokine that regulates growth of both normal and tumor cells although in a cell-type specific manner. OSM is a glycoprotein about 28 kDa in size that was originally isolated on the basis of its ability to inhibit the growth of A375 melanoma and other tumor cells while stimulating the growth of normal human fibroblasts. The oncostatin M receptor (OSMR) is coupled to the JAK/STAT signal transduction pathway. The OSM gene is located on chromosome 22q12. OSM belongs to the same family of cytokines as interleukin-11 (Il-11), leukemia inhibitory factor (LIF), interleukin-6 (Il-6), ciliary neurotrophic factor (CNTF), and cardiotrophin-1 (CT-1).

ontogenetic Of or pertaining to ontogeny, the complete life cycle or process of development of an organism.

oocyte The diploid germ cells of the female that generate gametes (eggs) by meiotic division.

oogamy The union of gametes to produce an embryogenic cell, as in fertilization of an egg by fusion with sperm.

oogenesis The process by which the mature egg(s) is generated from an oocyte.

oogonium (oogonia, pl.) The female reproductive organ in which the eggs are formed in thallophyte plants.

ooplasm The cytoplasm of an egg cell, or oocyte.

open reading frame (ORF) Any nucleic acid segment whose codons specify a continuous polypeptide; a nucleic acid sequence with a start codon.

operator A portion of the promoter in an operon that acts as a regulator of expression of the operon by serving as a site for the binding of a repressor protein.

operon A cluster of contiguous bacterial genes all under the control of a single promoter. The genes in an operon generally code for enzymes that catalyze steps in one biosynthetic pathway, for example, the synthesis of an amino acid.

opiate Any of the chemical derivatives of opium.

opines Unusual amino acids synthesized in plants infected by the T DNA portion of the Ti plasmid of the parasitic bacterium *Agrobacterium tumefaciens*. The opines include octopine, nopaline, and mannopine.

opsonization The process by which an antibody is taken up by a phagocyte in the presence of complement.

optical density The property of absorption of light by a solution of any given substance. The decrease in intensity of a light beam at a certain wavelength as it passes through a solution over a certain distance is proportional to the molar concentration of the solution. See BEER-LAMBERT LAW.

organelle Any subcellular, membrane-enclosed structure in the cytoplasm of a eukaryotic cell that carries out a specific cellular function, for example, mitochondria, chloroplasts, endoplasmic reticulum, and Golgi.

ornithine An intermediate in the urea cycle, the series of reactions in which nitrogen in the form of urea is formed. Ornithine is derived from the amino acid arginine through the loss of urea.

orotic acid A pyrimidine base, derived from carbamoyl phosphate and aspartate, that is the common precursor of CTP and UTP, the triphosphate nucleotides of cytosine and uracil.

Orphan Drug Act An act of Congress directed toward rare human diseases (defined as having a prevalence of less than 200,000 cases) that grants, as an incentive, a seven-year period of marketing exclusivity to the developer(s) of therapeutic drugs.

orphons Genes that are members of a gene family but are in distant locations.

ortholog Genes in different species that were all derived from a common ancestral gene. Normally, orthologs retain the same function in the course of evolution. For this reason identification of an unknown gene as an ortholog of one whose function is known is an important means of prediction of gene function.

orthophosphate Any salt of orthophosphoric acid (H_3PO_4); the name given to phosphoric acid that has been stripped of one or more of its hydrogen atoms as the result of its being placed in aqueous solu-

$$HO - \overset{\overset{\displaystyle O}{\|}}{\underset{\underset{\displaystyle OH}{|}}{P}} - O$$

Orthophosphate

tion. Orthophosphate is the form of phosphorous that is present in most important phosphorous-containing biomolecules, for example, nucleic acids.

osmolality The concentration difference between osmotic compartments as measured by molality; a 1-molal solution is defined when one mole of the solute is dissolved in 1,000 grams of the solvent.

osmosis The spontaneous diffusion of a substance from a compartment of relatively high concentration to a compartment of relatively low concentration where, generally, the two compartments are separated from one another by a semipermeable membrane. Many nutrients and other substances of biochemical importance enter into or pass out of cells by osmosis.

osmotic pressure A pressure produced on the side of a membrane with a higher solute concentration caused by the passage of water across the membrane by osmosis.

osteoblast The cells that initiate the formation of bone by secretion of the bone matrix, a material composed largely of collagen that is hardened into bony bone by the deposition of calcium phosphate crystals.

osteogenesis imperfecta A genetic disease caused by mutations in collagen genes that results in bone fragility, making them prone to fracture. Other symptoms include loose joints, bruising, hearing loss (caused by problems with the bones of the middle ear), and scoliosis. The affected genes are the collagen genes COL1A1 and COL1A2. The mutations in the COL1A1 gene tend to reduce the amount of collagen produced, while, less commonly, mutations in either COL1A1 or COL1A2 alter the structure of the collagen proteins making the collagen fibers weaker. The genes are located on human chromosome 17 at gene map loci 17q21.31-q22, 7q22.1.

oubain A toxic glycoside that specifically inhibits the Na^+–K^+ ATPase. The use of this inhibitor provided important information that helped elucidate the functioning of the ionic pump.

oxic Referring to an aerobic microbial habitat.

oxidative phosphorylation The formation of ATP from ADP (+ phosphate) using the energy of the electron transport process to drive the reaction.

oxidizing agent Any chemical agent that takes electrons, either as electrons or as electron-rich atoms, from another chemical with which it reacts.

oxidoreductase The class of enzymes that carry out electron transfers between two substrates. Oxidoreductases are the enzymes that are responsible for many of the electron transfers that occur in the electron transport chain in which energy from the electrons derived from the metabolism of sugars is used for oxidative phosphorylation.

oxygenases A group of enzymes that add oxygen across double bonds of the substrate molecule. This type of reaction is an essential step in the energy-producing metabolism of certain molecules, for example, fatty acids.

P

p19^ARF See INK4.

p53 A human phosphoprotein of 53 kilodaltons in size, originally discovered in studies of SV40 T antigen to which it is tightly bound. It was at first thought to represent the product of an oncogene but is now known to possess antioncogenic, or tumor-suppressing activity. p53 normally functions as a "gatekeeper" to block cells that contain potentially mutagenic DNA damage in the G1 phase of the cell cycle. Human cancers often contain mutations in p53, and the frequency of occurrence of such mutations in tumors is much higher than is true for any other known tumor suppressor.

packing ratio The length of a certain DNA divided by the length of the compartment into which it is packaged by folding. For example, if the DNA in the smallest human chromosome is 14 mm (4.6×10^7 bp) and the length of the chromosome is 2 μm then the packing ratio is: 14,000 μm/2 μm = 7,000.

paired-box homeotic gene (PAX5) A member of the paired-box (PAX) family of transcription factors that are important as regulators of genes involved in development; the gene family named for a highly conserved DNA binding motif called the paired box. The PAX5 gene encodes the B-cell lineage specific activator protein (BSAP) that is involved in differentiation of B cells but is also believed to play important roles in neural development and spermatogenesis. The PAX5 gene map locus is 9p13.

pair-rule mutants Mutants of the fruit fly, *Drosophila melanogaster*, in which features of normal development are missing in every other segment. See SEGMENTS, SEGMENTATION.

palindrome A nucleic acid-base sequence that is a "mirror image" of itself; for example, the base sequence GTGGCCGGTG is a palindrome because it consists of the sequence GTGGC and its "reflection," CGGTG. The recognition sequences of most restriction endonuclease enzymes are palindromes.

pandemic A worldwide epidemic.

papilloma virus A member of the papova group of DNA viruses that produces generally benign tumors (papillomas) of the epithelial cell layer in rabbits, cattle, and humans. Recently discovered members of the subclass representing human papilloma viruses (HPV) are now believed to cause some malignant genital cancers.

par A partitioning, functioning gene of some plasmids that ensures the proper segregation of plasmids into cells during cell division.

paralog Genes related to one another by duplication of an ancestral gene within a certain genome. Paralogs tend to develop new functions as the organisms that contain them evolve.

paranemic joint A side-by-side non-helical arrangement of DNA strands (as opposed to the usual plectonemic relationship) that occurs when a single-stranded circular DNA undergoes recA-mediated recombination with a linear double-stranded DNA.

parasegments An alternative scheme for labeling the segments of the fruit fly, *Drosophila melanogaster;* each parasegment begins in the middle of a segment. This scheme gains its utility from the fact that some mutants of *Drosophila* are more easily visualized in terms of parasegments. See SEGMENTS, SEGMENTATION.

parasexual The recombination of genetic material from different individuals that differs from sexual reproduction in that the genetic material is not derived from specialized meiotic cell types, for example, sperm and egg. Parasexual reproduction is characteristic of yeast.

p-arm The short arm of the human chromosome.

parthenogenesis The process of reproducing without fertilization, that is, asexually.

particle bombardment A technique used to transfer DNA into cells that are resistant to other methods of DNA transformation. This method has been used for certain plants and consists in coating the DNA with tungsten or gold particles and then projecting the DNA into the cells by an apparatus.

partition coefficient For some particular substance that is dissolved in two different solvents but that do not mix with each other, the partition coefficient is the ratio of the amount of the substance that remains dissolved in one solvent to the amount of the substance that remains dissolved in the other solvent when the two solutions are mixed and then allowed to separate from each other.

passive hemagglutination A test in which the presence (or absence) of an antibody is detected by the ability of the antibody to cause red blood cells to clump together (hemagglutination). Passive hemagglutination differs from the usual hemagglutination test because the surface of the red blood cells must be artificially modified by chemical linkage of a protein (the antigen) for the hemagglutination to take place.

passive immunity Immunity that is transferred from one individual to another by injection of blood components, for example, blood serum or cells.

passive transport Movement of a substance across a membrane from one side where the substance is at a relatively high concentration to the other side where the concentration is relatively low. See OSMOSIS.

Pasteur, Louis (1822–1895) French chemist who demonstrated the principle of sterilization, thereby destroying the idea that life could arise spontaneously from nonliving organic material (called spontaneous generation).

Pasteur effect The observation that when a microorganism living under anaerobic (oxygen-free) conditions is suddenly exposed to oxygen, sugar consumption as well as accumulation of the sugar breakdown product, lactate, drops.

pasteurization The process of destroying disease-causing microorganisms by heat. See STERILIZATION.

patch clamp A technique for measuring electric current flow across a neural membrane. In the patch-clamp technique, two electrodes, one inside and one outside the cell are used to record the voltage; simultaneously, a third electrode, usually inside the cell, is used to supply whatever current is needed to hold the membrane potential constant. The amount of current required to maintain a constant voltage provides a measure of the current passing through the membrane.

patent A document certifying an inventor or inventors to exclusive rights to an invention and any profits that may be derived from its use, sale, or license.

pathogen Any microorganism that causes disease or produces a pathological condition.

pathway A series of biochemical reactions occurring in a specified sequence by which a particular molecule (the precursor) is modified to become another; usually for purposes of synthesizing essential biochemicals or for degrading them.

pattern formation Generation of a particular three-dimensional body or structure during development of the embryo by the coordination of cell division, cell determination, and cell differentiation.

Pauling, Linus (1901–1994) American chemist who studied the behavior of electrons in chemical bonds. He won the Nobel Prize in chemistry in 1954 for his studies on the nature of structure of bonds in proteins (the peptide bond). Linus Pauling is also known for a wide variety of other contributions to the field of chemistry, including discoveries in the nature of quantum levels in chemical bonds, Van der Waals forces, electron resonance, and many others. Pauling's *The Nature of the Chemical Bond* is thought by many scientists to be one of the most influential scientific books of the 20th century.

pBR322 A commonly used plasmid for cloning recombinant DNAs in the bacterium *E. coli.*

PCNA Proliferating *c*ell *n*uclear *a*ntigen; a protein that functions during DNA replication in eukaryotes. PCNA binds DNA polymerase during synthesis of the leading strand and increases processivity of the polymerase.

PCR See POLYMERASE CHAIN REACTION.

pectin A jellylike substance released by plants. Pectin is made up of chains of the sugar derivative, galacturonic acid.

pectinase An enzyme that degrades pectin by breaking the links between the sugar units in pectin.

P element(s) A type of transposable element found in the fruit fly, *Drosophila.*

pellet The sediment portion of a biological extract after the extract is subjected to centrifugal force.

pendred syndrome (PDS) An autosomal recessive disorder that is characterized by deafness and thyroid goiter. The syndrome is caused by mutations in the PDS gene that encodes pendrin, an anion transporter localized at the apical membrane of thyroid follicular cells. Pendrin is believed to mediate iodide transport into the lumen of the follicle. There are at least 50 mutations in pendrin associated with PDS. The gene for pendrin is located at gene map locus 7q31.

penicillinase An enzyme that inactivates penicillin by breaking a key bond in the penicillin molecule by the process of hydrolysis.

penicillins Products of the *Penicillium* molds that act as an antibiotic by destroying bacteria by interfering with the cross-linking of proteins in the bacterial cell wall, causing the bacterium to break open, or lyse. The penicillins are all derived from a common chemical backbone (the lactam ring) by substituting various chemical groups (by convention the position of the substituent groups are given the general designation *R* in diagrams) at a certain position on the ring. See 6-AMINOPENICILLIC ACID and LACTAM ANTIBIOTICS.

pentose Any sugar with a five-carbon-atom backbone.

penicillin derivatives
obtained by substituting
groups for "R"

The penicillin backbone

N-terminal end · peptide bond · C-terminal end

amino acid 1 amino acid 2 amino acid 3 amino acid 4 amino acid 5 amino acid 6 amino acid 7

Peptide

peptidases A class of enzymes that breaks down proteins by cleaving the peptide bonds between the individual amino acids that make up the protein, for example, the digestive enzymes, chymotrypsin, trypsin, and pepsin. The breaking of peptide bonds by peptidases occurs by a process known as hydrolysis.

peptide A group of amino acids covalently linked by peptide bonds in a linear chain.

peptide antibiotic A short peptide with antimicrobial properties, for example, gramicidin A.

peptide bond A covalent linkage between the $-NH_2$ group of one amino acid and the $-COOH$ group of another amino acid. This type of linkage is known as an amide bond when applied to links between molecules that are not amino acids.

peptide hormone A short peptide secreted into the bloodstream that induces biological activity in a distant target gland or organ; for example, the pituitary hormones—the growth hormone (GH), the thyroid stimulating hormone (TSH), and the adrenocorticotropic hormone (ACTH).

peptidoglycan A peptide covalently attached to chains of sugars or sugar derivatives. Peptidoglycans are structural components of bacterial cell walls. In Gram-positive bacteria, the peptidoglycan portion of the cell wall is present in many layers. The peptidoglycans are the targets of the penicillin antibiotics that are incorporated into the bacterial peptidoglycan, where they prevent critical peptide cross links that weaken the bacterial cell wall.

peptone The water-soluble portion of a protein that has been partially broken down (i.e., hydrolyzed) such as by boiling. See HYDROLYSATE.

perfusion culture A culture in which there is a continual inflow of fluid that carries nutrients or other substances.

pericentriolar material Material of unknown composition surrounding the centriole of the chromosome. This material serves as the anchor points for the microtubules that pull the chromosomes apart during mitosis.

perinuclear space The space between the inner and outer nuclear membranes.

periodicity In a structure that has a regularly repeating subunit, periodicity refers to the distance that represents one complete subunit.

periplasmic space The area between the cell membrane and cell wall of Gram-negative (see GRAM STAIN) bacteria, such as *Escherichia coli*.

PERL Practical extraction and report language; a high-level, open-source computer language that was released in 1987. PERL is optimized for scanning and extracting information from arbitrary text files and, for this reason, is often used in bioinformatics. PERL is also optimized for ease of use and combines features of other popular languages such as PASCAL, C++, and BASIC.

permissive host A cell that, when infected by a virus, allows the expression of a particular viral function(s), usually replication of the virus.

peroxidase An enzyme that acts to promote the breakdown of hydrogen peroxide (H_2O_2) into water according to the reaction

$$\text{peroxidase}$$
$$H_2O_2 + \text{substrate–}H_2 \longrightarrow 2H_2O + \text{substrate}_{ox}$$

peroxidase labeling The attachment of a peroxidase enzyme (e.g., horseradish peroxidase) to a probe so that the presence of the probe can be visualized by a colorimetric reaction based on the activity of the enzyme.

peroxisome Small, self-replicating cytoplasmic organelles that contain no DNA but are composed largely of the peroxidase enzyme, catalase.

peroxisome proliferator-activated receptors (PPARs) PPARs are transcription factors that are activated by receptor-mediated signaling pathways that activate COX enzymes. Three different PPAR isotypes are known: α, β and γ. The gene for PPAR-γ codes for a product that regulates the development of fat cells (adipocytes). The PPARs also function as receptors for two classes of drugs: the hypolipidemic fibrates and the thiazolidinediones.

PEST motif A sequence of amino acids (Pro-Glu-Ser-Thr), discovered in the Notch protein but also present in other developmentally important proteins, that functions as signal for rapid proteolytic degradation. A number of developmental processes are believed to be regulated in part by the destruction of regulatory proteins at critical times in development.

petite mutant A microorganism (especially yeast and euglena) lacking mitochondria. In yeast, such mutants form tiny colonies when grown on a nutrient source low in sugar.

P factors DNA sequences carried on various chromosomes in the male fruit fly that bring about hybrid dysgenesis in matings with females of certain strains (referred to M strains for maternal contributing). See HYBRID DYSGENESIS.

PFGE See PULSED FIELD-GEL ELECTROPHORESIS.

pH A measure of the acidity of an aqueous solution; if the pH value of a solution is below 7, the solution is considered acid; solutions with a pH value above 7 are considered alkaline.

phage Short form of *bacteriophage* (literally meaning "bacteria eater"), a virus that infects and then usually destroys the bacterium that it infects. The destruction of the infected bacterium (the host) with the release of progeny viral particles is the last step in the virus life cycle.

phage display A technique used to search for proteins or peptides that will interact with a target protein. A microtiter plate is bait-coated, or coated with target protein. A library of possible interacting proteins or peptides is prepared by cloning DNA, which may encode those sequences into a vector that places the DNA in a gene for a phage head structure. When the phage is grown, millions of phage particles will be produced, each displaying a different fusion product on its surface. These phages are then portioned out into the microtiter plate wells. When the plates are repeatedly washed the phages displaying an interacting peptide to the bait will stick to the well, and all other phages will be washed away. The phages can then be eluted from the well and grown up to isolate the peptide of interest.

phagemid A cloning VECTOR constructed with components from plasmids and bacteriophages so that it can replicate either as a plasmid or phage.

phagocyte Any cell that normally carries out phagocytosis; usually applied to certain white blood cells, for example, macrophages that carry out phagocytosis as part of their function in the immune system.

phagocytic index An assay for the detection of phagocytic activity in a blood specimen. Among the tests used for this purpose are staining by the dye nitro-blue tetrazolium (NBT) that turns blue when particles containing the dye are phagocytized or uptake of latex beads by phagocytic cells.

phagocytosis The process by which a particle or cell becomes engulfed and ultimately devoured by another for purposes of sustenance or defense.

phagosome Following phagocytosis the engulfed particle is found in a membrane-enclosed vesicle in the cytoplasm of the phagocyte referred to as a phagosome.

phalloidin An alkaloid derived from the toadstool *Amanita phalloides* that binds to the actin filaments in a cell, thereby preventing cell movement.

pharmacology The study of the action of drugs, particularly as it relates to their therapeutic uses.

phase-contrast microscopy A type of microscopy in which the image of the specimen being viewed is enhanced by a technique involving manipulation of the light that is deflected by the specimen. In normal microscopy, only the light passing straight through the specimen is used to create the image of the specimen.

phase variation A mechanism of gene regulation in which a gene is switched on in response to an environmental condition in some of the cells in a population leading to a mixture of expressing (phase ON) and non-expressing (phase OFF) cells. An example of this type of regulation is seen in the expression of type 1 fimbriae (an adhesin) in *E. coli*. Transcription of a group of genes (fim genes) is controlled by a short invertible element in the promoter. Expression of the gene cluster is either ON or OFF, depending upon the orientation of the invertible element.

phenotype The set of characteristics that make a living organism distinct from others.

phenylalanine One of the 20 amino acids that make up proteins; designated by the three letter code *Phe* or by the single letter code *F*. Phenylalanine is also used in the body to make the neurotransmitter dopamine as well as adrenaline.

phenylketonuria (PKU) A genetic disease based on an inability to convert the amino acid phenylalanine into the amino acid tyrosine. This results in the accumulation of a toxic substance (phenylpyruvate) that causes severe mental retardation if the condition, which is manifest in newborns, is not treated by adherence to a diet low in phenylalanine. Because the disease has been traced to a deficiency of a particular enzyme (phenylalanine hydroxylase), prevention of the disease in susceptible individuals is a goal of modern genetic engineering.

pheromone A chemical signal secreted by an animal that brings about a specific behavior (e.g., mating) in an animal of the same species.

pheromone-responsive element (PRE) A nucleotide sequence in the promoters of certain yeast genes that binds the transcription factor STE12 in response to the binding of mating factors (pheromones) to specific cell-surface receptors. It is believed that genes required for mating are, in this way, activated by pheromones.

Philadelphia chromosome A type of reciprocal translocation between chromosomes 9 and 22 that is seen in patients with chronic myelogenous leukemia (CML).

phloem A type of plant vascular tissue surrounding the xylem that makes up the vessels that conduct fluids downward along the stem or trunk of the plant toward the root.

PHO5 The gene for acid phosphatase. The promoter for this gene has been incorporated in expression vectors for eukaryotic (see EUKARYOTE) genes. The advantage of this promoter is that it can be regulated because it is induced by the removal of phosphate from the culture medium. Many cloned genes are toxic to the host cell producing them and can only be expressed when the cells have reached a maximum biomass. Thus regulation allows the cells to grow to the maxiumum density before the toxic protein is expressed.

phorbol esters A class of compounds that act as tumor promoters. See TPA.

phorbol myristate acetate

Phorbol esters

phosphatide A type of phospholipid made up mainly of glycerol, fatty acids, and phosphate. Phosphatides are the type of lipid that make up the bulk of the phospholipids found in cell membranes.

phosphatidylethanolamine (PE) A type of phospholipid common to many cell membranes. PE is also a common constituent of many artificial membranes, such as those used to construct LIPOSOMES.

phosphatidyl inositol kinase (P13 kinase) An enzyme that adds phosphate groups onto the inositol portion of the membrane phospholipid, phosphatidyl inositol using ATP. The product, phosphatidylinositol-4,5-biphosphate, is an important intermediate in a signal transduction pathway in which the phosphoinositol group is cleaved from the phospholipid and is released into the cytoplasm, where it acts as a second messenger.

phosphodiesterase An enzyme that catalyzes the breakage of phosphodiester bonds.

phosphodiester bond A covalent bond that attaches a phosphate group to any other group by an oxygen-atom bridge. Phosphodiester bonds are the linkages that join the sugar molecules to one another in the backbone of nucleic acids.

$$(deoxy-)ribose$$
$$|$$
$$(deoxy-)ribose-O-P=O$$
$$|$$
$$O^-$$

phosphoglycerate kinase (PGK) An enzyme used in the process of glycolysis, or the breakdown of glucose, to form energy in the cell. Because the gene for PGK is highly expressed, the promoter for this gene has been incorporated in many expression vectors for eukaryotic genes.

phospholipase Any of a class of enzymes that acts to break down phosphatides by breaking the bonds between the glycerol portion of the phosphatide and the attached fatty acid(s) or the bonds between the glycerol portion and the phosphate. Phospholipases are the major mediators of phosphatide turnover in the membrane. Phospholipases also play a role in membrane signaling by liberating membrane molecules used in the synthesis of eicosanoids and PIP3 signaling.

phospholipase A2 An enzyme that catalyzes the hydrolysis of the fatty acid ester linkage at the second carbon of the glycerol backbone of membrane phosphatides, thereby releasing the esterified fatty acid as a free molecule. Phospholipase A2 is the enzyme responsible for releasing arachidonic acid from the plasma membrane for subsequent eicosanoid synthesis.

phospholipid The general class of lipids that are made up of fatty acids and phosphate and that are the main component of all cell membranes. Phosphatides and sphingosines are the two major phospholipid subclasses.

phosphomycin A phosphate-containing antibiotic produced by streptomyces.

phosphoramidite A chemically modified nucleotide used in the synthesis of oligo-nucleotides that contain an activated phosphoester group at the 3′ carbon and a DMT blocking group at the 5′ carbon.

Phosphoramidite

phosphoribosyltransferase An enzyme that is necessary to make nucleotides from free purine and pyrimidine bases. The action of this enzyme is responsible for a variety of important biomedical and research applications, for example, the labeling of nucleic acids or the killing of cancer cells with lethal analogs of the normal purines and pyrimidines.

phosphoric acid See ORTHOPHOSPHATE.

phosphorylation Any of a variety of biochemical processes by which a phosphate group is added to an organic molecule. However, the term usually applies to the phosphorylation of nucleosides, particularly adenosine, resulting in the formation of adenosine mono-, di-, and tri- phosphates (AMP, ADP, and ATP, respectively). Phosphorylation resulting in ATP formation is the major means of storing energy for all forms of biological activity. ATP formation produced during photosynthesis that is therefore light requiring:
- oxidative phosphorylation—oxygen-dependent formation of ATP; this means of generating ATP is coupled

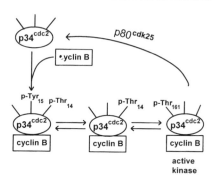

Phosphorylation

to the process of sugar oxidation (i.e., breakdown) in animal cells.
- substrate level—generation of ATP that occurs as part of the oxidation of sugars but does not require oxygen.

phosphotyrosine A form of the amino acid tyrosine in which the –OH group on the side chain is covalently attached to a phosphate group. Phosphotyrosine residues in proteins are of significance because signal-transduction pathways that effect cell growth regulation and oncogenesis involve phosphorylation of specific tyrosine residues by protein kinases.

phosphotyrosine-binding domain (PTB domain) A region on certain signal transduction proteins that binds to a portion of the cytosolic regions of receptor tyrosine kinases (RTKs) that contain phosphorylated tyrosine residues. While the PTB binds the RTK, a second domain on the same protein, called the src homology (SH2) domain, binds to another protein in the signal transduction chain. These PTB-containing proteins are components of signaling pathways often involved in regulating cell growth from ligand-activated receptors.

photoaffinity labeling The use of light to activate certain light-sensitive molecules so that they spontaneously bond to a protein, a nucleic acid, or another type of molecule. This is a technique for labeling biologically important substances if the activated molecule provides a highly visible color or generates some other type of signal with high detectability.

photoautotroph A photosynthetic organism capable of living on only minimal nutrients, that is, capable of making all its necessary biomolecules from simple organic molecues.

photoheterotroph A photosynthetic organism that is deficient in the ability to make one or more of its essential biomolecules and therefore requires nutritional supplements in its growth medium.

photolithography An automated technique for synthesizing oligonucleotides at specific locations (spots) on a microarray. In photolithography individual nucleotides are added one at a time to preexisting nucleotide chains by shining light through a screen. The light activates special light-sensitive nucleotides, causing them to form a covalent bond with the free end of the nucleotide chain(s). An ordered sequence of screens and activated nucleotides programmed by computer can generate thousands of oligonucleotides with defined sequences at specific locations on the microarray.

photolyases A group of light-activated, direct DNA repair enzymes that catalyzes the repair of pyrimidine dimmers resulting from ultraviolet radiation. Photolyases use $FADH_2$ and a folate derivative (MTHFpolyGlu; methenyltetrahydrofolylpolyglutamate) as coenzymes to repair the dimers by reducing the pyrimidine-pyrimidine bonds.

photomultiplier A photosensitive device that makes use of the phenomenon of photoemission and secondary electron emission to detect low levels of light. Electrons emitted from a photosensitive material by incident light are accelerated and focused onto a secondary-emission surface (called a dynode). Several electrons are emitted from the dynode for each primary electron produced. The secondary electrons are then directed onto a second dynode where more electrons are released. This process is repeated a number of times to amplify the initial photocurrent so that extremely low levels of light can be detected.

photon The unit of light that represents one discrete packet of light energy.

photophosphorylation The process in which sunlight is used to produce ATP.

photoreactivating enzyme See DNA PHOTOLYASE.

photorespiration A salvage process that occurs in plants under conditions of high O_2 and low CO_2 concentrations, where the enzyme responsible for fixing CO_2 instead takes up O_2 and releases CO_2.

photosynthesis The process by which plants utilize light energy to create sugars and produce oxygen from carbon dioxide and water.

photosystem A cluster of chlorophylls and other pigments that functions to capture the light energy that is used to carry out photosynthesis.

phototaxis Movement toward light.

phototroph An organism that is wholly dependent upon light for nourishment via photosynthesis.

phragmoplast The enlarged football-shaped spindle that is seen toward the end of mitosis in plant cells; the structure in which the cell plate forms.

phycomycetes A class of primitive fungi that shares many features in common with fungi.

phylogeny The construction of evolutionary trees based on relatedness of organisms. DNA sequencing and the field of bioinformatics are making large contributions to the understanding of the evolution of genes and organisms.

phytochrome A pigment-protein that is believed to play a role in the initiation of plant development when activated by light in the red or near-red part of the spectrum.

phytohemagglutinin A class of proteins that causes clumping of red blood cells (hemagglutination) by binding to certain sugar chains on the cell surface; also referred to as lectins. Examples are concanavalin A and ricin.

phytotoxin Any of a number of highly poisonous substances produced by plants.

picogram 10^{-12} or 0.000000000001 grams.

picornaviruses A class of RNA viruses originally termed *enteroviruses* because they were initially discovered in the intestinal tract. Currently, the picornaviruses are classified into two subclasses: the enteroviruses (poliovirus, coxsackievirus, echovirus, and enterovirus) and the rhinoviruses (rhinovirus). The name is derived from *pico-* (small) and *rna* to denote RNA.

pilus A hairlike structure, found on donor type *Escherichia coli,* that is used in the attachment of donor-type cells (F$^+$ and Hfr) to recipient cells (F$^-$) to mediate the transfer of DNA during mating. Also called the sex pilus.

pinocytosis A variation of phagocytosis in which the engulfed particle is taken into the cell in small vacuoles representing pinched-off pieces of the original particle.

piperidine A chemical that causes breakage of the sugar-phosphate backbone in nucleic acids at any point along a nucleic acid strand at which the purine or pyrimidine rings have been partially oxidized or completely removed. Piperidine treatment is used to create subfragments of a larger nucleic acid in the Maxam-Gilbert procedure for sequencing nucleic acids.

pituitary gland A small endocrine gland located at the base of the brain that secretes a number of important polypeptide hormones including follicle stimulating hormone (FSH), leutinizing hormone (LH), and prolactin, all of which play a role in stimulation of the female reproductive organs. The pituitary gland also produces adrenocorticotropic hormone (ACTH), somatotropin, and thyrotropin.

pK, pKa Terms that represent the "strength" of a chemical reaction; the degree to which some reaction will proceed in the direction written, for example, A + B → C + D. The negative logarithm of the equilibrium constant (–log[K], where K = [C][D]/[A][B]) for the chemical reaction.

plankton The small floating plant and animal life in a body of water.

plaque A clear area in an immobilized carpet of bacteria that is produced by local destruction of the bacteria in that area by bacteriophages.

plaque assay A means of determining the number of bacteriophage in a suspension by counting the number of plaques produced in a certain amount of the suspension. The results are usually expressed as plaque forming units per milliliter of suspension (PFU/ml).

plaque hybridization A process in which a labeled probe is annealed to the DNA from bacteriophages in a plaque. Plaque hybridization is used to identify plaques containing bacteriophage-carrying recombinant DNA.

plasma The liquid portion of blood.

plasma cell An antibody-secreting white blood cell.

plasma gel The protoplasm of a protozoan that is in the gel form, for example, the protoplasm in the pseudopodium of *Amoeba proteus.*

plasma membrane The membrane surrounding the cytoplasm of a eukaryotic cell. The plasma membrane is similar in structure to the cell membranes of prokaryotes and consists of a phospholipid bilayer and an overlying extracellular layer of glycoproteins.

plasma sol The protoplasm of a protozoan that is not in gel form.

plasmid A piece of DNA in the cytosol of bacteria that replicates independently from the bacterial chromosome. Naturally occurring plasmids have been found to carry a number of genes; perhaps most important are the genes that confer resistance to a number of antibiotics. Genetically engineered plasmids are important vectors for carrying recombinant DNAs.

plasminogen activator An enzyme that derives its name from the ability to catalyze the conversion of plasminogen to plasmin, which then catalyzes the breakdown of fibrin, a major component of blood clots. The secretion of plasminogen activator is a marker of cell transformation to a cancerous or precancerous state. Plasminogen activator is used as a therapeutic agent to dissolve blood clots associated with blockage of the coronary arteries.

plastid An organelle found in plant cells that contain its own genome; chloroplasts are a type of plastid.

platelet A subcellular particle in blood that is actually a fragment of a megakaryocyte cell that is formed in the bone marrow. Platelets bound to fibrinogen initiate the formation of a blood clot at the site of a wound.

platelet-derived growth factor (PDGF) A growth factor that is present in the granules of platelets and that probably plays a role in wound healing. PDGF consists of two subunits, one of which was found to represent a slightly changed version of the sis oncogene.

plectonemic Pertaining to the standard double helical arrangement of double-stranded DNA. See DOUBLE HELIX.

plectonemic supercoiling A term used to describe a type of supercoiling in which the DNA supercoils are folded over themselves to form branches. This type of structure is believed to represent a mechanism by which DNA is compacted. In plectonemically supercoiled DNA, the length of the DNA mass is approximately 40 percent of the length of the uncompacted DNA.

pleiotropic Any agent, such as a hormone, having more than one effect or having an effect on more than one target.

pleomorphic Having variable form, for example, variations in shape, behavior, or other characteristics of organisms of the same species.

point mutation A change in a single nucleotide in a gene resulting in loss of function or altered functioning of that gene.

pol One of the three major genes of retroviruses. The pol gene encodes the protein for the viral enzyme reverse transcriptase.

polar body A small cell that is produced as a result of uneven separation of cytoplasm during meiois when an oocyte is produced; the larger of the daughter cells becomes the oocyte.

polar group A small group of atoms held together by dipole-dipole linkages.

polarimeter An instrument for determining the percentage of polarized light in a beam of light. A polarimeter is also used to determine whether polarized light is rotated after passage through crystals of a given compound.

polarity 1. As applied to a molecular bond between two atoms, the term *polarity* refers to a state in which the electrons in the bond are localized more to one atom than the other, giving that atom a partial negative charge (the other atom is partially positively charged). The presence of polar bonds confers a number of important chemical properties to the compound that contains them, including solubility in water.

2. As applied to cell microanatomy, polarity refers to specialization of the cell architecture at different parts of the cell, for example, the presence of cilia and secretory vesicles located at the apical, as opposed to the basal, end of the epithelial cells lining the gut and respiratory tracts.

3. As applied to nucleic-acid strands, the term refers to the fact that the two ends of any nucleic-acid strand are distinguishable from one another by whether the end is 5' or 3'. This gives the strand a directionality or polarity.

polar microtubules The microtubules extending between the polar bodies that have the apparent function of pushing the poles of a dividing cell apart during mitosis.

polar mutation A nonsense mutation that causes early termination of normal transcription in a gene and that therefore also prevents transcription of any subsequent genes in a polycistronic unit.

poliovirus A picornavirus that infects individuals via ingestion but then attacks the central nervous system, resulting in varying degrees of paralysis.

polyacrylamide A polymer of acrylamide:

$$NH_2$$
$$|$$
$$CH_2=CH-C=O$$

A gel is formed to which the polyacrylamide strands are cross linked. Polyacrylamide gels are used for electrophoresis of proteins and nucleic acids.

polyacrylamide-gel electrophoresis Separation of nucleic acids or proteins from a heterogeneous mixture on the basis of size or charge by placing the mixture in a polyacrylamide gel and then subjecting it to an electric field.

polyadenylated The general term for nucleic acids that have undergone polyadenylation.

polyadenylation The addition of long tracts of adenosine polymers to the tail ends (i.e., the 3' ends) of messenger RNAs in eukaryotic cells.

adenosine   adenosine   adenosine
| | |
...–phosphate–ribose–phosphate–ribose–phosphate–ribose–...

polyadenylation signals Certain sequences of bases in an RNA molecule that are required for polyadenylation to-occur; for example, the sequence AAUAAA located in a region 11–30 nucleotides from the end of an mRNA molecule.

polyamine Molecules formed from repeating hydrocarbon chains separated by amino groups. The chain is always terminated at each end by a positively charged amino group. Because of their positive charge, the polyamines function to stabilize nucleic acids by neutralizing the strong negative charge of the nucleic acid phosphate backbone. Putrescine, spermidine, and spermine are the common polyamines.

poly-A polymerase The enzyme responsible for polyadenlylation of an RNA strand.

polycistronic A region of a nucleic acid that contains sequences representing multiple genes (cistrons) in an end-to-end tandem arrangement.

polycistronic mRNA The messenger RNA transcribed from a polycistronic DNA.

polyclonal antibody A set of antibodies that is secreted by a corresponding set of antibody-producing white blood cells. Although each of the antibodies carries a unique specificity, the set of antibodies as a whole reacts with a variety of antigenic molecules.

polyelectrolyte A large molecule that is highly charged under biological conditions.

polyethylene glycol (PEG) A long polymer of $-\overset{OH}{\underset{|}{C}}-$ groups. PEG is used in inducing cell fusion, precipitating microscopic particles, dehydrating samples of

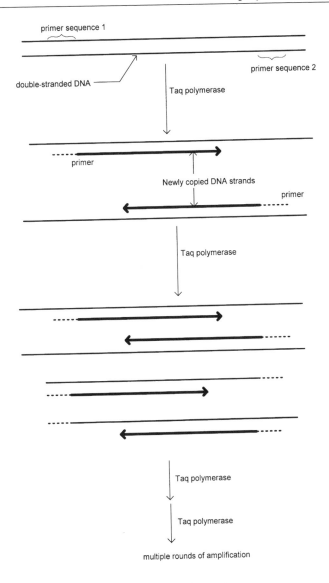

primer sequence 1

double-stranded DNA

primer sequence 2

Taq polymerase

primer

Newly copied DNA strands

primer

Taq polymerase

Taq polymerase

Taq polymerase

Taq polymerase

multiple rounds of amplification

Polymerase chain reaction

biological materials, and stabilizing certain enzymes.

polylinker A synthetic polynucleotide containing the sequences representing the restriction sites of certain specified restriction enzymes.

polymer A chain composed of a single molecule or small number of similar molecules that are linked together.

polymerase chain reaction (PCR) A technique for rapid amplification of extremely small amounts of DNA, using the heat-stable Taq I DNA polymerase enzyme. PCR has found wide application in forensic medicine because analyzable quantities of nucleic acid can be obtained even from microscopic tissue samples.

polymerases A class of enzymes that catalyze the formation of long nucleotide polymers, particularly as a means of making template-driven copies of nucleic acids. See DNA and RNA POLYMERASES.

polymorphism A naturally occurring variation in the normal nucleotide sequence within the individuals in a population.

polymyxin A group of antibiotics derived from the bacterium *Bacillus polymyxa*, with activity primarily against Gram-negative bacteria.

polynucleotide A polymer consisting of a long chain of nucleotides.

polynucleotide kinase (polynucleotide phosphorylase) An enzyme that catalyzes the transfer of a phosphate group from ATP to the free 3′ end of a polynucleotide.

polynucleotide ligase An enzyme that links two polynucleotides together. The enzyme catalyzes the formation of a covalent bond between a phosphate group on the 5′ end of one polynucleotide and the free 3′ hydroxyl group at the end of the other polynucleotide.

polyoma A member of the papova group of viruses, which normally infects rodent cells.

polypeptide A polymer of amino acids; the term usually applies to a peptide chain of fewer than 100 amino acids.

polyploid Having more than the normal diploid number of chromosomes.

polyploidy The state of being polyploid.

polyribosome (polysome) A number of ribosomes attached to the same messenger RNA.

polysaccharide A chain of sugar molecules linked together end-to-end.

polyspermy Fertilization of a single egg by more than one sperm.

polytene chromosomes Giant chromosomes found in insect salivary gland cells. They are actually made up of thousands of copies of the DNA normally found in one chromosome.

polyteny The state of a cell containing polytene chromosomes.

poly U Poly uridylic acid; a polymer of uridine of undefined length.

P/O ratio The amount of phosphate incorporated into ATP divided by the amount of O_2 taken up by the mitochodrion during the process of respiration; generally taken as a measure of the efficiency of energy production when sugars are oxidized.

porins Protein channels in the lipopolysaccharide layer of Gram-negative (see GRAM STAIN) bacteria that serve as portals for the flow of small molecules.

porphyrin An organic molecule made up of four nitrogen-containing rings called pyrrolles. Modified porphyrins are the basic constituents of the active sites of hemaglobin, myoglobin, chlorophylls, and cytochromes.

position effect The influence of the position of gene on its activity such as is seen when a gene that is moved (e.g., by translocation) to a new chromosomal location becomes inactive.

posttranscriptional processing Certain specific changes in the RNA that occur before the RNA leaves the cell nucleus in mature form. Polyadenylation, capping, and splicing are examples of posttranscriptional processing.

posttranslational import A process by which a certain class of proteins is brought into the interior of the endoplasmic reticulum (ER) either following or during its synthesis on ribosomes that are bound to the ER; polypeptides that are to be imported are recognized on the basis of the fact that they contain a small sequence of amino acids known as a signal peptide at one end.

posttranslational modification Some alteration in the structure of a polypeptide, for example, addition of a polysaccharide chain (glycolsylation), after it is synthesized and usually after it is imported into the interior of the endoplasmic reticulum. Such modification is required for the polypeptide to take on its biological activity.

posttranslational processing Removal of a specific end piece of a polypeptide known as the signal peptide, following its synthesis on ribosomes; one part of the process of posttranslational import.

posttranslational transfer The import of a polypeptide into the membrane of the endoplasmic reticulum or an organelle after synthesis of the polypeptide (i.e., translation) is completed.

poxvirus A class of DNA viruses that produces transient inflammatory skin lesions, for example, chicken pox and smallpox.

Prader-Willi syndrome (PWS) A disorder caused by chromosomal deletion and/or disomy involving genes on the proximal arm of chromosome 15. The syndrome is characterized by obesity, hypotonia, mental retardation, short stature, hypogonadotropic hypogonadism, and strabismus. About 70 percent of the PWS cases show a deletion in the chromosome 15 region, 15q11.2-q13. Several genes possibly involved in the disorder have been mapped to this region, including a small ribonucleoprotein gene (SNRPN), type II oculocutaneous albinism (P gene), and a ubiquitin-protein ligase (UBE3A).

PWS was the first human disease based on the phenomenon of genomic imprinting in which genes are differentially expressed depending upon the parent from which the disease originated. DNA methylation at cytosine bases may be the mechanism of this type of imprinting.

precursor A substance from which another substance is made by a series of sequential changes in molecular structure.

prednisone A synthetic steroid hormone used to reduce chronic inflammation such as occurs in arthritis.

pre-mRNA The general term given to that subclass of RNAs present in the nucleus that will be processed to become mature messenger RNA (mRNA) but that has not yet undergone that processing. See POSTTRANSCRIPTIONAL PROCESSING.

preproinsulin A precursor of insulin. Proinsulin, which is the immediate precursor of insulin, is produced from preproinsulin by cleavage of preproinsulin, resulting in the removal of a short polypeptide portion.

preprotein The polypeptide precursor of a membrane-bound protein prior to its actual insertion into a membrane. Because the leader sequences of membrane-bound proteins are removed on insertion into the membrane, preproteins have leader sequences.

prenatal diagnosis The diagnosis that a disease exists in a developing fetus made on the basis of examination of cell or tissue samples taken from the fetus in the womb.

presenilins Two proteins (PS1 and PS2) found on the surface of neurons that have been found to be involved in the development of familial Alzheimer's disease. The presenilins act together with another protein (gamma-secretase) to cleave the amyloid precursor protein that leads to the accumulation of amyloid plaques, causing rapid deterioration of the central nervous system at an early age (<60 years).

Pribnow box   A sequence of bases in the DNA that makes up part of the promoter of a prokaryotic gene. The Pribnow box occurs at 10 base pairs from the site at which transcription starts and consists of the sequence TATAAT or a close variation.

primary culture   The cell culture that arises from a tissue specimen when it is first placed into culture.

primary response   The elicitation of an immune response to a foreign antigen after an animal is first exposed to the antigen.

primary stucture   The sequence of amino acids that make up a polypeptide.

primase   The enzyme that catalyzes the formation of short RNA primers that are required to copy the DNA strand, starting from the 5' end.

primeosome   A complex of prepriming proteins and DnaG primase with a hairpin fold of single-stranded DNA in the bacteriophage φX174. This complex is the structure in which synthesis of the RNA primers in Okazaki fragments occurs.

primer   A short oligonucleotide that anneals to a specific region on a DNA or RNA strand and is used by a polymerase as a place to begin synthesis of a complementary nucleotide strand.

primer extension   A technique of mapping genes in which a primer is annealed to a DNA or RNA fragment and then extended, using an RNA or DNA polymerase (e.g., the Klenow fragment of DNA polymerase I) and the four nucleoside triphosphates to copy the nucleic acid to which the primer is annealed. Primer extension is most commonly used to detect mRNAs that contain the primer sequence.

priming   The process of annealing a primer.

prion   A type of protein particle that is responsible for a number of neurodegenerative diseases in humans and animals called *transmissible spongiform encephalopathies* (TSEs). The TSEs caused by prions includes scrapie (in sheep), kuru (afflicting the cannibalistic Foré tribe in Papua New Guinea), Creutzfeldt-Jakob disease (CJD), Chronic Wasting Disease, Fatal Familial Insomnia (FFI), Gerstmann-Sträussler-Scheinker syndrome (GSS), and bovine spongiform encephalopathy (mad cow disease). The name *prion* is derived from the descriptor *proteinaceous infectious particle*. Prions are unique as infectious agents because they contain no nucleic acids. The first prion protein was discovered by Stanley B. Prusiner, who was awarded the Nobel Prize in physiology or medicine in 1997. The infectious agent was named PrP (prion-related protein) and it is believed that the disease is caused by a change in shape of this protein (the altered version of the protein was called PrPSc) and that the misshapen protein can be "transmitted" by inducing the same change in shape to other nearby normal protein molecules.

probe   Any oligonucleotide that contains a chemical label allows the oligonucleotide to be traced when the oligonucleotide is annealed by hybridization to some target nucleic acid. To a lesser extent, any biomolecule including a protein, lipid, or polysaccharide that binds to some target molecule and that bears a chemical label that can be traced after binding has occurred.

processed pseudogenes   Pseudogenes that show a close similarity in nucleotide sequence to the mRNA for their active counterparts. The existence of processed pseudogenes has been taken as evidence that some pseudogenes were somehow originally derived from mRNAs.

pro-dynorphin/pro-enkephalin/pro-opiomelanocortin   Precursor molecules that give rise to various opioid peptides by proteolytic processing:

• Pro-opiomelanocortin is processed into γ-MSH, adrenocorticotropic hormone

(ACTH) and β-lipotropin. The latter two peptides are further processed into α-MSH and CLIP (from ACTH) and γ-lipotropin, β-endorphin.

- β-MSH and met-enkephalin (from β-lipotropin)
- Pro-enkephalin A gives rise to met-enkephalin and leu-enkephalin and some other larger peptides. Pro-enkephalin A contains four copies of met-enkephalin, one copy of leu-enkephalin, a heptapeptide (met-enk-arg-phe), and an octapeptide (met-enk-arg-gly-leu).
- Pro-dynorphin (pro-enkephalin B) gives rise to dynorphin A, dynorphin B, and β-neoendorphin as well as the smaller peptides dynorphin A1-8 and leu-enkephalins, three of which are contained within the pro-dynorphin peptide.

profilin A protein that complexes with actin proteins, preventing the polymerization of these proteins into actin filaments.

progeria A rare disease characterized by a collection of symptoms that give the appearance of premature aging, including baldness, prominent scalp veins, thin limbs with prominent joints, short stature, joint stiffness, hip dislocations, and heart and artery disease. At least two forms of the condition are known: Werner syndrome and Hutchinson-Gilford progeria. Hutchinson-Gilford progeria, the more severe form, occurs in children at a frequency of about one in 8 million and is caused by a point mutation in the gene for a nuclear lamin, known as lamin A (LMNA; gene map locus 1q21.2). Werner syndrome is caused by mutations in a gene called Wrn, which is a homologue of the *E. coli* helicase RecQ and is located on chromosome 8 at gene map locus 8p12-p11.2.

progesterone A steroid hormone produced by the ovary that prepares the uterus for reception of the fertilized egg.

prokaryote A term for the family of all primitive organisms, for example, bac-

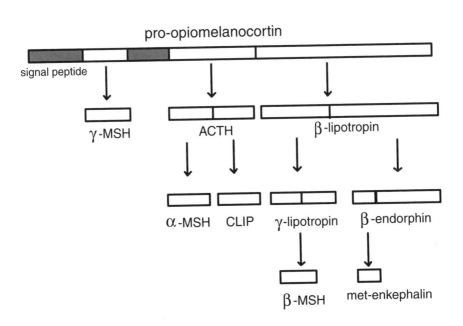

Opioid peptides

teria, in which the cellular DNA is not enclosed in a nucleus.

prolactin A hormone that stimulates the production of milk by the mammary glands following pregnancy.

proline-rich activation domain A characteristic region in a certain class of transcriptional activator proteins that contains a large number of proline residues. Transcriptional activators that contain proline-rich activation domains can stimulate transcription by binding to sequences that are either close to, or far away from, the TATA box unlike the glutamine-rich or acidic activators that tend to stimulate transcription from locations either very close to (glutamine-rich activators) or far away from (acidic activators) the TATA box. Examples of transcription factors containing proline-rich domains are AP-2 and CTF/NF1.

promoter A sequence of bases in a nucleic-acid strand that serves as a signal for the start of transcription of a given gene.

prometaphase The stage preceding metaphase in which chromosomes that have newly formed in the cytoplasm begin to migrate to the center of the cell where homologous chromosomes will pair with one another during metaphase.

proofreading The ability of the DNA polymerase to remove a mismatched base in the replicated strand with an exonuclease that cuts 3′–5′ and then to reinsert the correct base in the new strand with its polymerase activity. Organisms that lack this activity through mutation show a high rate of mutagenesis.

prophage The genome of a bacteriophage that has become integrated into the chromosome of the host bacterium.

prophase The first phase of mitosis characterized by the appearance of chromosomes from the amorphous chromatin.

prophylactic The treatment with a therapeutic agent to protect an individual from future disease.

prostaglandins A class of hormonelike chemicals derived from fatty acids. There are more than 16 prostaglandins that are classified into nine different groups. Stimulation of muscle contraction and inflammation are believed to be caused by prostaglandins. The anti-inflamatory effects of aspirin are related to inhibition of prostaglandin synthesis.

prostate-specific antigen (PSA) A glycoprotein originally isolated from semen that is now used as an indicator of the presence of prostate cancer. The PSA glycoprotein is a protease of 33 kDa called Kallikrein 3 (KLK3) that is believed to function to keep semen in a state of liquefaction. KLK3 is located on chromosome 19q13. As a diagnostic indicator, PSA levels in blood are reported in terms of ng/milliliter of blood: 0–2.5 ng/ml is low, 2.6–10 ng/ml is considered slightly to moderately elevated, 10–19.9 ng/ml is considered moderately elevated, and 20 ng/ml or more is significantly elevated.

prosthetic group An organic molecule often containing metal atoms that is tightly bound to an enzyme and which is required for the enzyme function.

protease The class of enzymes that catalyze the cleavage of a polypeptide or protein into smaller polypeptides.

protein Any polypeptide or cluster of polypeptides that has a defined biological function.

protein chip A protein microarray. In a protein chip, many individual proteins of a certain type are spotted onto a solid substrate such as a glass slide for the purpose of determining whether cells are producing factors that interact with the proteins on the chip. For example, some protein chips contain arrays of transcription factors, and these chips can be used to test particular cells for the presence of proteins that interact with particular transcription factors. In this example the proteins in a cell lysate could be labeled with a fluorescent molecule, and the presence of a protein that interacts with a certain transcription factor would then be determined by a fluorescent signal at certain spot(s) on the chip.

protein hydrolyzate The partial breakdown product produced by heating a protein mixture or subjecting it to treatment with acid or proteases.

protein kinase The class of enzymes that catalyzes the transfer of a phosphate group from one compound onto a protein.

protein kinase C A protein kinase embedded in the cell membrane that acts as a signaling molecule by virtue of its ability to phosphorylate certain cytosolic proteins on serine and threonine residues. Protein kinase C is activated by ca^{++} ions together with diacylglycerol and tumor promoters such as phorbol esters. Activated protein kinase C is then believed to cause transformation to a cancerous state by phosphorylation of as yet unidentified protein(s).

protein synthesis The process by which amino acids are assembled into peptides on ribosomes using the information supplied by a messenger RNA. Proteins and nucleic acids are held together by bonds that are susceptible to hydrolysis, and their assembly is accomplished by a reversal of the hydrolytic reaction. See TRANSLATION.

proteoglycan A large aggregate of protein that forms the core, and long polymeric saccharide chains that constitute the bulk. Hyaluronic acid, chondroitin sulfate, heparan sulfate, dermatan sulfate, and keratan sulfate are polysaccharides commonly found in proteoglycans. Proteoglycans are major components of the extracellular matrix and other extracellular structural components such as cartilage. See GAG.

proteolysis Breakdown of proteins by cleaving the bonds between amino acids by the process of hydrolysis. See PROTEASE.

proteome A bioinformatics term used to describe the proteins produced by all genes in a genome. The human genome is currently thought to contain a proteome of ca. 30,000 proteins.

proteomics The field of study devoted to the study of the proteins that are expressed in different cell types. A major technique employed in proteomics is the comparative analyses of two-dimensional protein gels from different cell types or from variants of a single cell type.

prothrombin A blood protein that is acted on to form thrombin.

prothymocytes Immature T cells that are formed from stem cells in the bone marrow prior to the time they enter the thymus. Prothymocytes are recognizable by the presence of an incomplete set of T-cell receptor proteins on the cell surface.

proton gradient An uneven distribution of protons caused by the accumulation of protons on one side of a membrane. Proton gradients are a means of storing energy for synthesis of ATP in mitochondria and chloroplasts.

proton-motive force The amount of energy stored in a proton gradient.

proton pump A cluster of membrane-embedded proteins that transports protons from one side of membrane to the other.

proto-oncogene A normal cellular gene that, when altered in a particular fashion (activation), acts to induce a cancerous state.

protoplast A plant cell or bacterial cell in which the cell wall has been removed, for example, by treatment with lysozyme.

protoplast fusion A technique for introducing foreign or genetically engineered DNA into a cell by fusion of that cell with a protoplast carrying the DNA of interest.

prototroph An organism with the same nutritional requirements as the parent organisms from which it was derived.

protozoa A phylum of single-celled organisms—for example, *Paramecium caudatum* or *Amoeba proteus*—representing the most primitive animals (from *proto* = first and *zoan* = animal).

provirus The name given to DNA representing the genome of a virus that has become integrated into the DNA of the host that it infects.

pseudogene A version of a gene that has become inactive over time as a result of accumulated mutations.

pseudouridine An unusual form of uridine found only in transfer RNAs (tRNA).

psoralens A type of organic molecule that spontaneously forms a variety of covalent bonds with nucleic acids in the presence of ultraviolet light.

psychrophile An organism thriving at low temperatures.

psychrotroph An organism that requires low temperature for normal growth.

Ptashne, Mark (b. 1940) Discoverer of the lambda bacteriophage cI repressor protein function. Ptashne was a cofounder of one of the first biotechnology companies, the Genetics Institute, and was a recipient of the prestigious Lasker Award.

pulsed field-gel electrophoresis A variation of agarose gel electrophoresis that allows the separation of extremely large (several thousand kilobases in length) DNA fragments by agarose gel electrophoresis.

pulvomycin An antibiotic that acts by blocking the elongation of a polypeptide chain as it is being synthesized. Pulvomy-

guanosine   adenosine

Purine ribonucleosides

cin interacts with an elongation factor (EFTu) and prevents the formation of an essential complex between GTP and an aminoacyl tRNA.

purine A nitrogen-containing, double-ringed organic molecule that is the parent compound for the purines found in nucleic acids.

purine bases in nucleic acids The purine derived molecules adenine and guanine. In nucleic acids, these molecules are attached to the sugar ribose or deoxyribose in the nucleic acid backbone.

puromycin An aminonucleoside antibiotic, derived from the soil bacterium *Streptomyces alboniger,* that acts by disrupting the process of protein synthesis. One part of the puromycin molecule resembles the 3′ end of a tRNA and binds to the ribosome during protein synthesis and then blocks the translocation step of protein synthesis. Puromycin inhibits the growth of Gram-positive bacteria and some animal and insect cells. The Pac gene which codes for a puromycin N-acetyl-transferase confers resistance to the antibiotic.

purple bacteria A type of bacterium that is capable of carrying out photosynthesis.

pyogenic The ability of a substance to induce the formation of pus and abscesses.

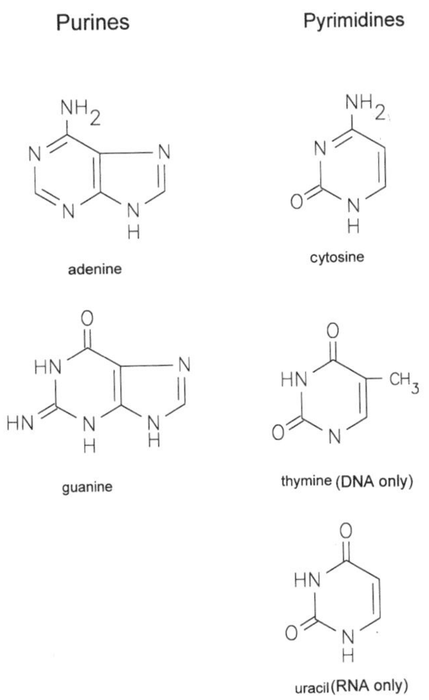

Purines   Pyrimidines

adenine

cytosine

guanine

thymine (DNA only)

uracil(RNA only)

Purine and pyrinidine bases found in nucleic acids

pyridine An organic molecule that contains five carbon atoms and one nitrogen atom in a ring. Used to dissolve otherwise difficult-to-solubilize biological materials.

pyrimidine A six-membered, nitrogen-containing, ringed molecule that is the par-

ent compound for the pyrimidines found in nucleic acids.

pyrimidine bases, in nucleic acids The pyrimidine-derived molecules thymine, uracil, and cytosine. Like the purine bases in nucleic acids, these molecules are attached to the sugar ribose or deoxyribose in the nucleic acid backbone.

pyrogen Any of a number of toxic, fever-causing agents that are usually of bacterial origin, such as endotoxins. The presence of pyrogens is a main concern when preparing solutions for injection because sterilization may destroy live bacteria but not their residual pyrogens.

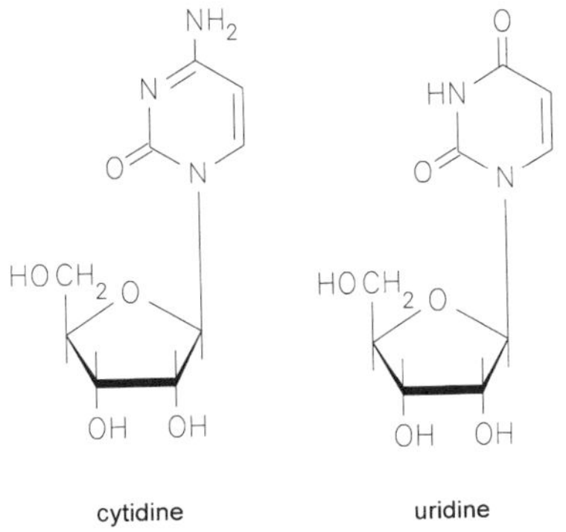

cytidine   uridine

Pyrimidine ribonucleosides

Q

Qa locus One of the genetic subloci within the mouse major histocompatibility locus (H2). Qa codes for an antigen that is only found on a subset of blood cells, and so it is considered to be a differentiation antigen.

q-arm The long arm of the human chromosome.

q banding The technique of staining chromosomes using quinacrine. This technique produces a unique pattern of chromosomal bands that can be used clinically for chromosome identification.

quadroma An antibody-producing cell that secretes antibodies made up of the random association of heavy and light chains from two different antibodies. This is an intermediate in the formation of a bispecific antibody, in which the two binding sites recognize different antigens. The quadroma is formed by the fusion of two monoclonal antibody-producing cells.

queuosine An unusual purine base found only in transfer RNA. Queuosine is formed by adding a pentenyl ring to 7-methylguanosine.

quinacrine A synthetic antimalarial compound that is also used as a fluorescent stain for chromosomes.

quinine An alkaloid drug derived from the bark of the cinchona tree that is found in South America and Indonesia. Quinine is an antimalarial drug that is also used to relieve fever and pain in other diseases. At one time, quinine was the only drug available for treatment of malaria, but it has been replaced to a large extent by synthetic drugs, such as quinacrine.

quinone A class of cyclic organic compounds such as phylloquinone, plastoquinone, and ubiquinone that is widely used to carry hydrogen atoms in certain critical steps in the process of energy production in both plants and animals. Chemically, quinones are characterized by the presence of two ketone groups (C=O) on the same hydrocarbon ring.

R

R1 particle An intermediate stage in the formation of the 30S ribosomal subunit. An R1 particle is formed from a strand of 16S RNA and 15 ribosomal proteins.

racemate A mixture of two different forms of a molecule that do not differ from each other chemically but that have a different physical arrangement of atoms that can be distinguished by methods using polarized light.

radial immunodiffusion An immunological test based on the reaction of an antibody with a protein that has been allowed to seep out of a central well into a slab of agar where the reaction takes place.

radioimmunoassay A sensitive test for a particular protein, based on the reaction of that protein with an antibody specific for it, where one of the reacting agents is radioactively labeled.

raf A protein intermediate in the signal-transduction-cascade pathway initiated by receptor tyrosine kinase activation. In this pathway a ras-GTP complex binds to the N terminal end of cytosolic raf, whose C terminal end has serine/threonine kinase activity that acts to phosphorylate and thereby activate MEK, a MAP-kinase kinase (MAPKK).

raf oncogene An oncogene that is found in murine sarcoma virus and that is associated with fibrosarcoma tumors in both rodents and humans. The normal homologue of the raf oncogene encodes the protein raf. raf is an acronym derived from *rat fibrosarcoma*.

RAG proteins Recombination *a*ctivating *g*ene proteins; two proteins (RAG1 and RAG2) that catalyze the cleavage of DNA segments within the immunoglobulin genes that is the first step in immunoglobulin gene rearrangements whereby different V (variable) segments are linked with different J (joining) segments to create a large spectrum of immunoglobulin molecules with many different specificities. The RAG proteins generate double-strand breaks at sites called recombination signal sequences (RSS) that are present in both the V and J segments.

Ramachandran plot For a given peptide, a graph that plots the angle of rotation around the bond between the alpha carbon and the carbonyl carbon (ψ) versus the angle of rotation around the bond between the alpha carbon and the amide nitrogen (Φ). This type of plot is used to give an idea of the combinations of bond angles that occur in the peptide. This information can be used to help determine how the peptide is folded.

Raman spectroscopy A type of spectroscopy that measures the wavelength of "inelastically" scattered photons (i.e., scattered photons whose wavelength is different from that of the incident photons). The scattering of light occurs at wavelengths that are shifted from those of the incident light by the energies of molecular vibrations (bond stretching). Like infrared spectroscopy, Raman spectroscopy gives information about the type of bonds present in a compound but is different from infrared spectroscopy by being able to see signals from bonds that are perfectly symmetrical and therefore have no dipole. Raman spectroscopy is used in making structure determinations of biomolecules.

random primer labeling A technique for labeling DNA that is based on the

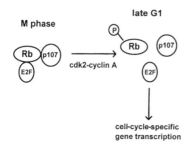

late G1

M phase

cdk2-cyclin A

cell-cycle-specific
gene transcription

E2F transcription factors induce the expression of genes required for cell-cycle progression, but these transcription factors are maintained in an inactive state in a complex with the retinoblastoma tumor suppressor, Rb, and an Rb-related protein, p107. E2F is released in active form from the complex in late G1 after phosphorylation of Rb by cdk2-cyclin A. Inactivation of Rb by phosphorylation begins in late G1 and continues throughout the S phase.

annealing of a mixture of short primers with randomly determined sequences to the DNA strand. Labeling takes place by extension of the primers (see PRIMER EXTENSION) using labeled nucleotides.

ras oncogene The oncogene that is found in rat sarcoma virus and that is associated with sarcomas in rodents and carcinomas in human. The name is an acronym derived from *rat* sarcoma. In mammals the ras proto-oncogene has a GTP binding activity that is believed to be homologous to the action of G proteins.

Rb An abbreviation for the antioncogene protein product of 110,000

molecular weight that is coded for by the gene associated with the familial form of retinoblastoma. (See figure at left.)

R banding A characteristic pattern of bands produced when chromosomes are stained by various dyes, for example, olivomycin. The basis of the pattern of R bands is the abundance of DNA rich in guanine-cytosine (G-C) base pairs.

reactive oxygen species (ROS) Any of a group of small molecules containing a highly active form of oxygen, including superoxide (O_2^-), hydroxyl radical ($OH\cdot$), hydroxyl ion (OH^-), or peroxides R-O-O-R$'$. ROS are produced as a consequence of ionizing radiation such as X-rays and gamma rays but are also produced naturally as a consequence of enzymes present in neutrophils and macrophages or normal respiratory chain reactions, especially those involving ubiquinone. Because of their reactivity, ROS are damaging to biomolecules, particularly DNA, and unrepaired DNA damage can lead to cancer-causing mutations. ROS are inactivated by antioxidants such as vitamin C and also by certain enzymes such as superoxide dismutase (SOD) and catalase.

reading frame Any one of the three possible ways of reading a sequence of amino acids from a nucleic acid using the genetic code; the three different reading frames are determined by which base in any group of three consecutive bases is chosen as the start point.

reassociation kinetics A technique for estimating the number of copies of a

amino acids

reading frame 1 → GlyProPheValCysSerProPheCysSerPheProIleValAlaProIlePheGluValHisTrpGluAla
reading frame 2 → GlyProLeuCysAlaLeuHisSerAlaProSerProSerLeuProProPheLeuArgCysThrGlyArgLeu
reading frame 3 → AlaLeuCysValLeuSerIleLeuLeuLeuProHisArgCysProHisPhe---GlyAlaLeuGlyGlySer
GGGCCCTTTGTGTGCTCTCCATTCTGCTCCTTCCCCATCGTTGCCCCCATTTTTGAGGTGCACTGGGAGGCTCC

DNA strand

start of reading frame 3

start of reading frame 2

start of reading frame 1

Reading frame

particular nucleic-acid base sequence in a sample by measuring the rate at which denatured nucleic acid in the unknown sample anneals to strands of a known nucleic acid with the same or similar sequence to that being determined.

reassociation of DNA Reannealing of the complementary strands of DNA after the paired strands have been separated by heat or strong acid or alkalai.

RecA A bacterial protein that is responsible for the major steps in recombination following the introduction of strand nicks by RecBCD: strand invasion, branch migration, and formation of the Holliday structure. The RecA protein also plays a role in DNA repair by regulation of the SOS response following DNA damage.

RecBCD A bacterial enzyme that mediates certain critical events in the process of recombination including DNA strand unwinding (helicase activity) and the creation of single-stranded nicks at specialized sites (chi sequences).

receptor A specialized cell surface molecule or complex of molecules that serves as a site of attachment for a specific effector molecule (the ligand), for example, a hormone. The receptor may also function to produce a biological response in the cell to which the ligand is bound via its receptor.

receptor His kinase A type of signal-transduction system found in plants and bacteria that transmits signals from membrane receptors by causing a phosphate group to be transferred to a histidine residue on a transduction protein. In these systems the phosphate is usually rapidly transferred from the histidine to an aspartate residue on another protein (his-to-asp phosphorelay). This type of signaling is involved in the sensing of certain environmental stimuli such as the presence of phytohormones, in *Arabidopsis thaliana*. Histidine kinase signaling is also involved in the ETR1 family of ethylene receptors, the CKI1 cytokinin-sensor, the ATHK1 osomo-sensor, and in the bacterial chemotaxis system mediated by the histidine kinase CheA.

receptor-mediated endocytosis The process by which many ligands that bind to receptors on the cell surface enter into the cytosol. During receptor-mediated endocytosis, the plasma membrane undergoes invagination in the region surrounding a receptor that has bound a ligand to form a vesicle (an endosome) that eventually separates from the plasma membrane and migrates into the cytosol.

recessive A form of a gene whose effect on the phenotype of an organism is masked by an alternate (dominant) form of the gene.

recessive allele The term for a recessive gene on a chromosome.

reciprocal translocation An exchange of material between chromosomes, usually by breakage of each of the participating chromosomes at a specific site. A translocation involves movement of a physical portion of a whole chromosome.

recombinant DNA A DNA that has become joined to another unrelated or foreign segment of DNA.

recombinant proteins Proteins produced by cloning technology, usually by splicing the DNA sequence for the protein into a vector, transferring that vector into a host organism, usually a bacterium or a yeast, and harvesting the protein after growth of the host.

recombinase An enzyme that catalyzes the joining of immunoglobulin gene segments during the recombination event involved in immunoglobulin gene switching.

recombination The process by which DNA from a gene on a large genetic unit; for example, a chromosome becomes exchanged with the corresponding DNA on a complementary genetic unit such as another chromosome that is an allele.

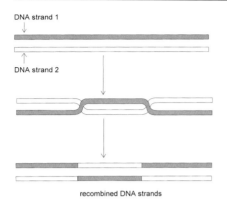

DNA strand 1

DNA strand 2

recombined DNA strands

Recombination

recombinational repair A mechanism for repairing thymine dimers based on recombining an undamaged piece of DNA from the undamaged strand into the damaged region during DNA replication.

reconstituted viral envelopes (RVEs) Viral envelopes whose contents have been removed or replaced with other substances. RVEs are made by a two-step process in which the whole virus is first completely disassembled and then the components of the envelope portion are allowed to reassemble. RVEs are used to deliver substances of biological interest to various types of animal cells. See DELIVERY SYSTEM and FUSOGENIC VESICLES.

recoverin A small (23 kDa) calcium-binding protein found mainly in the photoreceptors of the vertebrate retina. Recoverin plays a role in regulating the process of phototransduction by binding to rhodopsin kinase (GRK1), which prevents the inhibition of rhodopsin through phosphorylation. This in turn prolongs the light response. There is one mammalian gene (rcv1; gene map locus 17p13.1) and there are orthologues of recoverin present in the photoreceptors of most vertebrate species, beginning with amphibians.

redox potential A measure of the affinity that an atom or a molecule has for electrons. The redox potential is usually given by the symbol Eo′, a number that represents the extent to which the atom or molecule in question donates or accepts electrons to or from hydrogen as a standard reference.

redundancy Existing in more than one copy, for example, in repetitive DNA.

refractory phase A period of time required after emission of a nerve impulse for regeneration of the ability of a neuron to emit a new nerve impulse.

Refsum disease An autosomal recessive genetic disease that is one of the leukodystrophies, characterized by damage to the white matter of the brain and impaired movement. The disease is caused by a lack of the enzyme that breaks down phytanic acid, which as a result causes toxic levels of phytanic acid to build up in the brain, blood, and other tissues. The symptoms include increasing night blindness, loss of the sense of smell (anosmia), and over time, deafness, ataxia, peripheral neuropathy, and cardiac arrhythmias. Refsum disease can result from mutations in any of three genes: phytanoyl-CoA hydroxylase (PAHX or PHYH, chromosomal location 10pter-p11.2) or the gene-encoding peroxisome biogenesis factor, peroxin-7 (PEX7, gene map locus 6q22-q24).

regeneration The ability to grow an organ, a structure, or a whole organism from one or a few cells.

regulatory enzyme An enzyme that is part of a biochemical pathway, usually the first enzyme in the series, and that serves as a regulator of the chemical reactions in the pathway by either speeding up or slowing down the chemical reaction it controls in response to some environmental condition(s).

regulatory gene A gene whose product controls the expression of another gene or genes. The repressor proteins of the *lac* operon and the lambda bacteriophage cI regions are examples of regulatory gene products. See *LAC* REPRESSOR PROTEIN.

regulatory sequence A nucleic-acid sequence that serves as a site at which the protein product of a regulatory gene attaches; attachment of a regulatory protein to a regulatory sequence is the mechanism by which a regulatory gene controls the expression of another gene. See ENHANCERS and SILENCERS.

regulon A set of spatially separated genes under the control of a single repressor-operator system. See OPERATOR.

relaxed The state of a large molecule, for example, a long DNA molecule or a protein, being in a loose conformation or loosely folded over on itself. Change in biological activity is often related to the degree of twisting, folding, or compression of a molecule.

relaxin A peptide hormone produced by the corpora lutea of ovaries during pregnancy. The hormone causes softening of the cervix and plays a role in regulating contractions during the birth process. The relaxin family is a member of the insulin superfamily and contains seven peptides: relaxins 1, 2, and 3, and the insulin-like (INSL) peptides INSL3, INSL4, INSL5, and INSL6. The relaxins are now believed to be involved in a variety of functions other than parturition. These putative functions include regulation of pituitary hormone secretion, renal vasodilatation, enhancement of coronary flow, and promotion of nitric oxide biosynthesis that may affect smooth muscle function.

release factor A protein that causes the process of translation to terminate when a termination signal on the mRNA is present.

rel oncogene The oncogene product carried by the avian reticuloendotheliosis virus strain T (v-rel); the cellular product relA is the p65 subunit of the NF-kappa B transcription factor, a member of the rel/NF-kappa B family of transcription factors; these transcription factors are critical mediators of immune and inflammatory responses. In most cells NFkB is associated with IkB, an inhibitory protein. A number of stimuli,

including tumor necrosis factor, interleukin 1, T-cell activation signals, bacterial endotoxins, viral-transforming proteins, growth factors, and reactive oxygen species induce phosphorylation of NFkB, which leads to its translocation into the nucleus where IkB is degraded. In the nucleus NFkB acts as a transcription factor for genes encoding cytokines, cytokine receptors, cell adhesion molecules, proteins involved in coagulation, and genes involved in cell growth control. NFkB is also believed to be an important transcriptional regulator for HIV. rel is located on chromosome 2p13-p12.

renaturation The process of a molecule assuming its native shape or conformation after disruption of the native state, for example, the reassociation of DNA.

reovirus A group of RNA containing viruses that infect both the gut and respiratory tracts, usually without causing observable disease. The name is an acronym derived from *r*espiratory *e*nteric *o*rphan.

repair synthesis Synthesis of new DNA to replace a defective segment that has been removed; repair synthesis usually involves creating a complementary DNA strand from the remaining, nondefective DNA strand. See EXCISION REPAIR.

repetitive DNA A class of eukaryotic DNA sequences that are present in many, sometimes thousands or millions, of copies throughout the genome. See ALU ELEMENTS.

replacement sites DNA nucleotide bases that, when changed, for example, by mutation, result in a change(s) in the amino acid(s) that the DNA codes for.

replica plating A procedure by which bacterial colonies growing on one bacterial plate are reproduced on a second bacterial plate in the exact relative positions to one another as they were in the original plate.

replication eye The opening created between DNA strands as a result of unwinding of the DNA helix during the

rel oncogene

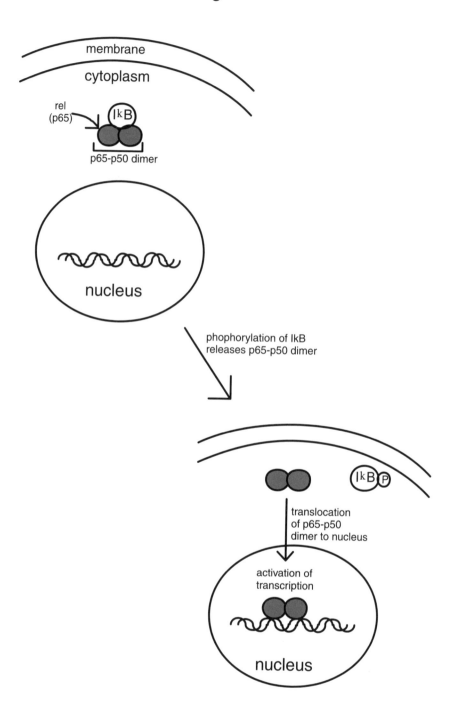

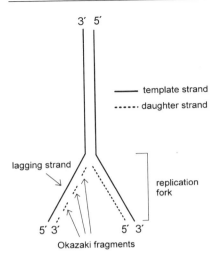

3′ 5′

——— template strand
······ daughter strand

lagging strand

replication fork

5′ 3′   5′ 3′
Okazaki fragments

Replication fork

process of DNA replication (replication forks).

replication fork The portion of the partially replicated DNA consisting of the separated DNA strands plus the newly synthesized copies. See ANTIPARALLEL, LAGGING STRAND, and OKAZAKI FRAGMENT.

replication of DNA The process in which DNA is copied by using each of the strands as a template for a new strand. Because with each round of DNA synthesis the new DNA consists of one newly synthesized strand and one parental strand, the process of DNA replication is called semiconservative. See DNA POLYMERASE(S).

replication origin A sequence of nucleotide bases that provides a signal for the start of DNA replication.

replicon The exact site on the DNA at which replication is actively taking place. See REPLICATION ORIGIN.

replicon fusion The meeting of two replicons approaching each other from opposite ends of replicating DNA.

replisome The complex of factors that are active at a DNA replication fork. The replisome includes DNA polymerase, primase, DNA ligase, DNA helicase, single-strand binding (SSB) protein, and topoisomerase.

reporter gene A gene whose expression is linked to the expression of another gene or biochemical process.

repressible enzyme An enzyme whose expression is governed by repression of gene transcription.

repressor protein A regulatory protein that exerts control over gene expression by binding to a regulatory sequence and thereby preventing transcription of the gene.

reptation The process by which a nucleic acid strand is threaded through the pores in an agarose gel matrix in a "headfirst" fashion during pulsed field-gel electrophoresis.

resolvase An enzyme that catalyzes the site-specific recombination event that results in the integration of some transposable elements. Resolvase is coded for by a gene on a transposable element.

resorufin-β-D-galactopyranoside (RG) A synthetic substrate for the enzyme, beta-galactosidase, the product of the lacZ gene that produces a fluorescent product in the presence of the enzyme. RG is used to detect the expression of the lacZ gene in individual cells by fluorescence-activated cell sort-ing (FACS).

respiration The oxygen-dependent process of generating energy in the form of ATP from sugars. See ELECTRON TRANSPORT.

response coefficient A number that gives a measure of how the rate of flow through a given biochemical pathway changes as a result of some outside influence such as a hormone or change in the concentration of an ion. The response coefficient (R) is made up of two components: the sensitivity of the pathway to the enzyme (C) and the sensitivity of the enzyme to the outside influence (ε), R = C·ε. If response coefficients are known for each enzyme in the pathway, then it is possible to predict how the rate of flow through the pathway will be affected by a particular outside influence.

resting potential The electrical potential across the membrane of a neuron in between nerve impulses.

restriction endonuclease (restriction enzyme) An enzyme, produced by bacteria, that cleaves DNA at a place defined by a specific sequence of nucleotide bases.

restriction fragments The DNA fragments that are produced by cleavage of DNA by a restriction endonuclease.

restriction mapping A technique of mapping DNA by determining the location of sites for different restriction enzymes.

restriction site The nucleotide base sequence on DNA that specifies the site where a given restriction endonuclease will cleave.

reticulo-endothelial system (RES) All the phagocytic cells of the body except the circulating leukocytes. Cells of the RES remove particulate matter, for example,

foreign antibody-agglutinated antigens from the bloodstream.

retinoblastoma A cancer of the cells of the retina that occurs in small children. Retinoblastoma was one of the first cancers that was shown to run in families (familial retinoblastoma) and was therefore genetic in origin. See RB.

retinoic acid receptors (RAR, RXR) Transcription factors that are activated by binding to retinoids in the cytosol. The activated factors are responsible for inducing the expression of genes seen in retinoid-treated cells. There are two types of retinoid receptors, RAR and RXR, which bind the trans form of retinoic acid and the cis form, respectively; each type of receptor has α, β, and γ isoforms. Because they serve a function in the nucleus (transcription), RAR and RXR are referred to as nuclear receptors similar to receptors for thyroid hormone, vitamin D_3, and steroids. Retinoic acid receptors play an important role in the growth and differentiation of epithelial tissues, the generation of new blood cells (hematopoiesis), and in central nervous system development. The RARα gene is a common target in chromosomal translocations in acute promyelocytic leukemia (APL).

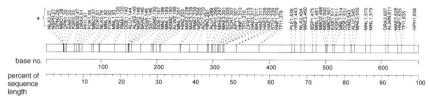

*standard abbreviation of the restriction enzyme name followed by the base number at which the restriction site is located

input sequence:

```
CACCATAGTTCTAATTTTTCCCACATGCGATCAGGAAGAGTAGTCCACCAAGTGGAAATAGAATTCTTCATCCTCCATGT
CCTCTATATCCATCTCTCTTTTCCATCCCACTCCACCCTAGTTTGGCTCTTTCTTGTCTGAGCTCTTGCGAGACGGCTCT
TCCTGAGTTTCCCTCCTCCAGTCTCTCCTCTACTCCGTTGACTGCCAGATTGTCTTACAGCATAGATGAAACCACGTGAC
TTCTGTGCCCCAAGACTTTGATGTCTACAGAATAAAGTTCAAGCTTCTCAACGTTGTCACCTGGAATCTGGCCACAATTG
ATCTTTTCAGGCCTATCTCTCCCTATCTCCTTTTCTAATTACACATTTTTGATTCTTGCCCAACCCAACCACTCACTATT
TCCAAGCACACCCTATACTTTCCCACGCCTTTGACTCCCACATGACCTTTGTCACACCTGCCTCCCTTCTGCTCTGCCTC
ACAAATTTTAACCTCTTCTTCAAACACCAGCCCAAATGCTCAGTTCCATAAAGCTTCTGTGACCTTGCCTCGCCTGCCTC
AGAGAGAAGTAATTTGCTTTTAGAGTTCACACAGTGCCTGTGAATACTTGTTGAGTGACTGAATCAACTTGCTCATAGCA
ATTTCATATT
```

Computer-generated restriction map

retrograde transport The process by which materials move from the plasma membrane to lysosomes or to the Golgi apparatus and then to the endoplasmic reticulum; the direction of transport is opposite from that taken by newly synthesized membrane proteins (anterograde transport). Retrograde transport is part of the process by which "worn-out" membrane components are recycled. This process is particularly evident in nerve cells, where parts of membranes from synaptic vessels move along the axon toward the cell body for lysosomal degradation and recycling. Various toxins, such as ricin, pertussis toxin, and tetanus toxin, make use of the retrograde transport system for their delivery to intracellular targets.

retroposon A type of transposon that has a structure similar to a retroviral genome in that it contains LTR-like sequences flanking a coding region and it replicates via an RNA intermediate that is reverse transcribed into a double-stranded form that integrates into the genomic DNA.

retrovirus A class of viruses characterized by having an RNA genome and carrying the enzyme reverse transcriptase within the virus capsid. The name was originally an acronym derived from *reverse transcriptase* = *retravirus,* which later became *retrovirus.*

retrovirus vector A genetically engineered DNA for cloning recombinant DNA that utilizes certain control elements from retroviruses.

reverse genetics The term used to describe the type of genetic analysis in which the structure of a gene is determined from the protein that it codes for. In the more common type of analysis, the protein structure is determined from the structure of its corresponding gene.

reverse mutation A mutation that reverses the effects of a previous mutation. The reverse mutation may or may not be localized near or at the site of the mutation whose effects it suppresses. See SUPPRESSOR MUTATION.

reverse transcriptase The enzyme, made and used by retroviruses during their life cycle, that catalyzes the synthesis of DNA copied from an RNA template. The enzyme is widely used in genetic engineering and molecular biology to make so-called complementary DNA (cDNA) from various RNAs so that base sequences in RNA can be cloned and manipulated by recombinant DNA technology.

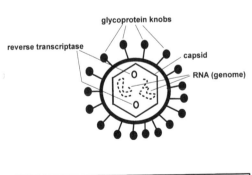

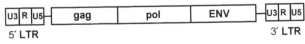

retroviral genes

Retrovirus

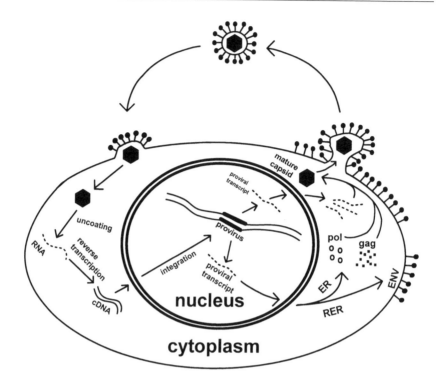

Retrovirus life cycle

reverse transcription The term that describes the action of the enzyme, reverse transcriptase.

rhesus blood groups Classification of blood cells according to whether or not they react with antibodies to the blood cells of rhesus monkeys.

rheumatoid factors Certain antibodies (IgM) present in the blood of some individuals with rheumatoid arthritis that react against other antibodies. Since it was discovered that different rheumatoid factors are specific for only certain subgroups of antibodies, rheumatoid factors became a means of classifying an individual's antibodies into subclasses (allotypes).

Rh factor An antigenic substance on the surface of the blood cells of any individual that carries the Rh trait (Rh+).

The presence of the Rh factor in a fetus whose mother is Rh− may provoke a life-threatening agglutination of the fetal blood cells. The term is named for the rhesus monkey, the organism that was used to demonstrate the presence of the antigen. There are at least 30 distinct subtypes of Rh factor.

rhinovirus A picornavirus that infects the nasal cavity and causes many of the symptoms of the common cold, particularly the nasal symptomatology.

rhizobium A leguminous plant that is a rich source of nitrogen-fixing bacteria that live in large nodules attached to the plant roots.

rho factor A small bacterial protein that is responsible for causing transcription to terminate when a ribosome encounters an appropriate termination signal on the mRNA.

riboflavin Vitamin B_2; an important cofactor for enzymes involved in the metabolism of sugars for ATP production. Riboflavin acts to transport electrons derived from the oxidation of sugars in energy production.

ribonuclease A class of enzymes that breaks down RNA by breaking the bonds between the phosphate and the ribose molecules in the RNA backbone.

ribonucleic acid See RNA.

ribonucleotide A molecule consisting of ribose that is bound to phosphate and with a purine or pyrimidine base attached to the ribose molecule. Ribonucleotides are the building blocks of ribonucleic acid (RNA).

ribose A five-carbon sugar normally in a ring conformation; the sugar used in ribonucleotides. See CARBOHYDRATE.

ribosomal protein Any one of many proteins that, together with a strand of RNA, form a ribosomal subunit.

ribosomal RNA (rRNA) A long strand of RNA that, together with ribosomal proteins, is a component of a ribosomal subunit. In eukaryotic cells there are two rRNAs denoted as 18s (found in the small ribosome subunit) and 28s (found in the large subunit). rRNA plays a role in binding mRNA in the process of translation.

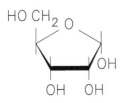

Ribose

ribosome A small organelle in the cytoplasm that is the site where protein synthesis takes place. Ribosomes are made up of two subunits. The subunits are assembled together with a strand of messenger RNA to begin protein synthesis.

ribozyme An RNA with enzymatic properties. The idea that RNAs could function as enzymes was first suggested by Carl Woese, Francis Crick, and Leslie Orgel in 1967. Thomas Cech described the first ribozyme, a self-splicing RNA in *Tetrahymena thermophila*. The bacterial 23S ribosomal RNA has also been shown to be a ribozyme that catalyzes the peptidyl transferase step in the process of protein synthesis. Synthetic ribozymes are being studied as possibly better alternatives to protein enzymes. In 1989 the Nobel Prize in chemistry was awarded to Thomas R. Cech and Sydney Altman for their discovery of catalytic RNAs.

ricin A potent, poisonous protein derived from the beans of the castor plant, *R. communis*, from which castor oil is derived. Its lethality and ease of extraction make ricin an attractive choice as a biological warfare agent. Ricin is composed of two hemagglutinins and two toxins (RCL III and RCL IV) that are dimers of approximately 66 kDa. The toxins are composed of an A and B chain. The B chain mediates entry of the toxins into the cells after binding to the cell surface glycoproteins. The A chain acts to block protein synthesis by attaching to the 60S ribosomal subunit and blocking the binding of elongation factor-2, which results in cell death.

Rickettsia Small bacteria which, unlike most other bacteria, are obligate parasites that live within, instead of outside of, the cells of the infected host. *Rickettsia* are the agents that cause typhus fever and Rocky Mountain spotted fever.

rifampicin An antibiotic that blocks transcription by inhibiting the action of RNA polymerase in bacteria, specifically by inhibiting the initiation of the process of transcription.

siRNA

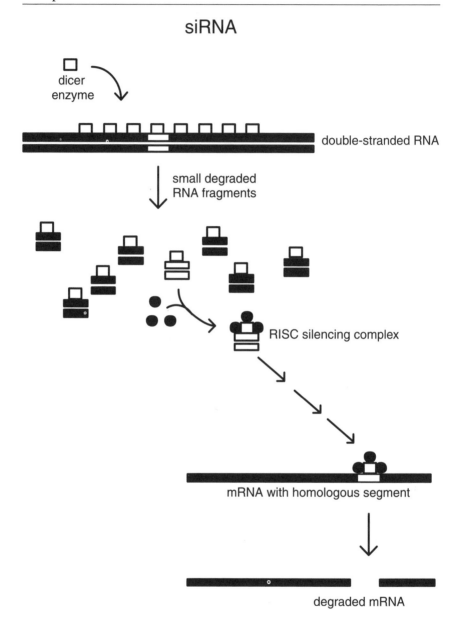

R loops R loops are the segments of DNA-representing introns that are seen as single-stranded loops in electron micrographs of heteroduplexes between a eukaryotic mRNA and the genomic DNA from which the mRNA was transcribed.

RNA Ribonucleic acid. A polymer of ribonucleotides, the purine and pyrimidine base sequence that generally is complementary to a DNA base sequence. There are four major classes of RNA that perform different functions in the process of protein synthesis: messenger RNA (mRNA), transfer RNA (tRNA), ribosomal RNA (rRNA), and small nuclear RNA (snRNA).

RNA-DNA hybrid(s) A double-stranded hybrid molecule in which RNA is based paired with a complementary strand of DNA.

RNA interference (RNAi) An experimental technique for silencing expression of a specific gene(s) through the use of small double-stranded RNA (dsRNA) to specifically target an homologous mRNA. In this technique dsRNA, which is introduced into a cell, is cleaved into small (ca. 23 bp) fragments by an enzyme called Dicer. The small RNAs (called short interfering RNAs; siRNAs) are "trigger" molecules for the siRNA-Dicer complex to recruit other factors to form an RNA-induced silencing complex (RISC). The siRNAs in the RISC can base pair with mRNAs that have complementary sequences which are then also cleaved and degraded by RISC.

RNA maturase An enzyme involved in the splicing of transcripts in the yeast mitochodial cytochrome b gene. The RNA maturase gene is unusual in that part of the gene is found in an intron of the cytochrome b gene itself.

RNA polymerases The class of enzymes that catalyzes the synthesis of a strand of RNA using DNA as a template to guide the assembly of ribonucleotides so that the order of the purine and pyrimidine bases in the DNA template is precisely copied, in complementary fashion, in the newly synthesized RNA.

RNA secondary structure The hairpin folding of an RNA molecule caused by internal base pairing of complementary stretches of purine and pyrimidine bases.

RNase D An exonuclease that removes nucleotides from the $3'$ end of an RNA in one-at-a-time fashion. RNase D is involved in the maturation of tRNAs that are synthesized in large precursor

strands that are shortened into functional tRNAs.

RNase H An enzyme that cuts RNA chains from within the chain, creating nicks in the phosphodiester backbone. This enzyme is used during production of cDNA. After the first strand of DNA is made from the RNA template, Rnase H is used to nick the template to create primer ends for second DNA strand synthesis.

RNA tumor virus A subclass of retroviruses that produces cancers by activation of oncogenes.

RNP Ribonucleoprotein.

ros oncogene An oncogene that is found in a stain of avian sarcoma virus and that is associated with sarcoma tumors in birds. The name is an acronym derived from Rochester 2 sarcoma virus.

R_0t In the annealing of RNA to DNA, a variable equal to the molar concentration of the RNA multiplied by time allowed for RNA-DNA annealing. R_0t values are generally used in plots of the annealing of RNA to complementary DNA sequences. See C_0T VALUE.

rotavirus A class of RNA-containing viruses that infects the intestinal tract and is responsible for epidemic gastroenteritis and infantile diarrhea.

rough ER Endoplasmic reticulum covered with attached ribosomes. Proteins synthesized by the ribosomes on the rough ER are destined to be transported out of the cell via vesicles that are derived from the endoplasmic reticulum.

Rous sarcoma virus (RSV) A retrovirus that produces sarcoma tumors in chickens. RSV, discovered and named for Peyton Rous, was the first RNA tumor virus discovered.

rRNA Abbreviation for ribosomal RNA, the RNA strands that are, together with the ribosomal proteins, the basic components of ribosomes. In eukaryotic cells

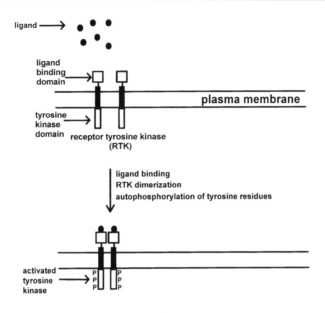

ligand

ligand binding domain

plasma membrane

tyrosine kinase domain   receptor tyrosine kinase (RTK)

ligand binding
RTK dimerization
autophosphorylation of tyrosine residues

activated tyrosine kinase

RTK

there are two major rRNAs denoted as 18s (found in the small ribosome subunit) and 28s (found in the large subunit).

RTK *Receptor tyrosine kinase*; a class of transmembrane proteins whose extracellular domain functions as a cell surface receptor; the cytosolic domain acts as a tyrosine kinase. Binding of a ligand to the receptor domain initiates the process of signal transduction by activating the tyrosine kinase domain that, in turn, brings about a cascade of subsequent protein phosphorylations, ultimately leading to induction of transcription of genes involved in cell-growth regulation.

rubella An RNA-containing virus (togavirus) responsible for German measles.

rumen bacteria Bacteria that live in the rumen of ruminant animals such as cows. The rumen bacteria utilize urea that would otherwise be excreted to make amino acids, that are then returned to the circulatory system of the animal.

Runting syndrome A pathological condition, characterized by skin lesions, diarrhea, and death, that results when the lymphocytes from a mature animal are placed in and then attack the tissues of a newborn.

S1 nuclease An enzyme that catalyzes the breakdown of any single-stranded (or single-stranded region of a) nucleic acid.

S1 nuclease mapping A technique of determining where, on a segment of DNA, the precise location of the sequences from which a given RNA is transcribed.

S-100 (calgranulin) Pro-inflammatory cytokines expressed by types of leukocytes called monocytes and granulocytes under conditions of chronic inflammation; elevated levels of calgranulins are found in patients with cystic fibrosis. The calgranulins are calcium-binding proteins that consist of at least two different polypeptides, designated A and B, coded for by genes on human chromosome 1q12-q21. A cell-surface receptor for S100A12, known as RAGE, interacts with a factor called ENRAGE (extracellular newly identified RAGE-binding protein) in endothelium, mononuclear phagocytes, and lymphocytes and triggers the generation of key pro-inflammatory mediators.

saccharide The biochemical term for a sugar.

Saccharomyces cerevisiae A yeast that is widely used as a vehicle for cloning extremely large segments of foreign DNA (see YEAST ARTIFICIAL CHROMOSOME) and for molecular studies on many animal genes that have homologues in yeast.

saline A solution of sodium chloride at a concentration exactly equivalent (eight grams per liter; 0.8 percent) to that found in bodily fluids.

Salmonella A group of Gram-negative, rod-shaped bacteria that is responsible for typhoid fever and a number of wide-ranging intestinal disorders.

saltatory movement The directed movements of organelles in the cell cytoplasm. This type of movement is thought to be controlled by microtubules.

saltatory replication Replication of a DNA sequence that produce extra copies of the sequence along the same DNA strand. This type of process is believed to have been responsible for the highly repeated, tandemly arrayed sequences seen in satellite DNA.

salting out The phenomenon of causing dissolved proteins or nucleic acids to precipitate out of solution by the addition of salts.

salt stabilization A phenomenon whereby slow denaturation of proteins and nucleic acids in aqueous solution is prevented by the addition of salts.

Sanger, Frederick (b. 1918) Discoverer of the first means by which the amino acid sequence of a polypeptide could by determined. Sanger is famous for the discovery of the amino acid sequence of insulin in 1954; he was awarded the Nobel Prize in chemistry in 1956.

Sanger method A method for determining the sequence of a polypeptide based on determination of the identity of the terminal amino acids of small subfragments of the original polypeptide.

Sanger (dideoxy) sequencing A technique for determining the sequence of a segment of DNA that utilizes synthetic nucleotides (dideoxy nucleotides) to create small polynucleotides representing small subfragments of the DNA that are to be sequenced but that can be made to terminate specifically at any one of the four purine or pyrimidine bases. See DIDEOXY SEQUENCING.

saprotroph An organism that obtains nourishment from nonliving matter.

sarcoma-derived growth factor (SGF) A growth factor secreted by cells infected with murine sarcoma virus, an RNA tumor virus. Because noninfected cells treated with SGF undergo changes generally characteristic of cells transformed into a cancerous state, sarcoma derived growth factor is now referred to as transforming growth factor (TGF).

sarcoplasmic reticulum A membranous structure that surrounds the myofibrils in muscle tissue. The sarcoplasmic contains calcium pumps that regulate the level of calcium ion (Ca^{++}) in muscle tissue.

sarcosine A component of the antibiotic, actinomycin D, an inhibitor of transcription. Chemically, sarcosine is N-methyl glycine.

satellite DNA A type of DNA made up mostly of repeated sequences that are not transcribed into RNA and that are found near the chromosome centromere.

satellite RNAs See VIRUSOIDS.

scanning electron microscopy (SEM) A variation of electron microscopy in which the specimen is given a thin coat of metal so that the electron beam can be used to visualize details of the cell surface as opposed to internal structures.

scatter plot A graph that shows the relationship between two variables as a set of data points.

Schiff's reagent A chemical (fuchsin leucosulfonate) used in the periodic-acid Schiff (PAS) stain that is used to identify the presence of certain infecting microorganisms; for example, fungi.

schistosomiasis A group of diseases whose symptoms range from dermatitis to cirrhosis of the liver. The symptoms are caused by parasitic infection by one of the trematode worms of the genus *Schistosoma*. Schistosomaiasis is endemic in the populations of Africa, the Middle East, and South America.

schizonte A subgroup of protozoa (sporozoa) that reproduces asexually. Plasmodium is a schizonte that causes malaria.

Schwann cells A type of brain cell that encompasses the axon of a neuron, thereby forming a sheath of myelin around the axon. The myelin sheath is essential for proper transmission of nerve impulses between neurons. Multiple sclerosis is an example of a disease that induces loss of muscle control by causing demyelination of the axon.

scintillation counter A sensitive device for detecting single emissions of particles produced by radioactive decay.

screen A method developed to detect and/ or select a recombinant protein, mutant, interacting protein, drug, hybridoma, and so on.

SDS Sodium dodecyl sulfate; a detergent widely used to dissociate biological materials into their component molecules.

SDS-polyacrylamide-gel electrophoresis (PAGE) A variation of the polyacrylamide-gel electrophoresis technique in which SDS is dissolved in the polyacrylamide gel. This type of gel is widely used to separate proteins in mixture from one another on the basis of size.

secondary culture The cell culture that is derived from the original outgrowth of cells derived directly from a tissue specimen (i.e., the primary culture).

secondary structure The manner in which a linear polypeptide is folded, twisted, or otherwise bent. The most common types of secondary structure are the alpha-helix and the pleated-sheet structures.

second-order kinetics (bimolecular kinetics) A term describing the rate at which a chemical reaction involving two reacting molecules occurs.

secretion   The ability of the host cells that produce recombinant proteins products to release the products extracellularly. Large-scale production of recombinant proteins requires the secretion of the product into the culture medium for easy harvesting. Vectors have been developed that fuse recombinant DNA protein products with sequences that will direct the proteins to the surface of the host cells. In addition, bacterial hosts are being developed that more easily secrete proteins than *E. coli* hosts.

segment polarity mutants   Mutants of the fruit fly, *Drosophila melanogaster,* in which one of the halves of each segment (the P compartment) is replaced by the other half (the A compartment) so that each segment contains two mirror images of one of the normal halves.

segments, segmentation   A pattern that develops in the embryo of the fruit fly, *Drosophila melanogaster,* which is defined by indentations giving the embryo the appearance of stacked disks with each disk representing a segment. Various structures of the adult such as legs, antennae, wings, and eyes develop from specific segments. Each segment consists of two halves: the A (anterior) compartment and the P (posterior) compartment.

selection   The ability to detect a recombinant protein, mutant, interacting protein, hybridoma, and so on. Selection techniques may make use of selective medium or specific markers on cells to be detected.

selective medium   A growth medium that, either by the inclusion of a toxic substance or by the lack of an essential nutrient, promotes the growth of only certain variant organisms in a population, for example, the growth of penicillin-resistant bacteria on a nutrient agar that contains penicillin. See SYNTHETIC MEDIUM.

SELEX   *S*ystematic *e*volution of *l*igands by *ex*ponential enrichment; an iterative technique for selecting oligonucleotides that bind to certain target molecules from a mixture of random oligonucleotides. In the SELEX technique, a mixture of oligonucleotides is passed through a column containing a matrix to which the target molecule (for example, ATP) is attached. Oligonucleotides that bind to the target will be retained on the column, while nonbonding oligonucleotides will wash through the column. The retained oligonucleotides can be eluted and amplified and passed through the column again in order to select oligonucleotides with stronger binding affinities to the target. This is repeated multiple times to select a few candidate oligonucleotides with very high binding affinities to the target.

self-assembly   The spontaneous, unassisted assembly of the components of a complex structure, for example, the protein viral coat of tobacco mosaic virus.

self-protein   Any protein that, as the result of immunological screening in early life, is determined to be "self" and therefore not recognized as a foreign antigen that would be attacked by the immune system. Certain illnesses, referred to as autoimmune diseases, result from a failure of the immune system to recognize self-proteins.

self-tolerance   The lack of an immune response to a self-protein.

semiconservative replication   The mode of DNA replication in which each of the original parental DNA strands is based paired with one newly synthesized daughter strand. Experiments performed by Matthew Meselson and Franklin Stahl in the mid-1950s demonstrated that DNA replication was semiconservative as opposed to conservative. This finding laid the foundation for future experiments that ultimately elucidated the molecular details of the process of DNA replication.

semidiscontinuous replication   DNA replication involving the synthesis of many small fragments that occurs on the lagging strand of double-stranded DNA in the form of Okazaki fragments.

Sendai virus   A member of the paramyxoviruses that is used to induce cell fusion, a technique for creating hybrid cells (heterokaryons) for the study of genetics in cultured cells.

sensitization A lowering of the threshold for a nerve impulse to be generated as a result of strong and repeated stimulation of a neuron by another neuron. Sensitization results from the tendency of some neurons to trigger an action potential if stimulated by weaker-than-normal nerve impulses or with shorter-than-normal refractory phases if action potentials have been triggered in that neuron in the recent past.

Sephacryl A composite matrix of polyacrylamide and dextran for column chromatography. This composite improves on the traditional gel filtration materials based upon cross-linked dextran (see SEPHADEX) or polyacrylamide (see BIOGEL), in that it is a more rigid gel type, which allows for higher flow rates and is better suited to large-scale chromatography.

Sephadex A polysaccharide-derived gel (formed by cross-linking of dextran strands). Filtration of mixtures of biological molecules through Sephadex gels in columns is a widely used procedure for separation of molecules based on size. See GEL-EXCLUSION CHROMATOGRAPHY and GEL FILTRATION.

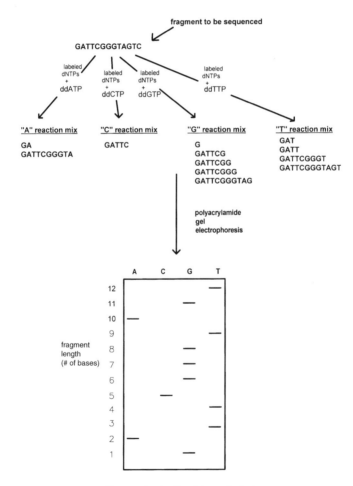

Sequencing by the dideoxy technique

Sepharose A form of agarose as small beads used in column chromatography for size separation of biomolecules similar to Sephadex and also a matrix for attaching antibodies and other ligands for various types of affinity chromatography.

sequenator A device for carrying out the automated sequencing of peptides by the Edman procedure.

sequence A term for the linear or end-to-end arrangement of biomolecules in a long polymeric molecule. Most often used to denote the order of purine and pyrimidine bases along the length of a nucleic acid, for example, AAGCTTCG..., where A=adenine, C=cytosine, G=guanine, T=thymine.

sequence conservation The tendency of certain DNA sequences to resist change in the course of evolution and therefore to be similar in dissimilar organisms.

sequence homology The degree of similarity between two nucleic acids as represented by the percentage of bases on one nucleic acid strand that match bases on the other nucleic acid strand when the two are aligned.

sequencing The process of determining the sequence of a polymeric biomolecule, for example, a nucleic acid or the amino acid sequence in a polypeptide, the sequence of sugars in a polysacharride, and so on. See DIDEOXY SEQUENCING.

SERCA pumps Sarcoplasmic and endoplasmic reticulum calcium ATPase; specialized pumps that transport Ca^{++} ions from the cytosol into the lumens of the endoplasmic reticulum and sarcoplasmic reticulum. In skeletal muscle sequestration of calcium is a mechanism for regulating muscle contraction; high cytosolic levels of calcium stimulate muscle contraction while low levels result in relaxation. In humans there are three SERCA genes that code for as many as 10 isoforms by alternative splicing.

serine An amino acid that, because it contains an hydroxyl group, can serve as a site for phosphorylation when serine is part of a protein.

serine proteases A family of proteolytic enzymes named for the fact that they always employ a serine residue in the catalytic site that is involved in the cleavage of a peptide bond at a specific site in a polypeptide or protein. In mammals serine proteases are particularly important in digestion, blood clotting, and the activation of factors in the complement system.

serodiagnostics A diagnosis based on the indirect evidence provided by serology indicative of a disease state or that an individual has been previously exposed to a pathogenic organism, for example, tuberculosis.

serologic reactions Any of several reactions based on the presence of specific antibodies in the blood serum. These reactions generally fall into three categories: bacteriolysis, precipitation, and agglutination.

serology A type of laboratory analysis based on the presence or absence of specific antibodies in the blood serum.

seropositive (seronegative) The finding, in a diagnostic test, that reactive antibodies to a given agent are (seropositive) or are not (seronegative) present in a sample of blood serum.

serotonin A monoamine neurotransmitter made from the amino acid tryptophan that regulates mood. A number of antidepressant drugs such as Prozac, Paxil, and Zoloft (selective serotonin reuptake inhibitors; SSRIs) act by inhibiting the reuptake of serotonin released at the synapse, which enhances the stimulatory effects of serotonin on mood.

Serotonin

serum The liquid part of blood from which the blood cells have been removed by clotting.

serum albumin One of the most abundant proteins in blood (albumin constiutes about 50 percent of the plasma protein). Albumin has at least two main functions: (1) to regulate water content of the tissues and (2) as a carrier of fatty acids in the blood stream.

serum globulins A group of abundant blood proteins with wide-ranging functions. The globulins are divided into three categories: alpha, beta, and gamma. Gamma globulins are the category that includes all the serum antibodies; the alpha and beta globulins form essential complexes with various substances, for example, lipids (these complexes are known as lipoproteins), carbohydrates (mucoproteins and glycoproteins), iron (transferrin), and copper (ceruloplasmin).

severe combined immunodeficiency (SCID) A group of inherited disorders in which an individual lacks an immune response due to a lack of infection-fighting lymphocytes. SCID is known in the popular media as the "bubble boy disease," for David Vetter, a boy with SCID who lived in a germ-free plastic bubble in the 1970s. There are several forms of SCID. One form is X-linked and so is most common in males. In another form the condition is caused by a deficiency of the enzyme adenosine deaminase (ADA). SCID mice are widely used in research to carry tissue xenografts from other animals, including humans, because their weakened immune systems allow the tissue to grow without being rejected. In this way the tissue can be studied while it is growing in an animal. See ADA.

sex chromosome See X CHROMOSOME.

sex-determining region Y (SRY) A region on the Y chromosome (gene map locus Yp11.3) that is responsible for development of the testis. SRY encodes a transcription factor of the high mobility group (HMG)-box family of DNA-binding proteins called the testis-determining factor (TDF), which initiates sex determination in males. Mutations in SRY give rise to XY females with a condition called gonadal dysgenesis (Swyer syndrome), in which there is gonadal degeneration leaving only "streak gonads" of fibrous tissue and ovarian stroma. In these patients there is no development of secondary sexual characteristics at puberty. Part of the Y chromosome containing SRY can also translocate to the X chromosome, which causes a condition known as XX male syndrome.

SH2, SH3 domains Domains of the GRB2 protein that function to mediate binding reactions of the signal-transduction protein, GRB2. The SH2 domain of GRB2 binds to the phosphorylatred tyrosine residues on the cytoplasmic domains of receptor tyrosine kinases and the SH3 domains bind to the protein, SOS. The SH prefix stands for *Src Homology* because of their homology to the src oncoprotein of rous sarcoma virus.

shadowing The process of coating a specimen with a thin layer of metal, such as platinum or palladium, by heat evaporation under a vacuum. Shadowing is necessary to view surface detail of the specimen under an electron microscope.

shaker mutation A mutation in a K+ channel in *Drosophila* that causes flies carrying the mutation to shake uncontrollably under anesthesia. The K+ channel was cloned from shaker mutants, and this allowed critical experiments to be carried out on how K+ channels function in the generation of action potentials.

shikimate pathway A major biochemical pathway by which all the "aromatic" amino acids (tyrosine, phenylalanine, and tryptophan) are synthesized from one parent chemical, shikimate.

Shine-Delgarno sequences Special sequences present on the 5′ region of each gene in a prokaryotic cell that are rich in the bases adenine and guanine and that help to align the ribosome on the mRNA so that translation can begin at the proper start site.

short-tandem repeat (STR) Sequences of DNA consisting of a core repeat of three to four bases, with an overall length of a few hundred bases. Such STR are used as markers in DNA profiling techniques because they are easily amplifiable, and STRs that differ from each other by one repeat unit can be easily resolved from each other on high-resolution sequencing gels.

shotgun-cloning method A technique of cloning a DNA sequence of interest based on mass ligation of a heterogeneous mixture of DNA fragments into a vector; the vector carrying the DNA of interest is then selected from mixture of cloned DNA fragments. This technique is useful when the DNA of interest is represented in low abundance or is difficult to purify.

shuttle vector A vector genetically engineered to permit the growth and/or expression of recombinant DNAs in both prokaryotic and eukaryotic cells

sialic acid A modified sugar found in the lipids of the membranes of neural cells that are part of the receptor for neurotransmitters.

sialophorin (SPN) A major sialoglycoprotein found on the surface of human T lymphocytes, monocytes, granulocytes, and some B lymphocytes that is important for immune function. Sialophorin is a component of a receptor-ligand complex involved in activation of T cells. The sialophorin gene is at gene map locus 16p11.2.

sickle-cell anemia A genetic condition involving a point mutation in the beta chain of the hemoglobin protein that results in a loss of ability to carry oxygen from the lungs to the tissues of the body. The disease derives its name from the fact that red blood cells carrying the mutant hemoglobin assume an elongated sickle shape.

sickle-cell disease The pathological condition caused by SICKLE-CELL ANEMIA that is characterized by an inability to handle exertion.

sigma factor A small protein that forms a complex with the RNA polymerase enzyme in prokaryotic cells. The formation of sigma factor–RNA polymerase is essential for the accurate intiation of transcription in bacteria.

signal peptidase An enzyme that catalyzes the cleavage of the signal peptide immediately after the polypeptide is inserted into the endoplasmic reticulum.

signal-recognition particle (SRP) A ribonucleoprotein comprised of six polypeptides and a small (7S) RNA molecule that mediates the binding of a signal sequence on a preprotein to its appropriate membrane receptor.

signal sequence A special sequence of amino acids on the amino terminal end of polypeptides that are destined to be exported from a eukaryotic cell. If the signal sequence is present, then the protein bearing that sequence is transferred into the endoplasmic reticulum where it is further processed for export.

signal transduction A process in which a substance binds to a receptor on the outside of a cell that then transmits a signal to induce a metabolic reaction. The chemical that acts as the signal is called a second messenger, the first messenger being the substance that bound to the receptor but cannot itself enter the cell to induce the metabolism. (See figure on next page.)

signature sequence A segment of a particular protein, generally between 10 and 50 amino acid residues, that is found only in one taxonomic group of organisms and not in others. For example, the elongation factor EF-Tu contains a 12 amino acid sequence near the amino terminal end that is found in archaebacteria and eukaryotes but not in other types of bacteria. Signature sequences have been subcategorized as superfamily signatures and motifs.

silencers Certain nucleotide sequences that act to suppress the activity of a promoter. Silencers may act at distances

ras-dependent signal-transduction pathway

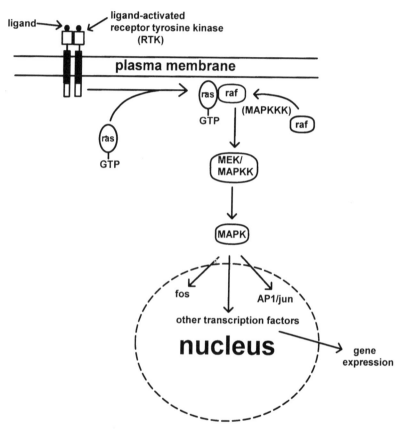

Schematic representation of the ras-dependent signal-transduction pathway. A membrane-bound receptor tyrosinase (TRK) becomes activated by ligand binding to its extracellular domain, and tyrosine residues on the cytosolic side of the membrane undergo autophosphorylation. An activated ras bound to GTP then binds to cytosolic raf, which then becomes activated. The raf C terminus is a protein kinase with dual specificity (serine/threonine) that acts to phosphorylate and thereby activate MEK, a MAP-kinase kinase (MAPKK). MEK, which is itself a dual-specificity kinase, activates MAP kinase (MAPK) via phosphorylation. Activation of MAPK leads to phosphorylation of a variety of transcription factors, including the AP1/jun complex, fos, and others, which results in gene transcription.

greater than one kilobase away from the promoter sequences on which they act.

silent mutation A mutation whose effect is not manifest either because it occurs in a nonessential region of DNA or because the effect of the mutation is masked.

silent sites DNA nucleotide bases that, when changed (for example, by mutation), do not result in any change in the amino acids in the polypeptide coded for by the DNA.

simian virus 40 A small DNA virus accidentally discovered as a contaminant in cultured African green monkey kidney cells that are used to grow polio virus for vaccine development. The virus was later found to be oncogenic in mice although not in humans or in its natural host.

simple sequence DNA DNA sequences that are extremely highly repeated throughout the genome of an organism. These highly repeated sequences that are generally very short in length are referred to as simple sequences.

sindbis virus A member of the family of RNA-containing togaviruses (alphavirus group). Infection in humans and other mammals is via mosquitoes where varying degrees of encephalopathy are produced.

single-locus probes (SLP) A technique used in DNA profiling in which the DNA from an individual, blotted to a membrane (see SOUTHERN BLOT HYBRIDIZATION) is mixed with a probe in a series of sequential tests that will detect a single sequence. Usually, only two bands are detected at each stage of the test.

sis oncogene An oncogene that is found in simian sarcoma virus and that is associated with sarcoma tumors in both monkeys and cats. The name is an acronym derived from *si*mian *s*arcoma. The sis oncogene protein is virtually identical to one of the subunits of platelet-derived growth factor (PDGF).

sister-chromatid exchange The exchange of material between the two daughter strands of a replicated chromosome (i.e., chromatids) during meiosis; recombination occurring at the chromosomal level.

site-directed mutagenesis The technique by which specific bases on a segment of DNA are experimentally altered.

site-specific drug delivery A technique for targeting drugs to certain tissues. Various strategies may be employed to accomplish this, for example, chemical linkage of antibodies to the drug molecule or attachment of the drug molecule to a ligand that is specific for a cell surface receptor. See FUSOGENIC VESICLES.

site-specific recombination Recombination between two DNAs that occurs at a specific site on each DNA. Site-specific recombination is exemplified by integration of lambda bacteriophage DNA in which recombination takes place at a site designated as attP on the bacteriophage DNA and the corresponding site (designated attB) on the bacterial host DNA.

Skatchard analysis (plot) A mathematical method for estimating both the number of receptors for a certain ligand and the affinity of the ligand for its receptor from a plot of the amount of unbound ligand versus (bound ligand)/(free ligand).

skeletal muscle The relatively more-striated muscle tissue associated with voluntary movement, for example, in the movement of the limbs.

ski oncogene The oncogene carried by a strain of avian sarcoma virus, it derives it name from *Sloan-Kettering Institute*, where it was discovered. The oncogene was identified on the basis of its ability to transform cultured cells to a cancer-like state. ski has subsequently been shown to cause non-muscle cells to differentiate into skeletal muscle. The proteins encoded by ski regulate transcription of genes by forming complexes with various transcription factors, including NF-I, Smad2, and Smad3. The ski proto-oncogene (c-ski) gene map locus is 1q22-q24.

Smads Signal transduction elements associated with signaling by TGF receptors. Binding of members of the TGF family of growth factors to their receptors causes Smads on the cytosolic side of the membrane to become activated by phosphorylation. Activated Smads form dimers that migrate into the cell nucleus, where they activate transcription of certain target genes. Smads fall into classes: R-Smads (those activated by phosphorylation by the kinase domains of receptors), I-Smads (inhibitory), and Co-Smads. The R-Smads are comprised of Smads 1, 2, 3, 5, and 8. Smads 2 and 3 are involved in signaling via the TGF-β family, and Smads 1, 5, and 8 respond signaling via the BMP subfamily of growth factors. Smad4 is a co-Smad. The R-Smads and Co-Smads contain two conserved structural domains: MH1 (MAD Homology domain) and MH2. Phosphorylated R-Smads form complexes with the Co-Smad, Smad4. The MH1 domain of the R-Smads is responsible for the DNA-binding activity of the complex, while the MH2 domain is involved in interaction with the receptor.

SMC proteins   Structural maintenance of chromosomes; a family of proteins that are involved in chromosome condensation, sister-chromatid cohesion, DNA repair, and recombination. There are six core SMCs in eukaryotes (SMC1–SMC6) that form functional complexes with other proteins. The cohesin complex contains SMC1 and SMC3 (together with cohesin proteins Scc1 and Scc3), which is needed for sister-chromatid cohesion during mitosis. The SMC1 and SMC3 also forms a complex with DNA polymerase ε and ligase III that mediate recombination. SMC2 and SMC4 are components of the condensin complex, which functions in the process of chromosome condensation during mitosis. The functions of SMC5 and SMC6 are not known, although SMC is believed to be involved in recombination-based DNA repair processes.

smooth ER   The endoplasmic reticulum that is not bound to ribosomes.

smooth muscle   The relatively less striated (i.e, smooth) muscle associated with involutary movement, for example, the heart muscle.

SNAP   Soluble NSF attachment proteins; cytosolic proteins that are required for NSF to bind the membrane of a Golgi vesicle and are therefore necessary for fusion of a transport vesicle with the Golgi.

SNARES   A family of proteins that mediate the fusion of synaptic vesicles with the synaptic membrane during the process of neurotransmitter release. The SNARES present on the surface of the synaptic vesicle are called v-SNARES, and those on the cell membrane are called t-SNARES. During synaptic fusion, the v-SNARES and t-SNARES bind to one another together with a protein called SNAP25 to initiate the fusion process that releases neurotransmitter. The SNARE-SNAP25 complex is targeted by the toxin of the bacterium *Clostridium botulinum*.

snRNA   Small nuclear RNA; a very short piece of RNA that complexes with a set of proteins (snRNPs) to form a structure whose function is to clip out loops in other RNAs, particularly for splicing of RNAs that are destined to become mRNAs

sodium-potassium pump   A specialized transmembrane protein that pumps sodium ions out of the interior of the cell and at the same time pumps potassium ions into the cell. Although sodium-potassium pumps are found in a variety of cell types, they especially abundant in nerve cells where they serve to establish an electric potential across the membrane that is the basis of nerve impulse transmission.

soma   A term for the entire body of an organism without reproductive cells.

somatic cell   A nonreproductive cell; any cell that does not generate either sperm or egg.

somatic cell hybrid   The product formed by somatic cell hybridization.

somatic cell hybridization   Combining the genetic material of two cells by cell fusion, such as that induced by Sendai virus or polyethylene glycol (PEG). See CELL FUSION.

somatic cell therapy A gene therapy based on the introduction of new genetic material or the alteration of existing genetic material in cells other than those that give rise to either sperm or egg, for example, the introduction of insulin genes into pancreatic cells.

somatic mutation Any mutation not affecting the reproductive cells. This type of mutation usually affects a particular tissue type and is not passed down to offspring in the form of a transmissible genetic defect.

somatomedin A polypeptide hormone, produced in the liver, that induces growth of bone and muscle.

somatostatin A polypeptide hormone, produced by the hypothalamus, that helps to regulate to blood sugar levels by inhibiting the release of glucogon and insulin by the pancreas.

somatotropin A polypeptide hormone, produced in the anterior pituitary, that simulates the liver to secrete somatomedin-1.

sorbitol An alcohol derived from glucose. In diabetes, sorbitol accumulates in the eye, the kidney, and the other tissues; this leads to osmotic swelling and eventual damage of critical cells such as the optic nerve.

SOS repair system A system of at least 15 different proteins that work to repair severe DNA damage in bacteria; the system appears to be induced by the presence of an excessive amount of single-stranded DNA as might be generated by DNA damage.

SOS response In bacteria, the induction of various proteins involved in DNA repair (such as the UvrA and UvrB proteins) and DNA synthesis (DNA polymerase III, UmuC, UmuD, and RecA) in response to the presence of high levels of DNA damage as might occur following exposure to a mutagenic agent.

Southern, E. M. (b. 1918) The discoverer of the Southern-DNA-blot-hybridization technique.

Southern blot hybridization In a complex mixture of DNA fragments separated by size on an agarose gel, a technique for identification of a DNA fragment(s) by first transferring the DNA fragments from the agarose gel to a special membrane and then hybridizing the DNA fragments to a specific probe.

spacer DNA Stretches of nontranscribed DNA that separate transcribed regions of DNA and that code for ribosomal RNAs.

species Different forms of an organism among the members of a genus that are incapable of producing offspring by interbreeding.

specific activity The activity of a substance that is present in some given amount of that substance, as defined for that substance by convention. For example:
- units of enzyme activity per microgram of protein
- units of hormone activity per milliliter of solution
- disintegrations per minute per mole of radiolabeled amino acid

specificity factors Proteins that act to alter the specificity of RNA polymerase to recognize a given promoter or a set of promoters, by making it more or less likely for the polymerase to bind to them. For example, vaccinia virus contains a protein that causes RNA polymerase to recognize the viral promoter and begin transcription of the viral genes.

spectrin A filamentous protein that comprises a cytoskeletal network that is attached to the cytosolic side of the plasma membrane in erythrocytes. The spectrin cytoskeleton largely accounts for the membrane rigidity that prevents deformity of red blood cells during their passage through the small vascular openings of capillaries. Spectrin also functions to anchor certain membrane-associated structural elements such as glyophorin and the membrane channel for chloride-bicarbonate exchange.

spermatids Immature sperm cells having the haploid number of chromosomes but lacking the morphological features of sperm, for example, the elongated acrosome-bearing head and the tail assembly that make spermi motile.

spermatocytes Cells representing stages in the formation of sperm: Primary spermatocytes are cells containing the diploid number of chromosomes but which, after dividing, form secondary spermatocytes that contain the haploid number of chromosomes. The secondary spermatocytes differentiate to form spermatids.

sperm cells (spermatozoa) The mature cells derived from the male reproductive cells (gametes) that is produced by meiosis.

SPF S-phase promoting factor; in yeast, a family of complexes between the cyclin-dependent kinase, cdc28, and G1 cyclins that mediate the transit through the S phase of the cell cycle.

S phase A part of the cell cycle during which the total complement of a cell's DNA is replicated.

sphingolipid A type of membrane lipid derived from the compound, sphingosine. Sphingolipids are subdivided into sphingomyelins, gangliosides, and cerebrosides, all of which are important components of the brain cell membranes. Altered metabolism of sphingolipids is the cause of the genetic syndrome, Tay-Sachs disease.

spinal muscular atrophy (SMA) A genetic neurodegenerative disease affecting motor neurons and characterized by wasting of the skeletal muscles. SMA is caused by progressive degeneration of the anterior horns of the spinal cord. There are several types of SMA:

• SMA type I (Werdnig-Hoffmann disease) manifest in utero or in neonates

• SMA type II, onset of symptoms between three and 15 months of age

• SMA type III (Kugelberg-Welander disease), onset of symptoms between two and 17 years of age

The disease is caused by mutations in two genes located on chromosome 5q13, SMN1 and SMN2 (SMN stands for survival of motor neuron). These genes code for proteins involved in RNA splicing. Over 90 percent of SMA cases lack part of, or all of, both copies of SMN1. A small percentage of SMA patients are missing one copy of the SMN1 gene and have small mutations in the remaining copy.

spindle apparatus The bundles of microtubules that are attached at one end to the centromere of chromosome and at the other to the centriole and are responsible for the movements that lead to segregation of the chromosomes during cell division. See MITOTIC APPARATUS.

spleen A large, ductless organ in the upper-left portion of the stomach; it plays a role in the maturation and differentiation of the antibody-forming blood cells.

splice, splicing A joining together of separated sections of an RNA molecule to generate new RNAs. In the process by which mRNAs are created from long RNA precursors in the nucleus, sections of RNA that represent introns are spliced out so that segments representing exons are joined together. Splicing is part of the process of RNA processing that takes place in the nucleus. See SPLICEOSOME and SPLICING JUNCTION.

spliceosome A complex that mediates the splicing of an RNA molecule during mRNA formation. The spliceosome contains the RNA precursor in which the ends of the regions that will be joined together are held in place by small ribonucleoprotein particles (snRNPs).

splicing junction The site on a spliced RNA where the ends of the spliced RNA segments meet.

spontaneous mutation A change in a nucleotide base in the DNA that occurs during the normal process of DNA rep-

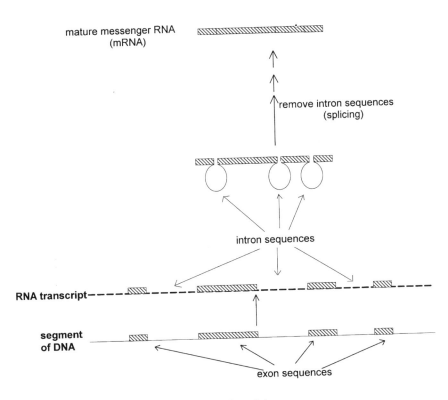

mature messenger RNA
(mRNA)

remove intron sequences
(splicing)

intron sequences

RNA transcript

segment
of DNA

exon sequences

RNA transcript splicing

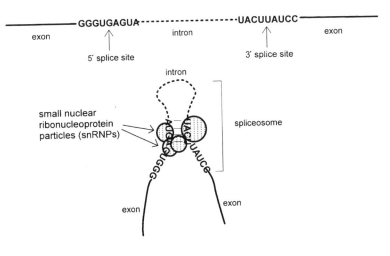

GGGUGAGUA - - - - - - - - - - - - - - UACUUAUCC

exon   intron   exon

5′ splice site   3′ splice site

intron

small nuclear
ribonucleoprotein
particles (snRNPs)

spliceosome

exon   exon

Spliceosome

lication and without the action of mutagenic agents.

src The oncogene that is carried by the Rous sarcoma virus that produces sarcomas in birds. The product of the src gene is a phosphorylated protein denoted as pp60. The src protein is a tyrosine kinase and is believed to cause transformation as a result of its ability to carry out phosphorylation of critical proteins.

SSCP Single strand conformation polymorphism; an analytical technique for determining the presence of changes in nucleic acid primary structure by observing the rate of migration of nucleic acid fragments in a gel where the nucleic acids are kept in a denatured state by chemicals such as urea or formamide. This technique is highly sensitive to changes in nucleotide sequence and is frequently used to analyze genes for the presence of small mutations.

staggered cut A term applied to the type of cleavage of DNA molecules produced by most restriction enzymes in which one end (usually the 5' end) protrudes past the cut end of the other strand.

standard deviation A statistical value given by the square root of the variance of a set of experimental values. This quantity is a measure of the average amount each experimental value or observation in a series differs from the mean of that series.

standard transformed constants The physical constants that, by international convention, are used in biochemical calculations to represent "standard" conditions in biological systems. The main standard transformed constants are: $[H^+]=10^{-7}M$ (pH 7.0), $T=298°K$ ($25°C$), 1 mM Mg^{++}, 55.5M H_2O, 1 atmosphere pressure, and 1M concentrations of all other products and reactants.

starch A complex polysaccharide used by plants as a means of storing glucose. Starch consists of long polymers of glucose that are joined to one another to form a compact, branched macromolecule similar to glycogen.

START A point in the G1 phase of the yeast cell cycle that represents the commitment of the cell to transit into S phase. See CLN1, CLN2, CLN3.

STAT Signal transducers and activators of transcription; a class of transcription factors that constitute one of the main components of the JAK/STAT signaling pathway. STATS are activated by phosphorylation catalyzed by JAKs. Once activated, STATS dimerize and move into the nucleus, where they bind to specific sequences in the promoters of target genes whose transcription is subsequently induced.

statins A class of drugs that lower the levels of cholesterol in the blood by inhibiting the pathway by which cholesterol is synthesized in the liver. The statins have structures similar to mevalonate, the normal substrate of the enzyme HMG-CoA reductase, which controls the key step in cholesterol biosynthesis. The statins therefore inhibit cholesterol biosynthesis by serving as competitive inhibitors of HMG-CoA reductase. Zocor, Pravacol, Mevacor and other pharmaceutical statins are chemically modified versions of statins that were originally derived from fungi.

stationary phase The point at which, in a bacterial culture, the cells become so numerous that the nutrient supply is exhausted and growth ceases. See GROWTH PHASES.

Zocor

STE genes A family of genes whose products function as part of a signal-transduction cascade to mediate mating in yeast; the STE designation is derived from the word *sterile* to indicate the fact that mutations in these genes result in sterility in yeast. STE2 and STE3 are receptors for mating pheromones in the α and a mating types, respectively. Other STE proteins are components of G proteins or function as protein kinases. STE12 is a transcription factor. See KSS1, FUS3.

stem cell Any cell that, in a tissue, is itself immature but gives rise, through cell division, to cells that become the mature form of the cells that characterize the tissue. The marrow in bone is a classic example of stem cells that give rise to the mature differentiated blood cells, including red blood cells, macrophages, and the antibody-producing cells of the immune system. However, only the completely undifferentiated stem cells derived from embryos (embryonic stem cells) are toti-

potent, that is, have the capability to give rise to all the different cell types characteristic of the different body tissues. Modern stem-cell research focuses on identifying the signals that can cause stem cells to differentiate into a particular cell type. If such signals can be found, it is believed that stem cells can then be used to replace damaged tissues seen in a number of conditions such as Alzheimer's disease, Parkinson's disease, spinal cord injuries, and others.

stem-loop structure A structure formed by nucleic acids, but particularly RNAs, in which a segment of the nucleic acid strand base pairs with a distant complementary sequence; the base-paired sequences form the "stem" and the sequences intervening between the base-paired regions form the "loop."

stereoisomer A form of a molecule involving different arrangements of atoms or molecules around a central atom, usually carbon in biomolecules. Stereoisomers are

a molecule containing an atom with four different substituents

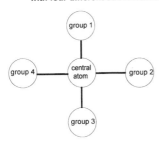

its optical isomer

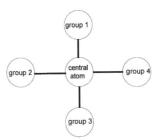

for example:

Stereoisomer

also referred to as optical isomers because crystals of stereoisomers cause polarized light to slant in ways that are characteristic for each stereoisomer. See DEXTROROTATORY ISOMER and LEVOROTATORY ISOMER.

sterile Completely free of living material.

sterilization The process by which objects or liquids are made sterile, usually for the purpose of preventing disease, infection, or contamination. Common methods of sterilization include heating to temperatures above 125°C and prolonged exposure to ultraviolet light.

steroid A class of potent hormones derived from cholesterol. Cortisone and the sex hormones, estrogen and testosterone, are examples of steroid hormones.

sticky ends The single-stranded ends of any two nucleic acids whose nucleotide base sequences are complementary to one another.

stimulatory neuron A neuron that, when stimulated, functions to enhance the effects of the nerve impulse from another neuron.

stock culture A culture of cells that serves as a common source of cells for experimental purposes.

stop codon A sequence of three nucleotide bases that do not represent the code for an amino acid but serve as signals for the termination of translation by the ribosome. There are three RNA stop codons: UAA, UGA, and UAG.

stop-transfer signal For preproteins that are to be inserted into, but not completely through, a membrane, a stop-transfer signal is a group of amino acids on the polypeptide that serves as a signal to stop its movement at a time when the polypeptide is properly positioned in the membrane.

strand displacement A variant of the normal mechanism of DNA replication in which replication of one of the DNA strands proceeds from opposite ends of a linear DNA molecule.

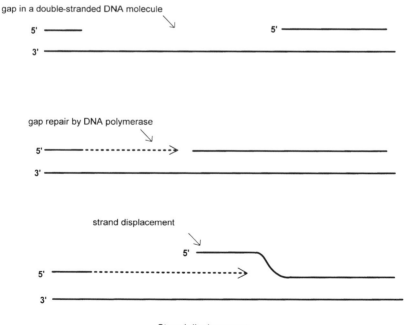

Strand displacement

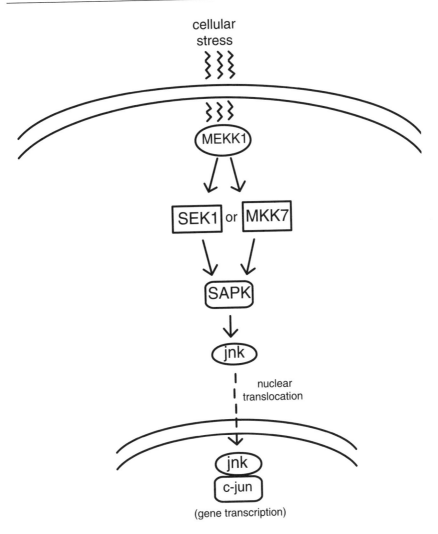

Stress-activated kinases

streptolydigin An antibiotic that inhibits the action of bacterial RNA polymerase. Streptolydigin binds to the beta-subunit of the polymerase and stops chain elongation after the addition of three or four nucleotides.

streptomycetes A funguslike bacterium found in soil. In addition to streptomycin, various isolates of streptomycetes have yielded more than 500 compounds of therapeutic value, including more than 90 percent of the clinically useful antibiotics.

streptomycin An antibiotic derived from molds that exerts its antibacterial effect by causing bacterial ribosomes to misread the codons on the mRNA, particularly with respect to the pyrimidines U and C where one is usually mistaken for the other.

stress-activated kinases (SAPKs) A family of protein kinases involved in signaling that are activated by a variety of environmental stressors such as ultraviolet irradiation, toxic chemicals, oxidative conditions, hypoxia and anoxia, heat shock, inhibitors of protein synthesis, and inflammatory cytokines. Activated SAPKs bind to, and phosphorylate, the N-terminal domain of the c-jun transcription factor, which subsequently translocates to the nucleus, where it upregulates the expression of genes involved in various stress responses. The SAPKs are activated by the upstream factors SEK1 and MKK7. There are three SAPK genes: α, β, and γ that code for 8–10 isoforms by alternative splicing mechanisms.

stress fibers   The fibrillar arrays that are seen on the surface of a cell that is oriented parallel to the direction in which a cell is moving. Stress fibers appear to be directly related to cell movement in that they are known to coincide with the actin filaments.

stress proteins   Proteins encoded by heat-shock genes and expressed when the cell undergoes stress conditions, such as a rise in temperature or exposure to certain chemicals. Many of these proteins are chaperons that aid in maintaining the structure of the protein under conditions of denaturation.

stringency   The conditions of temperature and ionic strength used during nucleic acid hybridization methods, for example, northerns or Southerns, to ensure proper binding of the probe to its target. Low stringency conditions allows for the probe to bind with less specificity to targets; high stringency conditions only will permit very specific binding of probe to its target.

stringent response   A bacterial response to conditions of nutritional deprivation in which expression of nonessential genes is shut down. A stringent response involves a rapid downregulating of certain bacterial biosynthetic pathways (e.g., synthesis of ribosomal and transfer RNAs) when the amino acid supply becomes limited. See ALARMONES.

stroma   1. The space between the grana and the chloroplast membrane that contains some of the enzymes of the dark reaction in photosynthesis as well as the chloroplast RNA and DNA.
2. The connective tissue underlying the epithelial cell layer, for example, in skin, the digestive tract, and the airways in lung.

structural gene   A gene in an operon that codes for the functional protein that is essential to the metabolism of the bacterial cell, for example, an enzyme, as distinguished from the genes for a repressor protein that controls the expression of a structural gene.

STS   *Sequence tagged site*; a means of cataloguing sequence data by recording only that part of the whole sequence necessary to create primers that can be used to amplify the entire sequence from a DNA sample by the polymerase chain reaction (PCR).

stuffer region   That part of the lambda phage that can be replaced by foreign DNA and still reproduce so that the phage can be used as a cloning vector.

subcutaneous   Just underneath the skin; as in subcutaneous injections.

substrate   Any one of the reacting chemicals in an enzyme-catalyzed reaction.

substrate analogue   A chemical that is similar in form to a particular substrate but that does not participate in the chemical reaction of the substrate. Substrate analogues used for various purposes such as inhibition of certain enzyme systems or for studying the mechanism of enzyme action. See COMPETITIVE INHIBITION.

substrate channeling   The direct transfer of intermediates from one enzyme to the next in multienzyme complexes. For example, the pyruvate dehydrogenase complex processes pyruvate into acetyl CoA by a series of enzymatic reactions catalyzed by five separate enzymes in a large complex.

The products of each reaction become substrates for the next, and these products/substrates are passed through the complex by substrate channeling.

subtilisin An proteolytic enzyme (protease) produced by the soil bacterium *Bacillus amyloliquefaciens*.

subunit One part of a complex biological molecule such as an enzyme or ribosome. The subunits combined together constitute the biologically active molecule.

subunit vaccine A vaccine created in the lab using recombinant DNA technology, in which a portion of the entire virus or bacterium is presented as an epitope. This method was used in the production of a vaccine against hepatitis B virus (HBV), a virus that cannot be cultured in the laboratory.

sucrose A disaccharide consisting of one molecule of fructose linked to one molecule of glucose. Common table sugar is sucrose.

sucrose-density centrifugation A technique that separates molecules in a mixture according to their density by using sufficiently high centrifugal force to cause the molecules to migrate through a solution of sucrose. In density-gradient separation, the sucrose solution increases in density the farther the molecules travel.

sudden-correction model The model that proposes that in gene clusters in which there are multiple copies of a gene (e.g., the genes coding for ribosomal RNA), the entire gene cluster is replaced "every so often" by a process that replicates the entire gene cluster from just one or a few copies. The sudden-correction model is actually an error-correcting mechanism that accounts for why mutational errors in some of the gene copies do not accumulate over time.

sugar Any compound that conforms to the general molecular formula $C_n.(H_2O)_n$, where n is any number between 3 and 7

Sugar

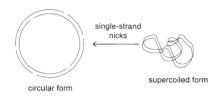

single-strand nicks

supercoiled form

circular form

Supercoiled DNA

and where the carbon atoms are linked together in a chain. See CARBOHYDRATE.

supercoiled DNA A circular double-stranded DNA molecule is itself twisted into a compact knot. This is the replicative form of many viral DNAs in their host cells.

suppressor gene Any gene that acts to suppress the effects of mutation.

suppressor mutation Any mutation that suppresses the effects of a previous mutation; for example, a mutation that suppresses the effects of frame shift mutation by reinstating the proper reading frame.

suppressor T cell A type of T lymphocyte that suppresses the antigenic response of antibody-forming B cells; that is, it inhibits the formation of antibody to a particular antigen.

suppressor tRNA A mutation in a transfer RNA that suppresses the effect of a previous mutation in a gene. The suppressor mutation allows the suppressor tRNA to read the first mutation correctly, thereby ensuring the process of translation. See AMBER SUPPRESSOR.

surfactant Any agent that lowers the surface tension of water. Soaps and detergents are the most common surfactants.

SV40 Simian virus 40.

Svedberg unit A measure of molecular size based on the rate of sedimentation of a molecule in a centrifugal field. The Svedberg unit is designated as s and is not directly proportional to size; for example s values of

nucleic acids vary with secondary structure, temperature, and salt concentration.

SW1/SNF A large complex of 10 proteins that acts as a transcriptional activator by indirect mechanism involving chromatin remodeling. The SW1/SNF complex uses energy from ATP derived from the ATPase activity of SW12 subunit to produce changes in chromatin structure, which include changes in nucleosome structure and positioning such that transcriptional activators that act by binding to DNA can have access to their recognition sequences.

symbiosis A state of two or more organisms living in permanent close proximity for the mutual purpose of supplying some essential nutrient or life function to one another.

synapse The specialized junction between the tip of the axon from a neuron and the dendrite of an adjacent neuron. The transmission of nerve impulses from one neuron to the next is carried out by neurotransmitters that cross the synapse.

synapsis A stage in the recombination process mediated by the RecA protein in which the RecA protein forms a complex with the single-stranded and double-stranded DNAs that will then align with each other before undergoing recombination.

synaptic cleft The space intervening between the axon and dendrite membranes in a synapse.

synaptic vesicle A membrane-enclosed vesicle that carries the NEUROTRANSMITTERS to the synapse where they are released by fusion of the synaptic vesicle membrane with the membrane at the axon terminus.

synaptonemal complex The structure that joins chromosome pairs when homologous chromosomes align during the process of meiosis.

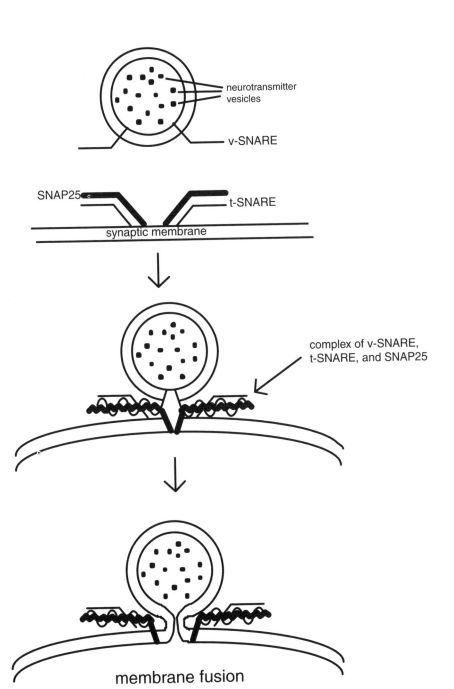

neurotransmitter
vesicles

v-SNARE

SNAP25

t-SNARE

synaptic membrane

complex of v-SNARE,
t-SNARE, and SNAP25

membrane fusion

Synaptic vesicle fusion

synaptophysin A polypeptide located in a transmembrane fashion in the membrane of a synaptic vesicle that is thought to mediate the fusion process between the synaptic vesicle membrane and the plasma membrane at the axon terminal.

synchronous culture A cell culture in which all cells are simultaneously at the same phase of the cell cycle. Experimentally, synchronization of cells can be achieved by techniques that transiently but specifically block in one phase of the cell cycle, for example, mitosis leading to accumulation of cells at the block; synchronous growth ensues when the block is released.

syncitium A multinucleated cytoplasm such as occurs when cells are fused by treatment with polyethylene glycol or Sendai virus.

syndrome A set of characteristics that are usually associated with the same cause. For example, genetic diseases caused by a single mutation can result in a number of different phenotypes.

synergism Facilitation of a response by multiple stimuli such that the magnitude of the response is greater than the sum of the individual stimuli. The principle of synergism is often exemplified by the facilitation of a nerve impulse when a single neuron is stimulated by several excitatory neurons.

syntaxin A transmembrane protein located in the active zone of the plasma membrane at the axon terminal that is believed to mediate the docking process between the synaptic vesicle and the plasma membrane. The docking process involves anchoring the synaptic vesicle in close enough proximity to the plasma membrane to allow fusion to occur. During this process, syntaxin is thought to bind to synaptobrevin, a transmembrane protein located in the synaptic vesicle membrane.

synteny The state of two or more genetic loci being present on the same chromosome. All the genetic loci that are present on one chromosome are said to be syntenic. The concept of synteny has been extended to include the organization of the genetic loci on a chromosome, and this has been used to create synteny maps that compare the arrangements of homologous genes on chromosomes from different species.

synthesis The process of creating a new substance from precursor molecules. Biosynthesis is the process by which living cells create new biomolecules, whereas the term *synthesis* is generally applied to processes used in the laboratory for creating biomolecules.

synthetic medium Solutions of nutrients are created for the purposes of growing cells of various types in culture. Most synthetic media formulations attempt to recapitulate the natural nutrient environment for the cell type being cultured as nearly as possible.

synthetic peptides The creation of peptides in the laboratory, using techniques of organic chemistry to link amino acids together according to some prescribed sequence so that a peptide of any given primary structure can be synthesized.

syntrophism Cross-feeding by organisms sharing a common growth medium, for example, bacterial colonies whose growth is dependent on a factor or factors secreted by a neighboring bacterial colony on a common agar plate.

syphilis A venereal disease caused by the spirochete *Treponema pallidum*. If left untreated, the disease may cause blindness and neuological symptoms, including a syndrome characterized by a loss of motor control known as general paresis.

systemic lupus erythematosis An autoimmune disease of the connective tissue that is characterized by a reddish skin rash (erythema) and a wide variety of conditions related to internal organ malfunction. Antibodies to a wide variety of self antigens are seen.

T4 RNA ligase An enzyme, isolated from bacterial cells infected with the bacteriophage, T4, that catalyzes the formation of a covalent bond between the phosphate group on the 5' end of either single-stranded DNA or RNA and the 3' hydroxyl end of either single-stranded DNA or RNA.

T7 promoter A sequence that forms the promoter for transcription of the genes of the T7 bacteriophage. The T7 promoter is widely used in synthetic cloning vectors where expression of recombinant DNAs is desired.

tachykinin A group of biologically active amidated neuropeptides that excite neurons and are potent vasodilatators and cause contraction of many smooth muscles. The tachykinins are found in both vertebrates and invertebrates. There are three tachykinins in humans that are encoded by two genes. They are made in the form of precursor peptides that are enzymatically converted to their mature forms, all of which are between 10 and 12 residues long and share a common carboxy-terminal sequence: Phe-X-Gly-Leu-Met-NH$_2$. One of the precursor peptides contains both substance P and neurokinin A, while the other encodes a precursor that contains only neurokinin B.

TAFs TBP *a*ssociated *f*actors; a set of at least eight proteins that assembles onto the DNA-bound TATA Binding Protein (TBP) to form the general transcription factor TFIID. TFIID binding to the TATA box in eukaryotic promoters is the first event required for subsequent assembly of other transcription factors and RNA polymerase II to form the fully functional transcription complex that initiates transcription.

tamoxifen An anticancer drug specific for breast cancer that acts as an antagonist of the estrogen receptor in low-grade cancers whose growth is estrogen-dependent.

Tamoxifen

tandem In general a group of objects arrayed in a line, one next to the other. As applied to molecular genetics, the term refers to genes arranged in tandem along a stretch of DNA. A number of viral and cellular genes (e.g., rRNA genes) that undergo amplification are tandemly arrayed.

T antigen(s) The products of the early genes of the papova viruses. In normal hosts, the T antigen(s) function to stimulate viral DNA replication and to regulate expression of the viral genes. However, in host cells that do not support virus replication, the T antigens are known to be responsible for transformation of the cells to a cancerous phenotype.

Taq polymerase   A DNA polymerase isolated from the thermophilic bacterium *Thermus acquaticus.*

tariquidar (XR9576)   An experimental drug used to inhibit the ability of cancer cells to become resistant to chemotherapeutic agents (multidrug resistance). Tariquidar acts by binding to a membrane glycoprotein known as P-gp (P-glycoprotein pump), a transmembrane protein that acts to pump administered anticancer drugs out of the tumor cell. The binding of tariquidar blocks the ability of P-gp to act as a pump, thereby allowing the chemotherapeutic agent to be retained in the cell.

tastin   A cytoplasmic cell-adhesion molecule that, in combination with bystin, forms part of the machinery that mediates the process by which the embryo attaches to the wall of the uterus.

TATA box   Another name for the Pribnow box.

tautomerism   The rapid and continual transition between different forms of a molecule based on delocalization of an electron(s) on different atoms of the molecule.

taxol   A plant alkaloid that stabilizes microtubules, thereby freezing cells in mitosis. Because cancer cells are rapidly dividing cells, taxol is currently being used as a chemotherapeutic agent.

taxonomy   The science of classification.

Tay-Sachs disease   A hereditary disease in which accumulation of a certain type of sphingolipid accumulates in the brain and the spleen, leading to degeneration of the nervous system and death at an early age.

T cell   A lymphocyte named for the thymus (i.e., thymus cell) where the majority of T cells mature. T cells are responsible for so-called cell-mediated immunity, the immune function directed toward detecting and destroying foreign cells rather than foreign proteins.

T-DNA   A term for the Ti plasmid carried by *Agrobacterium*, a parasite that induces various plant tumors. The tumors are a direct result of expression of the genes carried by Ti in the plant cells.

teichoic acid   A long polymer of glycerol or ribitol molecules linked together by phosphate groups. Teichoic acid is a structural component of the outer cell wall of Gram-positive bacteria. See GRAM STAIN.

telomere(s)   Special tandemly arranged, guanine-rich, repeated sequences that prevent loss of DNA at the end of a DNA strand during chromosome replication and so are required for faithful replication of the DNA in a chromosome.

telomeric sequences   Special sequences on the ends of DNA strands that are required for synthesis of the terminal segments of the lagging strand. Telomeric sequences are present on the ends of chromosomes (the telomeric region) and are used in the construction of yeast artificial chromosomes (YACs).

telophase   The stage of mitosis in which the new cell membrane that divides the daughter cells forms (the cell plate) and chromosomes reform into diffuse chromatin.

temperate phage   A bacteriophage that is capable of establishing lysogeny in a host rather than undergoing a normal lytic cycle.

temperature-sensitive mutant (Ts mutant)   Any organism that expresses a function with a temperature dependence, for example, bacteriophages that establish lysogeny at one temperature but not at another.

template   In general a pattern for creating a copy of something; nucleic acid strand whose sequence is used to create a complementary nucleic acid copy.

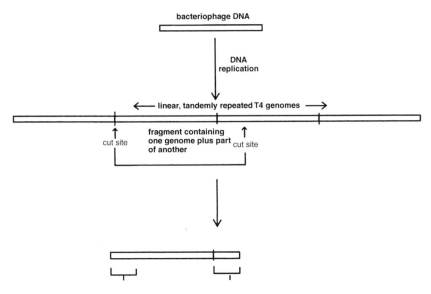

bacteriophage DNA

DNA replication

←— linear, tandemly repeated T4 genomes —→

cut site

fragment containing one genome plus part of another

cut site

Terminal redundancy

teratoma A type of tumor derived from a developing embryo.

terminal redundancy During the replication of bacteriophage T4, slighty more than one genome equivalent is cut from a long, linear DNA that represents T4 genomes repeated in an end-to-end head-to-tail fashion. This leads to the packaging of T4 genomes in which the ends are repeated.

terminal transferase An enzyme that catalyzes the addition of an unspecified number of deoxyribonucleotides from deoxyribonucleotide triphosphates (dNTPs) to a free 3′ hydroxyl end of double- or single-stranded DNA:

terminal
transferase
+dNTPs
3′OH—5′ -------> NNNNNNNNNNN—5′

Terminal transferase catalyzes the addition of long polymeric chains of whichever nucleotide whose triphosphate (dATP, dCTP, dGTP, or TTP) is used as a substrate. The addition of long nucleotide tails to DNA fragments is a tool for cloning DNA fragments into vectors for which no convenient restriction enzyme sites exist.

ternary initiation complex A three-part complex that is necessary to start the process of translation. The ternary initiation complex consists of met-tRNA, GTP, and an initiation factor (eIF2).

tertiary structure The overall, three-dimensional folding of a polypeptide; the folding, twisting, or conformation of the secondary structure of the polypeptide.

testosterone The steroid hormone produced by the testes that regulates sperm production and male sexual behavior.

tetanus A syndrome caused by infection by the anaerobic bacterium, *Clostridium tetani*. The disease symptoms (uncontrollable muscle spasms) are due to the presence of a potent neurotoxin produced by the bacterium.

tetracycline A broad spectrum (both Gram-positive and Gram-negative) antibiotic produced by *Streptomyces venezuelae*.

TGF See TRANSFORMING GROWTH FACTOR.

thalassemia A disease that results from a mutation, often a deletion of DNA within the gene, that causes a reduction or complete loss of expression of one or both of the globin proteins (alpha or beta), resulting in gross defects in hemoglobin function that may be fatal. The thalassemias are examples of genetic disease brought about by uneven crossing over.

thalassemia, β A type of thalassemia affecting the biosynthesis of the β globin chain of hemoglobin. Some β thalassemias have been found to be due to defects in gene regulation such as RNA processing in the nucleus and so have provided important insights into mechanisms of the control of gene expression.

thermogenin (uncoupling protein) A protein that forms a channel in the inner mitochondrial membrane for the passage of protons from the cytosolic side of the membrane into the mitochondrial matrix. Because the path of proton flow using this channel bypasses the FoF1 ATPase, energy from the oxidation of fats and sugars is released in the form of heat that is used to raise body temperature.

thermo-inducible Stimulated by heat. In the context of molecular genetics, the term is usually applied to genes and/or their products whose activity is rapidly increased when the temperature rises by several degrees higher than optimal for the growth of the organism.

thermophile An organism that thrives at high temperature.

theta structure The term used to describe the structure formed when a circular, double-stranded DNA molecule is engaged in replication proceeding in both clockwise and counterclockwise directions from the same starting point.

thiamine Also known as vitamin B_1. An important cofactor for the reactions involved in the O_2^- dependent oxidation of sugars (respiration) in energy (ATP) production.

thiazolidinediones A class of drugs that lowers the levels of fatty acids in the blood. Thiazolidinediones act by binding to and activating PPARγ, which, in fat tissue, leads to the induction of the enzyme phosphoenolpyruvate carboxykinase. This in turn diverts pyruvate away from fatty acid synthesis.

thin-layer chomatography A sensitive analytical technique for separating mol-

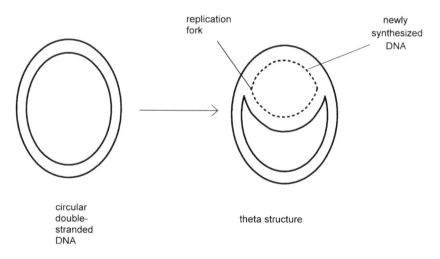

circular double-stranded DNA

theta structure

Theta structure

ecules on the basis of their differing solubility in various solvents. The sensitivity of the technique derives from running the sample on a thin, inert matrix to maintain the sample in a concentrated form.

thiol The sulfur containing analog of an alcohol (–OH) group.

thiostrepton An antibiotic that acts by blocking a critical step (translocation) in protein synthesis (i.e., translation) by binding to the large subunit of the ribosome.

6-thioguanine A purine derivative that is acted on by the enzyme hypoxanthine-guanine phosphoribosyl transferase (HGPRT) to form a toxic compound. For this reason, 6-thioguanine is used to select cells that contain low levels of HGPRT. See HAT SELECTION.

1-thiouridine An unusual pyrimidine base that is found only in tRNA. Thiouridine is derived from uridine by the replacement of an oxygen atom with sulfur. See TRANSFER RNA.

threonine An amino acid that, like serine and tyrosine, contains an –OH group on its side chain. For this reason, threonine serves as a site of phosphorylation in proteins.

thrombin In the blood-clotting pathway, the enzyme that produces fibrin from fibrinogen by cleavage of a portion of the fibrinogen molecule. Fibrin polymerizes to form a clot.

thromboxanes A class of eicosanoids produced in platelets that cause platelet aggregation during blood clotting and help reduce blood flow at the site of a clot.

Thy 1 A protein on the surface of T cells with homology to a portion of the Fc region of immunoglobulins. Like other such T-cell membrane proteins, Thy-1 is believed to play a role in the recognition of foreign antigens.

thylakoid disk A membrane-enclosed, coin-shaped structure that contains the light-reacting pigments and other components of the light reaction in photosynthesis; these components are contained in the membrane of the thylakoid disk.

thymectomy Removal of the thymus by surgery. A procedure frequently performed for the purpose of rendering an animal unable to mount an immune response to foreign cells, for example, tissue grafts or lymphocytes from another animal.

thymic nurse cells (TNC) Cells that engulf developing T lymphocyes to educate the lymphocytes. Once the T cells are released from the TNC, they possess the appropriate receptors to interact with foreign cells that invade the body.

thymidine A pyrimidine base attached to the deoxyribose sugar in deoxyribonucleotides.

thymidine kinase An enzyme that catalyzes the major step in the formation of TTP from thymine. Because this pathway is the only means by which thymine or thymine analogs can enter into nucleic acids, manipulation of this enzymatic step provides an important experimental tool for studying gene action by the incorporation of modified bases into DNA.

thymidine triphosphate (TTP) The thymine-containing nucleotide precursor of DNA.

thymine One of the nitrogenous bases found in nucleic acids. Thymine is a pyrimidine that forms hydrogen bonds with the purine adenine.

thymine dimers A type of mutation in which adjacent thymine bases in a DNA strand are covalently linked to one another, causing the two bases to be read as a single base during DNA replication or transcription. Thymine dimers are largely caused by exposure of tissue to ultraviolet light. See ULTRAVIOLET REPAIR.

thymosin A mixture of small naturally occurring peptides that acts to promote the appearance of the T-cell surface proteins that are seen in mature T cells, for example, Thy 1. Thymosin is thought to mimic the effects of the hormone(s) that normally induce maturation of prothymocytes.

thymus A structure comprised of lymphatic tissue located in the upper portion of the chest cavity in mammals. Some of the immature lymphocytes from the bone marrow migrate to the thymus where they develop into the mature lymphocytes that are responsible for cellular immunity. In mammals, the thymus is present in young animals but decreases in size or disappears in adults.

thyroxine (T4) One of two major hormones secreted by the thyroid gland in response to thyrotropin, a pituitary hormone. Thyroxine is made from iodinated tyrosine and has the effect of raising the basal metabolic rate, an indicator of the oxygen-dependent oxidation of sugars.

tight junction A structure in which the cell membranes of neighboring epithelial cells are brought into extremely close contact, preventing the seepage of even small molecules through the space between cells. Tight junctions are particularly evident between the epithelial cells lining the gut where they function as a barrier against unregulated diffusion of substances into the bloodstream from the digestive tract.

Ti plasmid See *AGROBACTERIUM* and CROWN GALL PLASMID.

tissue culture The general technique of keeping tissues and/or cells derived from tissues alive outside the organism from which they were derived by creating an artificial environment that provides the essential aspects of the natural setting. The development of sophisticated tissue culture systems has been a major factor in recent advances in biomedicine because tissue culture permits organ-specific cells to be studied and manipulated in an experimental environment.

tissue plasminogen activator (tPA) An enzyme that catalyzes the cleavage of the blood protein plasminogen to the active form of the blood-clotting protein, plasmin. tPA has recently been used as a therapeutic agent for destroying blood clots in the blood vessels. The tPA gene has been cloned, and the protein has been synthesized in large quantities using recombinant DNA techniques so that it may find widespread therapeutic use as a preventative agent for heart attack and stroke.

titer The concentration of live virus in a fluid; the number of plaque-forming units (PFU) or focus-forming units (FFU) per unit volume of fluid, for example, PFU per milliliter.

titration The process of determining the concentrations of substances experimentally by adding known amounts of chemical antagonists to the solution until the effects of the target substance are neutralized.

T-loop 1. A specialized structure for protecting the single-stranded 3′ end of telomeres. In a T-loop the end of the telomere is folded back so that the single-stranded end is base paired with complementary sequences in the preceding double-stranded region. The loop is held in place by the proteins TRF1 and TRF2. 2. A control region of the cyclin dependent kinase cdk7, where phosphorylation of critical serine and threonine residues occurs. After activation of cdk7 as a result of the phosphorylation, the kinase forms a complex with cyclin H.

Tn5 A type of insertion sequence that carries the gene for resistance to the antibiotic kanamycin. See TRANSPOSON.

Tn10 A type of insertion sequence that carries the gene for resistance to the antibiotic tetracycline. See TRANSPOSON.

tobacco mosaic virus (TMV) A large, filamentous, RNA-containing plant virus. TMV was one of the first viruses to be studied in detail; among the findings derived from studies on TMV was the spontaneous assembly of the viral coat from its component subunits.

Tonegawa, Susumu (b. 1939) An immunologist who discovered the process by which antibody-producing cells of the immune system rearrange segments of the antibody genes (translocation of the variable and constant regions of the immunoglobulin genes) to create novel antibody-encoding genes. This discovery showed how the wide range of antibodies present in the adult immune system were derived during the process of immune cell differentiation.

tonofilament A filament type that is characteristic of epithelial cells. Tonofilaments are approximately 8–10 nanometers in diameter by transmission electron microscopy and terminate as filament bundles at cell junctions that are characteristic of epithelial cells known as desmosomes. Tonofilaments have been shown to be identical to keratin filaments that make up the intermediate filament network that is characteristic of epithelial cells. See INTERMEDIATE FILAMENT.

topoisomerase A class of enzymes that catalyzes the relaxation of supercoiled DNA by creating transient nicks in the DNA strands that permit tightly wound DNA to uncoil. Type I topoisomerases cause breaks in only one of the DNA strands, while type II topoisomerases (also known as DNA gyrase) nick both strands. DNA gyrases are involved in the process of DNA replication, which requires the unwinding of the DNA helix.

topoisomerase inhibitors A class of drugs that acts as anticancer agents by inhibiting the activity of topoisomerase enzymes. The inhibition of topoisomerase activity blocks DNA replication and the ability to cause cell-cycle arrest at the G2/M interface, which is lethal to actively dividing cells. Two isoforms of topoisomerases exist, I and II, and anticancer drugs that act as topoisomerase inhibitors are classified according to the isoform that they inhibit. Some examples include apigenin, kaempferol, rebeccamy-

cin, campothecin (topoisomerase I inhibitors), aurintricarboxylatic acid (ATA), amsacrine hydrochloride, chromomycin, ellipticine, etoposide, novobiocin, sobuzoxane (topoisomerase II), and netropsin (topoisonerases I and II).

topoisomers Alternative forms of a circular DNA that differ from one another only in terms of linking number. Changes in linking number result from the action of topoisomerase enzymes.

totipotent The concept that a particular cell (e.g., the fertilized egg) has the capability to generate or differentiate into any cell type in the body of an organism. Because the DNA in all cells of the body was believed to be essentially equivalent, the concept of totipotency was originally thought to apply even to specialized cells of a highly differentiated structure such as the eye, but modern understanding of the fluidity of the genome now suggests that, in many cases, differentiation is accompanied by alterations in the DNA.

toxin A chemical poison secreted by one organism for purposes of defense against a competing organism. Because toxins normally target highly specific cell/organ systems, many toxins have been used to gain insight into normal biochemical mechanisms; for example, tetrodotoxin, a toxin secreted by the puffer fish, specifically paralyzes the sodium transport channel in nerves and has been used to study ion transport in the neural system.

TPA A plant-derived phorbol ester (12-O-tetradecanoylphorbol-13-acetate) that is a potent tumor promoter. TPA appears to exert its tumor promoting activity by activating protein kinase C in the cell membrane.

trans In general, on the opposite side of or across from. In organic chemistry, the term refers to a molecular configuration where groups are on the opposite side of a chemical bond from one another. In molecular genetics, the term is used to indicate changes in expression of a particular gene that are caused by an agent located on different DNA molecules, such as changes in gene expression caused by an agent acting on the gene from a distance (e.g., a hormone).

trans acting Pertaining to a genetic element exerting an effect on a target that is located on a physically separate unit. For example, a gene coding for a regulatory protein is said to be trans acting with respect to the genes it controls because the target genes may be located on DNA strands or even chromosomes at some distance from the regulatory gene.

transamination A type of biochemical reaction that allows amino acids and,

therefore, proteins to enter into the same biochemical pathway by which sugars are oxidized for energy (i.e., ATP) production. See TRANSAMINASE.

transcellular transport A mechanism for carrying certain substances from one side of a cell to the other. This type of transport is the means by which substances (for example, glucose) move across the epithelial cells that line the intestinal tract to the bloodstream.

transcription The process of making an RNA complementary to a strand of DNA. In transcription, an RNA polymerase, using the order of nucleotide

Initiation of transcription in eukaryotics

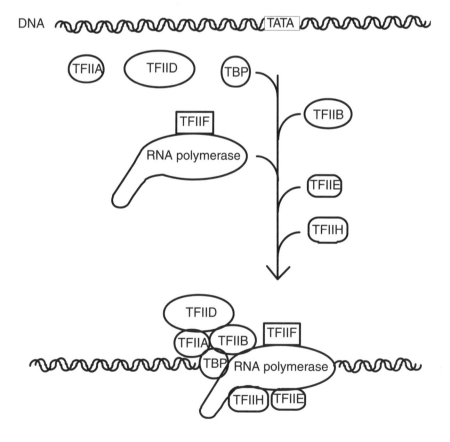

bases present in the DNA template as a guide, assembles nucleotides from the four ribonucleotides (ATP, CTP, GTP, and UTP) to create the RNA strand.

transcription factor A protein or hormone that binds to a certain sequence on the regulatory region of a gene at the PROMOTER or the ENHANCER and either turns on TRANSCRIPTION, enhances transcription (up-regulation), or inhibits transcription (down-regulation).

transducin An alternative term for G protein.

transducing phage A bacteriophage that, during its normal replicative cycle, occasionally packages some of the DNA from the host into the bacteriophage head, along with the normal bacteriophage DNA. The DNA so packaged can then be carried from the previous host and introduced into a new host that is infected by the transducing bacteriophage.

transduction The term for the process of carrying sections of DNA from one bacterial cell to another by a transducing bacteriophage.

transfection The technique of introducing DNA into eukaryotic cells. Transfection is the process homologous to transformation in bacteria. Transfection encompasses a number of techniques that utilize different principles to introduce the DNA including electroporation and precipitation by calcium phosphate.

transfer factor An as yet unidentified factor extracted from living T cells that, when taken from one human and injected into another, induces some of the cell-mediated immunity that was present in the donor.

transferase A class of enzymes that catalyzes the transfer of a chemical group from one substrate to another, for example, methyl transferases for transfer of methyl groups from one molecule to another.

transferrin A plasma protein that carries iron in the blood. Transferrin transfers the bound iron to the appropriate cells via a special cell surface receptor.

transfer RNA (tRNA) A type of RNA that recognizes the codon on the mRNA during the process of translation and brings the proper amino acid (attached to it) into close proximity to the end of the peptide chain being synthesized so that the amino acid can be added to the peptide chain. The tRNA molecule is folded so that a group of three nucleotides complementary to the codon in the middle of the molecule (the anticodon) is exposed, while the end of the tRNA is used for attachment of the amino acid that corresponds to the codon. See ADAPTOR MOLECULES.

transformation, cancerous or neoplastic The process by which a normal cell comes to attain the characteristics of a cancerous cell. Because the actual transformation process cannot be directly observed, steps in the process are inferred by the expression of certain properties that cells taken from tumors exhibit when grown in culture (e.g., the ability to grow without being attached to a solid surface and lowered dependence of growth on serum).

transformation, DNA The process of introducing foreign DNA into bacteria. See COMPETENCE.

transforming growth factor (TGF) Any of a group of proteins secreted by transformed cells that can stimulate the growth of normal cells. Transforming growth factor alpha (TGFα or TGF-A) binds the epidermal growth factor receptor (EGFR) and stimulates the growth of endothelial cells. TGFα is produced by macrophages and keratinocytes and is secreted at high levels by some human tumors. Transforming growth factor beta (TGFβ or TGF-B) has two subtypes β1 and β2 and is found in hematopoietic (blood-forming) tissue and initiates a signaling pathway that suppresses the early development of cancer cells. Bone morphogenetic proteins (BMP) are members of the TGFβ family. Overexpression of TGF can bring about renal fibrosis, leading to end-stage renal disease as well as diabetes. Certain types of TGF beta-receptor antagonists have been found to be effective in halting renal fibrosis. See SARCOMA GROWTH FACTOR.

transgenic animal A animal, typically a goat, a pig, a cow, or a horse, that has been modified by introduction of a foreign gene into its germline so that some specific aspect of phenotype, such as production of a human protein in the milk or resistance to disease, is conferred on the offspring.

transit peptide A preprotein destined for insertion into a mitochondrion.

translation The process of assembling amino acids together to form a polypeptide; the sequence of amino acids is specified by the codons on the mRNA being used as the template. Translation is carried out on ribosomes that carry sites for tRNAs carrying the appropriate amino acids.

translational domain One of the two major classes of binding sites on the ribosome. The factors that bind within the translational domain are directly involved in the translation of mRNA into proteins. The translational domain contains binding sites for peptidyl transferase, mRNA, EF-TU, EF-G, and 5S RNA. See ELONGATION FACTORS.

translocation 1. During translation the process of moving the tRNA carrying the growing polypeptide chain from one site on the ribosome to another to make room for an incoming tRNA carrying a new amino acid.
2. In biochemistry the process of actively transporting a molecule across a membrane.
3. The breakage of a chromosome followed by subsequent rejoining of one of the pieces to another chromosome. See RECIPROCAL TRANSLOCATION.

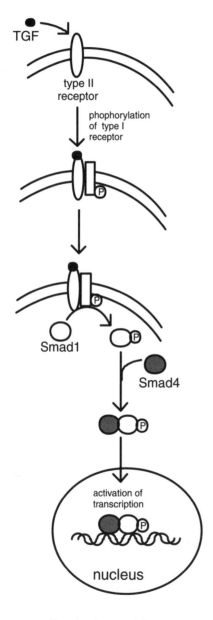

TGF

type II receptor

phophorylation of type I receptor

Smad1

Smad4

activation of transcription

nucleus

Transforming growth factor

transmembrane protein A protein that is inserted into and spans the cell membrane so that one end of the protein protrudes out of the cell (the extracellular domain) while the other end (the internal domain) remains in the interior of the cell. The two major functions of transmembrane proteins are: (1) to serve as channels for or transporters of specific molecules and (2) as devices for transmembane signaling. See INTEGRAL MEMBRANE PROTEIN.

transmembrane signaling A signal mechanism in which the binding of a specific molecule (the ligand) to the extracellular domain of a transmembrane protein (the ligand-binding domain) causes a physical change in the internal domain; this then sets in motion a series of chemical reactions, for example, phosphorylation(s) of certain proteins that then produces specific changes in the cell behavior. This is called signaling because the ligand never actually enters the cell. See G PROTEIN(S).

transmission electron microscope (TEM) A device that is similar in principle to a conventional microscope but that uses an electron beam instead of light, and a magnetic field instead of a glass lens to focus the beam on the specimen. The image of the specimen is seen as a pattern of greater or less electron intensity in the beam that emerges from the specimen. The great advantage of the electron microscope over the conventional light microcope is that, because electrons have a much shorter wavelength than photons, resolution of much finer detail is possible.

transplacement vector A vector that is designed to transfer a defined segment of DNA of interest to another vector by recombination. Transplacement vectors are useful in situations where one wishes to introduce a DNA segment into a particularly large vector (e.g., baculovirus or a yeast artificial chromosome) or another vector in which it is difficult to engineer a unique site for making the

recombinant by conventional restriction enzyme technology.

transplant The removal of a tissue or portion of a tissue from its natural location and its placement in a new location, either in the same organism or in some other organism.

transplantation antigens Certain proteins produced by the major histocompatibility locus that are found on the surface of all cells in the animal and that are responsible for provoking rejection of tissue grafts. See MAJOR HISTOCOMPATIBILITY COMPLEX.

transport In biochemistry the process of moving a molecule from one location to another, usually across a membrane. The term implicitly indicates that expenditure of energy is required for the transport. See ACTIVE TRANSPORT.

transport protein A transmembrane protein that mediates the transport of a molecule across a membrane. Transport proteins are often in the form of a channel spanning the membrane that allows molecules to pass through. See INTEGRAL MEMBRANE PROTEIN.

transposase An enzyme encoded by genes on a type of transposon, called an insertion sequence, that recognizes the terminal inverted repeat sequences and catalyzes the events in the transposition.

transposition immunity A term used to describe the observation that plasmids containing one copy of a transposon (such as Tn3) are resistant to insertion by another copy of the transposon.

transposon A mobile genetic element found primarily in prokayotic cells that carries genetic information from one site in the genome to another. Transposons may carry a variety of genes or other genetic units, but they always carry the information needed to carry out the transfer function, for example, short terminal-inverted repeat sequences that are

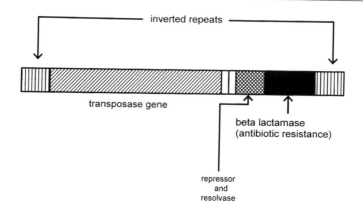

inverted repeats

transposase gene

beta lactamase
(antibiotic resistance)

repressor
and
resolvase

Transposon

needed for insertion of the transposon into their target sites in the genome.

transvection The influence of the synapsing of paired chromosomes on the expression of genes in the region of the synapse. The phenomenon was first described in *Drosophila melanogaster* within the bithorax complex (BX-C). Transvection usually involves enhancers acting on genes on the opposite chromosome (i.e., in trans) but can also involve silencers acting in trans.

trehalose A disaccharide of the sugar glucose. Trehalose is mainly found in insects where it is used as a source of energy.

tricarboxylic acid cycle A series of reactions beginning with the formation of citric acid (also known as the citric acid cycle or Krebs cycle) by which the majority of the energy from the oxygen-dependent oxidation of glucose is derived. The cycle consists of a critical series of reactions in the oxidation of sugars in which CO_2 and H_2O and electrons to be used in electron transport are produced from the intermediate, acetyl-CoA. The majority of CO_2 produced in the body from muscular activity is generated via this pathway.

trigger factor A bacterial protein which, by forming a complex with a pre-protein, holds the polypeptide in a specific conformation necessary for its insertion into the bacterial cell membrane.

triglyceride A type of lipid formed from glycerol and fatty acids. Triglycerides are the major form of storage for fatty acids and are the main constituents of body fat.

trimethoprim A folate antagonist antibiotic with activity against both bacterial and parasitic infections. Trimethoprim blocks the synthesis of folate, an essential nutrient, from para-amino benzoic acid (PABA). The effectiveness of trimethoprim depends upon the fact that mammals can obtain folate from food whereas bacteria and lower eukaryotes must synthesize it de novo.

triplet In general any group of three. In molecular genetics, the term generally refers to the triplet of nucleotide bases that makes up the genetic code for an amino acid. See CODON and WOBBLE HYPOTHESIS.

triploidy The state of having one extra haploid set of chromosomes, that is, 3n, where n = the haploid chromosome number.

triskelion A subunit of the clathrin coat of coated pits. The triskelion is named for its tripartite structure consisting of three light and three heavy polypeptide chains.

Triton X-100 A nonionic detergent used for the biochemical isolation of cell membranes and nuclei.

troponins A class of proteins that regulate muscle contraction by binding to tropomyosin and inhibiting of the interaction between the actin and myosin filaments. There are three troponin subunits in the troponin complex that assembles onto tropomyosin filaments: inhibitory (I), tropomyosin binding (T), and calcium binding (C). The inhibition of muscle contraction by the I subunit is relieved by calcium binding to the C subunit.

trypsin A protease that cleaves polypeptides specifically after the amino acids arginine or lysine.

tryptophan An amino acid containing an indole group in its side chain. Tryptophan is the precursor of the neurotransmitter, serotinin, in animals and the plant hormone, indole acetic acid (IAA).

TSH-releasing hormone (TRH) A short (three amino acids) polypeptide hormone produced in the hypothalamus that stimulates the anterior portion of the pituitary gland to secrete thyroid-stimulating hormone (TSH).

tuberous sclerosis complex (TSC) An inherited genetic disease of organ systems, particularly the brain, skin, heart, lungs, and kidneys. This can result in epilepsy, learning and behavioral difficulties, skin and renal lesions; kidney complications can include cysts, polycystic renal disease, and renal carcinoma. Tuberous sclerosis complex is caused by mutations in the tumor suppressors hamartin (located on chromosome 9q34) and tuberin (located on chromosome 16p13.3) that are encoded by the genes TSC1 and TSC2. Hamartin and tuberin

are critical regulators of cell growth that are believed to act as negative regulators of S6 kinase 1 (S6K1) and eukaryotic initiation factor 4E binding protein 1 (4E-BP1).

tubulins Two proteins (alpha and beta) of about 55,000 daltons each that constitute the subunit proteins of microtubules.

tumor An abnormal growth of a tissue. In benign tumors, the tissue growth eventually stops and the tumor remains confined to the site at which it began; in malignant tumors, growth is unlimited and may take place at sites distal from the site of origin, known as metastases.

tumorigenesis The process of tumor formation. Tumorigenesis is the end point in the process of transformation.

tumor necrosis factor (TNF) A factor originally isolated from serum and found to be produced by certain leukocytes (for example, macrophages) that appear to be selectively toxic to tumor cells. The gene for TNF has been cloned and expressed in quantities that are sufficient for ongoing clinical studies of its effectiveness as an anticancer agent.

tumor promoter A chemical that, by itself, does not cause tumors but that will induce the formation of tumors after exposure of a tissue to any of a class of other agents called tumor initiators.

tumor virus Any virus that induces the formation of a tumor in the tissue that it infects.

tunicamycin An antibiotic isolated from *Streptomyces lysosuperficus* that inhibits the synthesis of N-linked glycoproteins. Tunicamycin is a structural analog of UDP-N-acetylglucosamine (UDP-GlcNAc) that acts as a competitive inhibitor of the critical reaction between UDP-GlcNAc and Dolichol phosphate, the first step in the process of N-glycosylation. Tunicamycin is widely used as a research tool for studying the process of protein glycosylation.

turgor/turgor pressure Water pressure resulting from the diffusion of water into cells. The term is usually applied to the rigidity of plant structures, for example, leaves and stems resulting from osmotic pressure.

Turner's syndrome A genetic defect in which the cells of the afflicted individual contain one X chromosome but no Y chromosome. Although such individuals have the outward appearance of females, the sexual organs are incomplete or undeveloped.

turnover number For an enzyme-catalyzed reaction, the number of substrate molecules undergoing reaction per unit time. For example, in the reaction catalyzed by carbonic anhydrase,
$$CO_2 + H_2O \rightarrow H_2CO_3,$$
the enzyme has a turnover number of 600,000 molecules of CO_2 per second.

T vector A specialized vector, usually a plasmid, for cloning PCR products. Because the polymerases used in the polymerase chain reaction also possess terminal transferase activity, PCR products contain overhanging chains of polyadenylate residues on their 5' termini. T vectors are linearized DNAs that contain complementary thymidylate residues on their 5' termini that permit base pairing between the T vector and the PCR product to be cloned.

Tween 80 A nonionic detergent used for isolation of cell membranes and to reduce background signal interference in immunoblotting.

twin sectors If the process of transposition carried out by a transposon occurs just at the time of cell division such that only one daughter cell carries the transposition, then the descendants of such genotypically different daughter cells are called twin sectors.

twisting number For a given double helical DNA, a number representing the total number of helical turns. The twisting number is calculated as the total number of base pairs in the DNA under consideration divided by the number of base pairs per turn.

Ty 1 elements A type of transposon in yeast; Ty is an acronym for *t*ransposon *y*east. The Ty elements are about 6.3 kb long and contain genes similar to the gag and pol genes of retroviruses.

tyrosine An amino acid with a phenolic group in its side chain. Tyrosine is a precursor of adrenaline and the skin pigment melanin.

tyrosine kinase A type of enzyme that catalyzes the transfer of a phosphate group (phosphorylation) to a tyrosine amino acid in a protein. The oncogenes activated by a number of retroviruses have been shown to be tyrosine kinases. See KINASE.

U

ubiquitin A protein originally discovered in the nucleosomes of the fruit fly, *Drosophila melanogaster,* whose binding to A- and B-type mitotic cyclins leads to their destruction at the end of mitosis. Ubiquitin binding to the cyclin allows it to be proteolytically degraded by a multiprotein proteosome complex.

ultracentrifugation Centrifugation at speeds great enough to produce forces of greater than 300,000 times the force of gravity. The forces produced by ultracentrifugation are capable of separating molecules of different sizes from one another. The ultracentrafuge was invented by the Swedish physical chemist Theodor Svedberg in 1923.

ultrafiltration A technique using gas pressure to drive samples through an ultrafine meshed filter for the filtration of particles smaller than bacteria.

ultrastructure Features of cell architecture discerned in an electron microscope.

ultraviolet radiation Light with a wavelength in the range of about 200–400 nanometers. Ultraviolet light is readily absorbed by tissue that is exposed to it, for example, skin. Ultraviolet radiation produces numerous alterations in biomolecules, including mutations in DNA, that are linked to tumorigenesis. See THYMINE DIMERS.

ultraviolet repair A system used by bacteria to repair regions of DNA damaged by exposure to ultraviolet light. The ultraviolet repair system removes altered nucleotides (e.g., thymine dimers) and then recopies the region using the remaining strand as a template. See EXCISION REPAIR.

unequal crossing over A form of recombination in which a segment of DNA from one chromosome is transferred to a position adjacent to its own allele on the homologous chromosome:

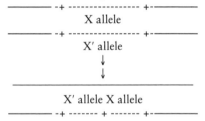

unicellular Composed of a single cell; for example, protozoa are unicellular organisms.

upstream activator sequences DNA sequences found in yeast that are similar to enhancer sequences in higher organisms in that they can stimulate the transcription of a gene from a long distance but only when they are located upstream (i.e., in the 5′ direction) from the gene.

uracil A pyrimidine base used instead of thymine in ribonucleotides and RNA.

urea A compound derived from ammonia and carbon dioxide. Ammonia produced from the breakdown of amino acids is excreted as urea in the urine.

uric acid A purine-like molecule formed as a degradative product of the purine nucleotides. Because uric acid is relatively insoluble in body fluids, overproduction of uric acid produces a condition known as gout, the accumulation of uric acid crystals in joints. Uric acid is the main vehicle for the elimination of ammonia by birds and reptiles.

uridine The nucleoside derivative of uracil, that is, uracil bonded to ribose.

uridine triphosphate (UTP) The triphosphate derivative of uridine; the uracil-containing molecule used in the synthesis of RNA. See NUCLEOTIDE.

urokinase A protease, found in urine and blood, that has the same activity as plasminogen activator; urokinase is therefore used therapeutically to dissolve blood clots that may accumulate in the coronary arteries.

urotensin II See GPR14.

Uridine

V

vaccination The capacity of a vaccine to induce an immune response to a pathogenic organism in an individual usually by multiple, direct injection of the vaccine at a body site favorable for exposure to the immune system. Vaccination is a prophylactic procedure that is generally ineffective in treating disease after onset.

vaccine A preparation, derived from inactivated or attenuated pathogenic organisms, that is used for vaccination.

vaccinia virus A member of the poxviruses that causes cytopathic destruction of epithelial cells (pocks); vaccinia is the agent responsible for cowpox. Vaccinia virus is closely related to variola virus that causes smallpox in humans. Vaccinia is easily grown in tissue-cultured cells for purposes of creating vaccines. For this reason, vaccinia is being researched as a vector for expressing recombinant DNA for other pathogens with a view toward using this as a system for creating vaccines to other agents.

vacuole A membrane-enclosed cytoplasmic organelle generally arising from phagocytosis and often containing enzymes involved in degradation of biologic material, for example, proteases.

valine An amino acid with an isopropyl group as its side chain; valine belongs to the group of nonpolar amino acids.

valinomycin An antibiotic composed of lactate, hydroxyisovalerate, and the amino acid, valine, joined together in a ring configuration that carries a potassium ion in its center. In this way, valinomycin acts as a vehicle for transporting potassium through the cell membrane, thereby destroying the delicately balanced ion concentration in the cell. See IONOPHORE.

vanadate (HVO_4^{-2}) A compound of vanadium that, in biological systems, functions as an analog of phosphate, which inhibits the action of transporters known as P-type ATPases.

Van der Waals forces Weak electrostatic forces between nonpolar hydrocarbon molecules such as those that make up paraffins and the lipids in membranes. Van der Waals forces are primarily responsible for the aggregation of waxy, oily, or fatty substances in water.

variable region The terminal portion of an antibody molecule that contains the site that binds the antigen that the immunoglobulin carries specificity for. The variable nature of this region is a reflection of the large number of antigens for which specific antibodies can be made.

variable surface glycoprotein (VSG) The major component of the cell surface of the trypanosome parasite, such as the sleeping-sickness-producing parasites that are carried by the tsetse fly. The VSGs, which are the only antigenic molecule on the trypanosome surface, change throughout the development of the organism. These rapid changes in the VSG allows the organism to evade the immune system of the host. The VSGs are coded for by a large family of genes, and the expression of different VSGs at various stages of development is due to the activation of different VSG genes from this gene family.

vascular endothelial growth factor (VEGF) A family of growth factors produced by cells at the site of a wound and by some tumors that stimulates the process of ANGIOGENESIS. The mammalian VEGF family consists of five members—VEGF, VEGF-B, VEGF-C, VEGF-D, and placenta growth factor (PlGF)—that are all encoded by separate genes. New anticancer drugs, such as Avastin, target VEGF or its receptor as a means of starving tumors by depriving them of blood supply from new blood vessel growth induced by tumor-derived VEGF.

vasoactive intestinal peptide (VIP) A neuropeptide hormone that is found to have effects on the immune system, including anti-inflammatory effects, inhibition of cytokine production, and inhibition of macrophage functions such as chemotaxis and phagocytosis. VIP has been found to exert some of these effects via a cAMP-mediated signal transduction system.

vasopressin A polypeptide hormone, produced by the posterior part of the pituitary gland, that causes an increase in blood pressure but also decreases the flow of urine. Previously known as antidiuretic hormone.

vav-1 oncogene A rho-type guanine nucleotide exchange factor (GEF) of the Dbl family. vav-1 plays a role in development and activation of T and B cells. vav-1 is the binding partner of the HIV-1 nef proteins. The vav-1/nef complex causes cytoskeletal rearrangements, activates the JNK/SAPK signaling cascade, and upregulates HIV transcription and replication. The vav gene map locus is 19p13.2

vector Any DNA that can propagate itself rapidly in a host and can also maintain this capability after insertion of foreign DNA into the vector. Although there are many types of vectors, the most common ones are derived from bacterial plasmids or the DNAs of both bacterial and animal viruses. Most vectors in use today have been subjected to genetic engineering for specific purposes such as the expression of foreign proteins in bacteria or animal cells (expression vectors).

vectorial discharge Secretion of a substance by a cell at only one location or area on the cell surface. Examples of vectorial discharge include the secretion of mucin at the apical surface of epithelial cells that line the gut and the discharge of neurotransmitters into the synaptic cleft.

Venter, J. Craig (b. 1946) Founder of The Institute for Genomic Research (TIGR) who, backed by venture capital, competed with the NIH-backed Human Genome Project in sequencing the entire human genome. Venter is known for employing an innovative "shotgun" approach that utilized robotic sequencing of short genomic fragments and advanced computer algorithms to assemble the sequence data.

veratridine A powerful neurotoxin, derived from the lily *Schoenocaulon officinalis,* that is used to study nerve function. Veratridine interferes with the action of sodium channels such that sodium ions can pass freely into the neuron.

very short patch repair (VSPR) The type of excision repair that involves mismatches between single bases.

vimentin A protein comprising one of the subclasses of intermediate filaments that make up a tissue-specific cytoskeleton in mammalian cells. Vimentin filaments are characteristically found in cells of mesenchymal origin such as fibroblasts.

vinblastine An antitumor drug isolated from the Madagascar periwinkle plant *(Vinca rosea).* Vinblastine acts to cause depolymerization of microtubules by binding to the tubulin protein subunits.

vincristine An antitumor drug isolated from the Madagascar periwinkle plant *(Vinca rosea);* a chemical variant of vinblastine that acts to cause depolymerization in the same manner as vinblastine. See VINBLASTINE.

vinculin A component of the adhesion plaque that, together with alpha-actinin, attaches to both the terminus of a stress fiber and a transmembrane integrin molecule to form an adapter complex that connects the integrin to the stress fiber. The phosphorylation of tyrosine residues in vinculin in transformed cells is believed to play a role in the altered cell-substrate interactions that is seen in transformed cells grown in tissue culture. See EXTRA-CELLULAR MATRIX.

viomycin An antibiotic that acts by blocking a critical step in protein synthesis (translocation), thereby causing the synthesis of a polypeptide to be blocked before completion. Viomycin is itself a peptide that binds to either the large or the small ribosomal subunit to block translocation.

virion A complete virus particle that includes the viral nucleic acid (either RNA or DNA) and, in some cases, enzyme molecules enclosed in a protein capsid.

viroid An unusual infectious agent that produces diseases in plants and consists only of a naked, circular strand of RNA.

virulent, virulence The property of rapid spread of a pathogenic agent (e.g., a virus or bacteria) through a susceptible population.

virus An agent that infects single cells but consists only of the components of the virion and does not possess the cellular machinery required for its own replication. For this reason, viruses are, of necessity, intracellular parasites that are not clearly classifiable as living organisms.

virusoids One of two classes of small infectious RNA molecules in plants. Virusoids do not contain genes for their own replication or packaging but require a second, helper virus to accomplish these functions. Virusoids are also referred to as satellite RNAs.

viscosity The property of resistance to flow exhibited by a substance in a fluid, semisolid state. In biochemical solutions, viscosity is an indicator of a solution containing large macromolecules.

vitamin An essential nutrient in the diets of mammals; any of a number of organic molecules that generally function as cofactors for specific enzyme-catalyzed reactions involved in energy production.

vitamin A Any of the various chemical relatives of retinol (e.g., retinoic acid). Vitamin A is the precursor of the visual pigment, visual purple, and also has profound effects on the differentiation of epithelial including anticancer effects. Derivatives of vitamin A have been used as therapeutic agents for a variety of skin conditions such as icthyosis, acne, and wrinkling.

vitamin B$_{12}$ A ring-shaped, cobalt-containing molecule also known as cobalamine. Vitamin B$_{12}$ is an essential cofactor for the entry of certain amino acids and fatty acids into the tricarboxylic acid cycle (Krebs cycle).

vitamin B$_6$ Any of various derivatives of pyridoxine (e.g., pyridoxal phosphate) used as a cofactor in the transamination reaction, the critical step by which amino acids enter into the tricarboxylic acid cycle (Krebs cycle).

voltage-gated channel A specialized type of transmembrane channel that opens only when there is an electrical potential of a certain value across the plasma membrane (the threshold). Voltage-gated channels for ions are present on a variety of cell types but are especially characteristic of neurons where they are responsible for the generation of an action potential.

von Willebrand disease (vWD) An inherited disorder in which blood fails to clot properly. Symptoms of vWD include bleeding from the gums, nose, and intestinal lining and prolonged or excessive bleeding from small cuts. vWD is caused by mutations in a gene coding for a substance known as von Willebrand factor, which causes platelets to stick to damaged blood vessels and which carries a clotting factor, called factor VIII. There are three types of von Willebrand disease:

- type 1, in which lower levels than normal of von Willebrand factor are present in the blood
- type 2, in which a defective von Willebrand factor is made
- type 3, in which von Willebrand factor is virtually absent and there are very low levels of factor VIII

The gene map locus for von Willebrand factor is 12p13.3.

W

Waldenstrim's macroglobulinemia A tumor of the lymphatic system characterized by oversecretion of IgM immunoglobins. Immunoglobulins derived from this tumor were used to derive the pentameric structure of IgM-type immunoglobulins.

Warren, Robin J. (b. 1937) A pathologist who, in collaboration with Dr. Barry Marshall, a gastroenterologist at the Royal Perth Hospital, discovered the bacterium *Helicobacter pylori* and showed that the microbe was the main cause of peptic ulcers. This stood in contrast to prevailing dogma of the time that ulcers were caused by stress. The discovery of the bacterium eventually led to a cure for ulcers and earned Warren and Marshall the Nobel Prize in physiology or medicine in 2005.

Watson, James D. (b. 1928) Along with Francis Crick, he demonstrated the double helical structure of DNA using the technique of X-ray crystallography. This discovery showed that DNA had the requisite characteristics of a macromolecule that could serve as the genetic material. Watson and Crick were awarded the Nobel Prize in medicine in 1962.

Wernicke-Korsakoff syndrome A genetic disease caused by mutations in the gene coding for the enzyme transketolase, which is critical for the metabolism of pentoses (five-carbon sugars). The disorder is characterized by severe memory loss, confusion, and partial paralysis. The disease-causing mutations result in lowered affinity of the enzyme for thiamine, which is a coenzyme for transketolases.

western blot A technique for identifying polypeptides that have been separated by polyacrylamide gel electrophoresis based on the reaction of specific antibodies to the proteins after they are transferred from the gel to an artificial membrane.

Williams syndrome A rare congenital disorder characterized by an "elfinlike" facial appearance, cardiovascular and blood vessel problems, dental and kidney abnormalities, and musculoskeletal problems. Individuals with Williams syndrome may show good language, music, and interpersonal skills, but their IQs are usually low. The genetic basis for the disease is a deletion on a segment of chromosome 7 that includes the gene that codes for elastin and the enzyme LIM kinase. Elastin is a key component of connective tissue and the loss of elastin leads to the vascular disease seen in Williams syndrome. LIM kinase is strongly expressed in the brain, and its absence is believed to account for neurodevelopmental brain abnormalities.

Wilson's disease A rare autosomal recessive disorder affecting the transport of copper that results in toxic accumulation of copper in the liver and brain. In children liver disease is the most prominent symptom, while neurological disease is seen mostly in young adults. Wilson's disease is caused by mutations in a gene called ATP7B, a P-type ATPase transporter located in the Golgi network. The gene map locus for ATP7B is 13q14.3.

white matter That portion of the brain consisting of myelinated nerve fibers (axons) serving to carry nerve impulses from the gray matter of the brain that give it a characteristic white color.

wild-type gene The normal, nonmutated version of a gene.

wobble hypothesis The idea that, for an amino acid for which there is more than one triplet in the genetic code, the first base will always be the same in the different triplets but that the second and third positions of the triplets will vary or "wobble," with the third base of the triplet exhibiting the most wobble. See CODON.

writhing number (**W**) A number representing the turning of the axis of a supercoiled DNA: W = L − T, where L = the linking number and T = the twisting number.

xanthine-guanine phosphoribosyl transferase (gpt) gene See HGPRT.

X chromosome One of the two sex chromosomes. The sex of the fetus is determined from the sex chromosomes present in the fertilized egg: Two X chromosomes in the fertilized egg will produce a female, and one X chromosome and one Y chromosome will produce a male.

xenograft A graft from a foreign donor, for example, human skin grafted onto a mouse.

xeroderma pigmentosum (XP) A family of autosomal recessive genetic diseases in which excision-repair mechanisms are faulty. This illness results in increased sensitivity to sunlight and, in particular, a much higher incidence of sunlight-induced cancers. At present nine XP-related genetic loci all involved in the process of excision repair of pyrimidine dimers and other bulky groups have been identified.

X-gal A synthetic substrate for the enzyme beta-galactosidase that produces a blue product when acted upon by the enzyme. In vectors that carry the gene for beta galactosidase, growth of bacteria that contain a recombinant DNA that is inserted into the gene (thereby inactivating the gene) by cloning can be detected by the color of the colony they produce. See INSERTIONAL INACTIVATION.

X-linked diseases Genetic diseases that are carried on one of the sex chromosomes.

X-ray crystallography A technique for deducing the physical dimensions of a molecule (sizes of the atoms, lengths of the bonds between them) by examining how the path of an X-ray beam is altered as it passes through a crystallized sample of the molecule of interest. See X-RAY DIFFRACTION.

X-ray diffraction A technique for determining distances between atoms in a molecule by analyzing the diffraction pattern produced when an X-ray beam passes through molecules in a crystallized form.

XRCC1 X-ray Repair Cross Complementing; a protein that plays a role in excision repair of DNA following ionizing irradiation. The C-terminal end of XRCC1 binds DNA ligase III and the N-terminal end binds DNA polymerase β to form a repair complex after damaged DNA is excised.

YAC See YEAST ARTIFICIAL CHROMO-SOME.

Yalow, Rosalyn (b. 1921) Inventor of the radioimmunoassy (RIA) technique for detection of minute quantities of protein using a specific radiolabeled antibody. By applying this technique to insulin she was able to demonstrate the existence of diabetic states that did not result from insulin insufficiency. She was awarded the Nobel Prize in physiology and medicine for this work in 1977.

Yanofsky, Charles (b. 1925) A biochemist who studied the regulation of the bacterial operon that governs the synthesis of the amino acid, tryptophan. His findings showed that the *trp* operon is regulated in at least two ways: (1) via a special regulatory protein termed the *trp repressor* and (2) by a unique mechanism termed *attenuation*. These were among the pioneering discoveries in the field of gene regulation.

yeast(s) A subclass of fungi whose members are single-celled. Yeasts display the major characteristics of higher cells including chromosomes, an endoplasmic reticulum, and sexual mating. Because they are among the simplest eukaryotic cells with these characteristics, they are used as convenient and easily manipulable model systems to study the molecular genetics of eukaryotic cells.

yeast artificial chromosome A type of vector that is used for cloning extremely large DNA fragments in yeast. The vector is constructed by combining those elements of the yeast chromosome necessary for chromosome replication with the foreign DNA. The recombinant DNA created in this way can then be grown in a yeast host for many generations.

yeast two-hybrid system A method for determining whether two proteins form a complex in vivo using genetically engineered DNAs. In this technique, two DNA constructs (hybrids) are created: (1) DNA coding for the DNA-binding domain of a particular transcription factor fused to a DNA segment coding for one of the test proteins and (2) DNA coding for the activation domain of a particular transcription factor fused to a DNA segment coding for the other test protein. When the two constructs are transfected together into yeast cells, an active transcription factor will be formed if, and only if, the two test proteins bind to one another; that is, a complex will be formed that contains both the DNA-binding domain and the activation domain. If the promoter that is normally activated by the transcription factor is itself fused to a reporter gene, expression of the reporter indicates that the test proteins do interact (i.e., bind to one another) in vivo.

yes oncogene A non-receptor cytoplasmic protein tyrosine kinase that is a member of the src family of tyrosine kinases. The human gene is the cellular homologue of the oncogene carried by the Y73 avian sarcoma virus (Yamaguchi sarcoma virus) and is a proto-oncogene located on chromosome 18p11.32. c-yes is highly expressed in neurons, spermatozoa, platelets, and epithelial cells and is believed to have a general role in growth control. Oncostatin M utilizes a signal transduction pathway in human endothelial cells involving the activation of the yes tyrosine kinase. yes also becomes activated after binding of colony stimulating factor (CSF) to its receptor.

z-DNA A form of DNA in which the two strands are twisted around each other in a left-handed helix as opposed to the right-handed helix found in the more common form (i.e., B-DNA).

Zellweger syndrome A rare, autosomal recessive disorder that begins in utero and is characterized by defective myelination of nerve tracts, mental and growth retardation, craniofacial malformations, glaucoma, seizures, cataracts, kidney cysts, and cardiac complications. The disease is caused by a reduction in, or an absence of, perioxisomes in the liver, kidney, and brain. Zellweger syndrome is caused by mutations in any of several different genes whose products are needed for peroxisome formation, for example, peroxin-1 (PEX1), peroxin-2 (PEX2), peroxin-3 (PEX3), peroxin-5 (PEX5), peroxin-6 (PEX6), peroxin-12 (PEX12), peroxin-14 (PEX14), and peroxin-26 (PEX26). The genetic loci map to chromosomes 1 (PEX14), 7q21 (PEX1), 8q (PEX2), 6q23 (PEX3), 12 (PEX5), and 6p (PEX6).

zinc finger A feature of many DNA binding transcription regulatory proteins (transcription factors) in which a zinc atom is bonded to four amino acids (generally cysteine and histidine residues) so as to hold the polypeptide in a loop which has been termed a *finger.* The finger is necessary for the DNA-binding properties of the protein.

Zinder, Norton (b. 1928) A geneticist who studied recombination in bacteria and bacteriophages. As a graduate student in the laboratory of Joshua Lederberg, he carried out the critical experiments involving mutants of the *Salmonella* bacterium that led to the discovery of transduction by bacteriophages that became an important tool in the mapping of bacterial genes.

zoo blot A Southern blot in which a probe from a DNA that is suspected to represent a gene in one species is tested for its relatedness to sequences in other species. If the probe is found to hybridize to DNAs from other species, then, because genes tend to be conserved in other species, this suggests that the probe represents a gene. See HYBRIDIZATION STRINGENCY and SOUTHERN BLOT HYBRIDIZATION.

zygotic-effect genes Genes that effect the segmentation pattern of the embryo of the fruit fly, *Drosophila melanogaster,* that are derived from both the maternal and paternal parent as opposed to those that are active in the egg even before fertilization and so are referred to as maternal-effect genes. See SEGMENTS, SEGMENTATION.

zymogen An inactive form of a proteolytic enzyme. The active enzyme is generated by the action of other proteolytic enzymes via cleavage of the zymogen peptide at specific sites. The term is derived from the phrase: en*zyme gen*erating. For example, the zymogen, chymotrypsinogen, is cleaved by the protease, trypsin, to generate the active enzyme, chymotrypsin.

zyxin (ZYX) A component of adhesion plaques that functions both as a regulator of actin filament assembly and as a component of the mitotic regulatory apparatus. During mitosis cytoplasmic zyxin forms a complex with a factor called h-warts/LATS1 (the human homologue of a *Drosophila melanogaster* tumor suppressor) on the mitotic apparatus. Complex formation between zyxin and h-warts/LATS1 is regulated by phosphorylation of zyxin h-warts/LATS1 by Cdc2 kinase. The zyxin gene map locus is 7q34-q35.

Appendixes

I. Acronyms (and Other Abbreviations)

A	adenine
ACP	acyl carrier protein
ACTH	adrenocorticotrophic hormone
ADP	adenosine diphosphate
AIDS	acquired immunodeficiency syndrome
AMP	adenosine monophosphate
APC	adenomatous polyposis coli
APH	aminoglycoside-3'-phosphotransferase
araA	arabinosyladenine
araC	arabinosylcytosine
ARC	AIDS-related complex
ARS	autonomously replicating sequences
ATP	adenosine triphosphate
AZT	3'-azido-3'-deoxythymidine
BAP	benzylaminopurine
BOD	biochemical oxygen demand
bp	base pair
5-BU	5-Bromuracil
C	cytosine
cAMP	cyclic AMP
CAP	catabolite activator protein
CAT	chloramphenical acetyl transferase
ccc DNA	covalently closed circular DNA
Ccrit	critical dissolved oxygen concentration
cdc	cell-division cycle
cDNA	complementary DNA
CF	complement-fixation
CFT	cystic fibrosis transmembrane conductance
CFTR	cystic fibrosis transmembrane conductance regulator
CFU	colony-forming unit
cM	centimorgan
CML	chronic myelogenous leukemia
CMP	cytidine monophosphate
CNBr	cyanogen bromide
CoA/CoASH	coenzyme A
con A	concanavalin A
CRP	catabolite repression protein
CsCl	cesium chloride
ctDNA	chloroplast DNA
CTP	cytidine triphosphate
d	deoxy
DAG	diacylglycerol
dATP	deoxyadenosine triphosphate
dCTP	deoxycytidine triphosphate
dd	dideoxy
ddNTP	dideoxyribonucleotide triphosphate
dGTP	deoxyguanosine triphosphate
DHFR	dihydrofolate reductase
DMSO	dimethyl sulfoxide

DMT	dimethoxytrityl
DNA	deoxyribonucleic acid (See also CTDNA, MTDNA.)
DNase	deoxyribonuclease
dNTP	deoxyribonucleotide triphosphate
DP	docking protein
EBV	Epstein-Barr virus
ECM	extracellular matrix
E. coli	Escherichia coli
EDTA	ethylenediaminetetraacetate
EF	elongation factor
EGF	epidermal growth factor
ELC	expression-linked copy
ELISA	enzyme-linked immunoabsorbent assay
EMBL	European Molecular Biology Lab
EMS	ethylmethane sulfonate
ER	endoplasmic reticulum
erb	erythroblastosis
ERK	extracellular receptor tyrosine kinase
EST	expressed sequence tags
FACS	flourescence-activated cell sorter
f-actin	filamentous actin
FAD	flavin adenine dinucleotide
FBJ	Finkel, Biskis, and Jinkins (discoverers of the FBJ murine osteosarcoma virus)
FCS	fetal calf serum
fes	feline sarcoma
FFU	focus-forming unit
FGP	fluorescent green protein
FISH	fluoresence in situ hybridization
5-FU	5-fluorouracil
FMN	flavin mononucleotide
FRA	fos-related antigens
FRAP	fluorescence recovery after photobleaching
FSH	follicle-stimulating hormone
FSV	feline sarcoma virus
ftz	fushi tarazu
G	guanine
G actin	globular actin
GABA	gamma amino butyric acid
GAG	glycosaminoglycan
gal	galactosidase
GALT	gut-associated lymphatic tissue
GAP	GTPase-activating proteins
GC	gas chromatography
GDP	guanosine diphosphate
GEF	guanine nucleotide exchange factor
GFAP	glial fibrillary acidic protein
GH	growth hormone
GLC	gas-liquid chromatography
GMP	guanosine monophosphate
gpt	guanine phosphoribosyl transferase
GST	Glutathione S-transferases
GTP	guanosine triphosphate

HAT	hypoxanthine-aminopterin-thymine
HCG	human chorionic gonadotropin
HDL	high-density lipoproteins
Hfr	high-frequency recombination strain
HGPRT	hypoxanthine-guanine phosphoribosyl transferase
HIV	human immunodeficiency virus
HLA	human leukocyte-associated antigens
HLTV	human T-cell leukemia virus
HMG	high-mobility group
HN	hemagglutinin-neuraminadase
hnRNA	heterogeneous nuclear RNA
HPFH	hereditary persistence of fetal hemoglobin
HPLC	high-performance liquid chromatography
HPV	human papilloma virus
HRP	horseradish peroxidase
HSE	heat-shock response element
hsp	heat-shock protein
HSR	homogeneously staining region
HSV	herpes simplex virus
IAA	indole acetic acid
IDL	intermediate-density lipoprotein
IF	initiation factor
IL	interleukin
IMP	inosine monophosphate
IPTG	isopropyl-β-D-thiogalactopyranoside
IR	infrared
IS	insertion sequence
IUdR	iododeoxyuridine
j	gene-joining gene
KD	diffusion coefficient/constant
Ki-MuSV	Kirsten sarcoma virus
k_M	Michaelis-Menten constant
LAV	lympho adenopathy virus
LDL	low-density lipoprotein
LH	lutinizing hormone
LINES	long-period interspersed sequences
LTR	long terminal repeat
MAPS	microtubule-associated proteins
MAR	matrix attachment regions
MAT	mating-type locus
MBP	maltose binding protein
MCP	methyl-accepting chemotaxis protein
mdr	multidrug resistance
5MeC	5-methylcytosine
MHC	major histocompatibility complex
MIF	migration-inhibitory factor
MMTV	mouse mammary tumor virus
MPF	M-phase promoting factor
mRNA	messenger RNA
mtDNA	mitochondrial DNA
MTOC	microtubule organizing center
MuLV	murine leukemia virus
MVR	minisatellite variant repeat

myb	myeloblastosis
myc	myelocytomatosis
NAD	nicotamide adenine dinucleotide
NADP	nicotamide adenine dinucleotide phosphate
NANA	N-acetylneuraminic acid
NBT	nitro-blue tetrazolium
N-CAM	neural cell adhesion molecule
NCBI	National Center for Biotechnology Information
NGF	nerve growth factor
NHGRI	National Human Genome Research Institute
NK cells	natural killer cells
NMR	nuclear magnetic resonance
NP40	nonidet P40
NTG	neomycin, thymidine kinase, glucocerebroside
oc	open circle
PAGE	polyacrylamide gel electrophoresis
PaPoVa	papilloma, polyoma, and vacuolating viruses
PAS	periodic acid-Schiff stain
PCR	polymerase chain reaction
PDGF	platelet-derived growth factor
PE	phosphatidylethanolamine
PEG	polyethylene glycol
PEP	phosphoenol pyruvate
PFU	plaque-forming unit
PITC	phenyl isothiocyanate
PKU	phenylketonuria
PML	progressive multifocal leukoencephalopathy
PMU	polymorphonuclear leukocyte
poly U	polyuridylic acid
PPLO	pleuropneumonialike organisms
PVP	polyvinylpyrrolidone
raf	rat fibrosarcoma
ras	rat sarcoma
RBC	red blood cell
RER	rough endoplasmic reticulum
RES	reticuloendothelial system
RFLP	restriction fragment-length polymorphism
RG	resorufin-β-D-galactopyranoside
RNA	ribonucleic acid (See also HNRNA, MRNA, RRNA, SNRNA, TRNA.)
RNP	ribonucleoprotein
ros	Rochester 2 sarcoma
rRNA	ribosomal RNA
RSV	Rous sarcoma virus
RTK	receptor tyrosine kinase
RVE	reconstituted viral envelope
SAM	S-adenosylmethionine
SAR	scaffold attachment regions
SDGF	sarcoma-derived growth factor
SDS	sodium dodecyl sulfate
SEM	scanning electron microscopy
sis	simian sarcoma
snRNA	small nuclear RNA

SPF	S-phase promoting factor
SRP	signal-recognition particle
STR	short tandem repeat
STS	sequence tagged site
SV40	simian virus 40
T	thymine, twisting number
Taq	*Thermus aquaticus*
TB	tuberculosis
Tc	cytoxic T cell
TCA	tricarboylic acid
TEM	transmission electron microscope
TGF	transforming growth factor
TI	tumor inducing
TMV	tobacco mosaic virus
TNF	tumor necrosis factor
tPA	tissue plasminogen activator
TPA	12-O-tetradecanoylphorbol-13-acetate
TRH	TSH-releasing hormone
tRNA	transfer RNA
TSH	thyroid stimulating hormone
Ts mutant	temperature-sensitive mutant
TTP	thymidine triphosphate
Ty	transposon yeast
U	uracil
UTP	uridine triphosphate
UV	ultraviolet
VLDL	very low-density lipoprotein
VSG	variable-surface glycoprotein
VSPR	very short patch repair
W	writhing number
XP	xeroderma pigmentosum
YAC	yeast artificial chromosome

II. The Chemical Elements

The Chemical Elements

element	symbol	a.n.	element	symbol	a.n.	element	symbol	a.n.	element	symbol	a.n.
actinium	Ac	89	erbium	Er	68	molybdenum	Mo	42	selenium	Se	34
aluminum	Al	13	europium	Eu	63	neodymium	Nd	60	silicon	Si	14
americium	Am	95	fermium	Fm	100	neon	Ne	10	silver	Ag	47
antimony	Sb	51	fluorine	F	9	neptunium	Np	93	sodium	Na	11
argon	Ar	18	francium	Fr	87	nickel	Ni	28	strontium	Sr	38
arsenic	As	33	gadolinium	Gd	64	niobium	Nb	41	sulfur	S	16
astatine	At	85	gallium	Ga	31	nitrogen	N	7	tantalum	Ta	73
barium	Ba	56	germanium	Ge	32	nobelium	No	102	technetium	Tc	43
berkelium	Bk	97	gold	Au	79	osmium	Os	76	tellurium	Te	52
beryllium	Be	4	hafnium	Hf	72	oxygen	O	8	terbium	Tb	65
bismuth	Bi	83	hassium	Hs	108	palladium	Pd	46	thallium	Tl	81
bohrium	Bh	107	helium	He	2	phosphorus	P	15	thorium	Th	90
boron	B	5	holmium	Ho	67	platinum	Pt	78	thulium	Tm	69
bromine	Br	35	hydrogen	H	1	plutonium	Pu	94	tin	Sn	50
cadmium	Cd	48	indium	In	49	polonium	Po	84	titanium	Ti	22
calcium	Ca	20	iodine	I	53	potassium	K	19	tungsten	W	74
californium	Cf	98	iridium	Ir	77	praseodymium	Pr	59	ununbium	Uub	112
carbon	C	6	iron	Fe	26	promethium	Pm	61	ununpentium	Uup	115
cerium	Ce	58	krypton	Kr	36	protactinium	Pa	91	ununquadium	Uuq	114
cesium	Cs	55	lanthanum	La	57	radium	Ra	88	ununtrium	Uut	113
chlorine	Cl	17	lawrencium	Lr	103	radon	Rn	86	unununium	Uuu	111
chromium	Cr	24	lead	Pb	82	rhenium	Re	75	uranium	U	92
cobalt	Co	27	lithium	Li	3	rhodium	Rh	45	vanadium	V	23
copper	Cu	29	lutetium	Lu	71	rubidium	Rb	37	xenon	Xe	54
curium	Cm	96	magnesium	Mg	12	ruthenium	Ru	44	ytterbium	Yb	70
darmstadtium	Ds	110	manganese	Mn	25	rutherfordium	Rf	104	yttrium	Y	39
dubnium	Db	105	meitnerium	Mt	109	samarium	Sm	62	zinc	Zn	30
dysprosium	Dy	66	mendelevium	Md	101	scandium	Sc	21	zirconium	Zr	40
einsteinium	Es	99	mercury	Hg	80	seaborgium	Sg	106			

a.n. = atomic number

III. Periodic Table

Periodic Table of Elements

Key:
1	← atomic number
H	← symbol
1.008	← atomic weight

Numbers in parentheses are the atomic mass numbers of radioactive isotopes.

1	2	3	4	5	6	7	8	9	10	11	12	13	14	15	16	17	18
1 H 1.008																	2 He 4.003
3 Li 6.941	4 Be 9.012											5 B 10.81	6 C 12.01	7 N 14.01	8 O 16.00	9 F 19.00	10 Ne 20.18
11 Na 22.99	12 Mg 24.31											13 Al 26.98	14 Si 28.09	15 P 30.97	16 S 32.07	17 Cl 35.45	18 Ar 39.95
19 K 39.10	20 Ca 40.08	21 Sc 44.96	22 Ti 47.88	23 V 50.94	24 Cr 52.00	25 Mn 54.94	26 Fe 55.85	27 Co 58.93	28 Ni 58.69	29 Cu 63.55	30 Zn 65.39	31 Ga 69.72	32 Ge 72.59	33 As 74.92	34 Se 78.96	35 Br 79.90	36 Kr 83.80
37 Rb 85.47	38 Sr 87.62	39 Y 88.91	40 Zr 91.22	41 Nb 92.91	42 Mo 95.94	43 Tc (98)	44 Ru 101.1	45 Rh 102.9	46 Pd 106.4	47 Ag 107.9	48 Cd 112.4	49 In 114.8	50 Sn 118.7	51 Sb 121.8	52 Te 127.6	53 I 126.9	54 Xe 131.3
55 Cs 132.9	56 Ba 137.3	57-71*	72 Hf 178.5	73 Ta 180.9	74 W 183.9	75 Re 186.2	76 Os 190.2	77 Ir 192.2	78 Pt 195.1	79 Au 197.0	80 Hg 200.6	81 Tl 204.4	82 Pb 207.2	83 Bi 209.0	84 Po (210)	85 At (210)	86 Rn (222)
87 Fr (223)	88 Ra (226)	89-103‡	104 Rf (261)	105 Db (262)	106 Sg (263)	107 Bh (262)	108 Hs (265)	109 Mt (266)	110 Ds (271)	111 Uuu (272)	112 Uub (285)	113 Uut (284)	114 Uuq (289)	115 Uup (288)			

*lanthanide series:

57 La 138.9	58 Ce 140.1	59 Pr 140.9	60 Nd 144.2	61 Pm (145)	62 Sm 150.4	63 Eu 152.0	64 Gd 157.3	65 Tb 158.9	66 Dy 162.5	67 Ho 164.9	68 Er 167.3	69 Tm 168.9	70 Yb 173.0	71 Lu 175.0

‡actinide series:

89 Ac (227)	90 Th 232.0	91 Pa 231.0	92 U 238.0	93 Np (237)	94 Pu (244)	95 Am (243)	96 Cm (247)	97 Bk (247)	98 Cf (251)	99 Es (252)	100 Fm (257)	101 Md (258)	102 No (259)	103 Lr (260)

The periodic table as it looks today.

IV. The Genetic Code

UUU ⎫ phenylalanine
UUC ⎭
UUA ⎫ leucine
UUG ⎭

UCU ⎫
UCC ⎬ serine
UCA ⎪
UCG ⎭

UAU ⎫ tyrosine
UAC ⎭
UAA ⎫ STOP*
UAG ⎭

UGU ⎫ cysteine
UGC ⎭
UGA ⎰ STOP*
UGG ⎱ tryptophan

CUU ⎫
CUC ⎬ leucine
CUA ⎪
CUG ⎭

CCU ⎫
CCC ⎬ proline
CCA ⎪
CCG ⎭

CAU ⎫ histidine
CAC ⎭
CAA ⎫ glutamine
CAG ⎭

CGU ⎫
CGC ⎬ arginine
CGA ⎪
CGG ⎭

AUU ⎫ isoleucine
AUC ⎬
AUA ⎭
AUG ⎰ methionine

ACU ⎫
ACC ⎬ threonine
ACA ⎪
ACG ⎭

AAU ⎫ asparagine
AAC ⎭
AAA ⎫ lysine
AAG ⎭

AGU ⎫ serine
AGC ⎭
AGA ⎫ arginine
AGG ⎭

GUU ⎫
GUC ⎬ valine
GUA ⎪
GUG ⎭

GCU ⎫
GCC ⎬ alanine
GCA ⎪
GCG ⎭

GAU ⎫ aspartate
GAC ⎭
GAA ⎫ glutamate
GAG ⎭

GGU ⎫
GGC ⎬ glycine
GGA ⎪
GGG ⎭

*STOP = translation stop signal

V. Purine and Pyrimidine Bases Found in Nucleic Acids

Purines

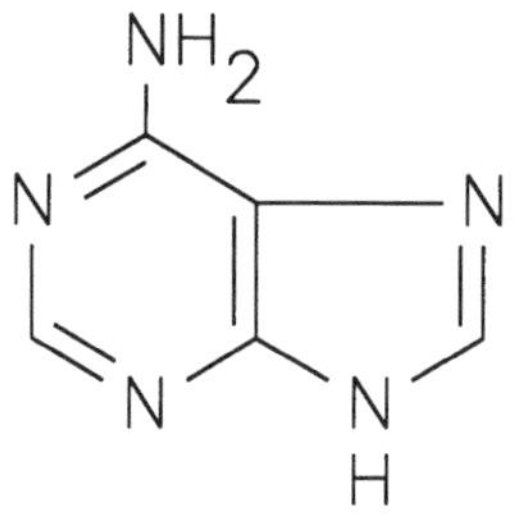

Pyrimidines

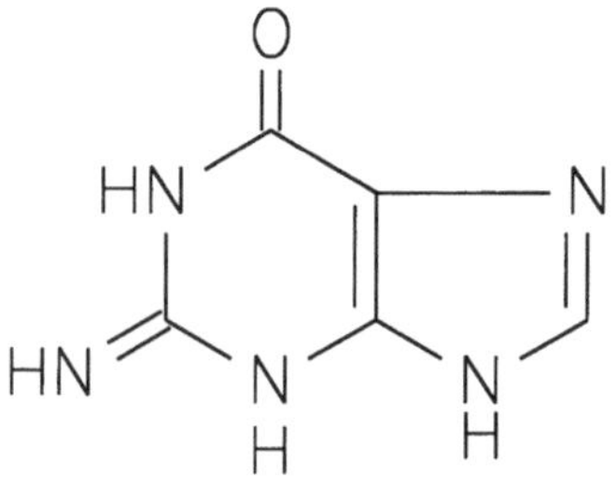

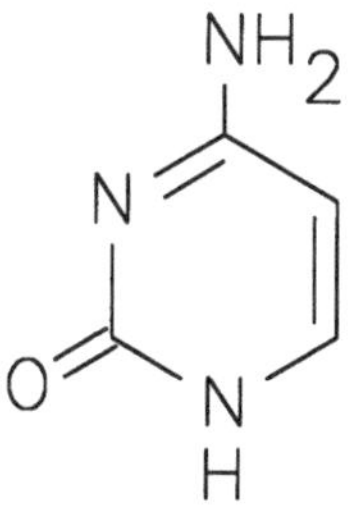

adenine

cytosine

guanine

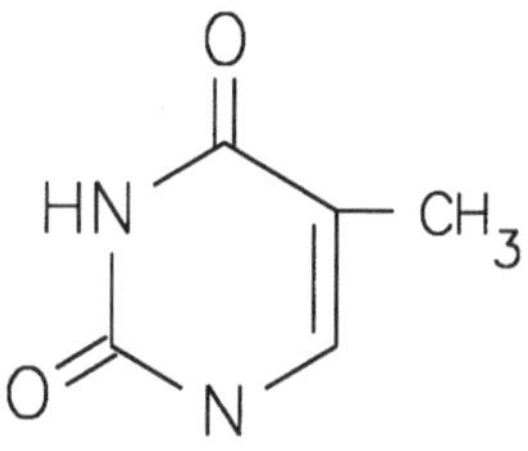

thymine (DNA only)

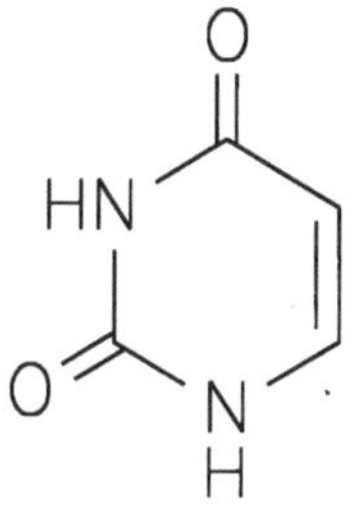

uracil(RNA only)

VI. Side Chains (R Groups) for Individual Amino Acids

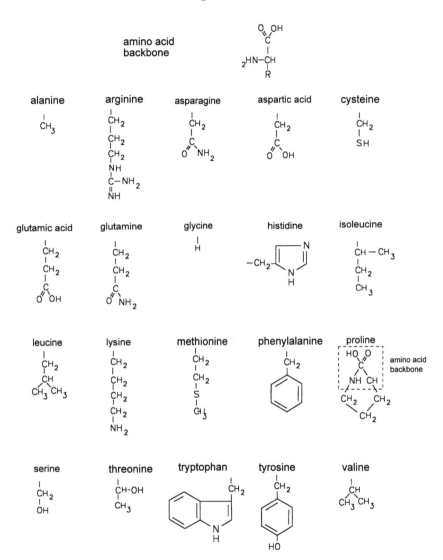

amino acid
backbone

alanine

arginine

asparagine

aspartic acid

cysteine

glutamic acid

glutamine

glycine

histidine

isoleucine

leucine

lysine

methionine

phenylalanine

proline

amino acid
backbone

serine

threonine

tryptophan

tyrosine

valine

VII. Bioinformatics Web Sites

AlignACE

A Web site for identifying motifs by alignments of multiple sequences. The AlignACE software can be downloaded. http://atlas. med.harvard.edu/cgi-bin/alignace.pl. Accessed on September 12, 2005.

The European Bioinformatics Institute

Access to the EMBL nucleotide sequence, PDB protein, and microarray databases. Searches of Medline literature database and patent abstracts. ClustalW for multiple sequence alignment and tools for depicting protein three-dimensional structure. http://www.ebi.ac.uk. Accessed on September 12, 2005.

ExPASy Proteomics Server

The ExPASy (Expert Protein Analysis System) proteomics server of the Swiss Institute of Bioinformatics (SIB) is devoted to the analysis of protein sequences and protein structure. This Web site also has tools for analysis of two-dimensional polyacrylaminde protein gels. Tools are available for primary, secondary, and tertiary structure analysis, including protein subfragment mass prediction, and programs for modeling structures. Access to the Swiss Protein database. http://www.expasy.org. Accessed on September 12, 2005.

The GENSCAN Web Server at MIT

Scans large nucleotide sequences for various gene features such as exons and splice sites in genomic DNA. http://genes.mit. edu/GENSCAN.html. Accessed on September 12, 2005.

MEME

On the MEME Web site, searches of protein or DNA sequences can be carried out for motifs that are present in different sequences. MEME will analyze multiple sequences for similarities among them. http://meme.sdsc.edu/meme/website/ meme.html. Accessed on September 12, 2005.

Mitomap

A database of variant human mitochondrial genomes detailing known polymorphisms and mutations and literature references. http://www.mitomap.org. Accessed on September 12, 2005.

National Center for Biotechnology Information (NCBI)

A major bioinformatics Web site maintained by the National Institutes of Health. The NCBI is the repository of the GenBank nucleotide and protein sequence database. This Web site also provides BLAST for alignment searches of the databases. There is a wide variety of useful tools for data analysis, such as ORF finder, Entrez Gene, Model Maker, CD Search, Open Mass Spectrometry Search Algorithm, ProtEST, Cn3D, VAST Search, and CD Search. The NCBI also provides access to the PubMed database of biomedical literature. http://www. ncbi.nlm.nih.gov. Accessed on September 12, 2005

National Human Genome Research Institute

Basic information on topics related to the Human Genome Project, including ethics and legal issues. Links to all major sequence databases. http://www.genome. gov. Accessed on September 12, 2005.

Online Analysis Tools

This Web site contains a variety of tools for carrying out sequence manipulations and analyses, including alignments, DNA motifs, PCR primer design, phylogenetic trees, restriction mapping, and open reading frames and motifs. http://molbiol-tools. ca/Restriction_endonuclease.htm. Accessed on September 12, 2005.

Pfam

A database of protein families derived from multiple sequence alignments. In the Pfam Web site, known protein structures and domain architectures can be viewed. Distribution of protein domains among species. Links to other databases. http://www.sanger.ac.uk/Software/Pfam. Accessed on September 12, 2005.

Regulatory Sequence Analysis Tools
Algorithms to scan genomes from various organisms for nucleotide pattern representing putative regulatory sequences. http://rsat.ulb.ac.be/rsat. Accessed on September 12, 2005.

The Repeat Masker Server at the University of Washington.
Web-based utility that scans DNA sequences for interspersed repeats (such as Alu and L1 elements) and low-complexity short repeats (such as dinucleotide repeats).

http://repeatmasker.genome.washington. edu. Accessed on September 12, 2005.

REPFIND
REPFIND finds clustered, exact repeats in nucleotide sequences. Output is in the form of a graphical display of the repeats found. http://zlab.bu.edu/repfind. Accessed on September 12, 2005.

Wadsworth Center
Run by the New York State Department of Health, the Wadsworth Center provides news and information on health and health-related research. Image analysis using SPIDER (System for Processing Image Data in Electron Microscopy and Related Fields). http://sfold.wadsworth.org/index.pl. Accessed on September 12, 2005.

VIII. Enzymes Used in Gene Cloning

Enzyme	Activity	Applications
alkaline phosphatase	removes 5′ terminal phosphates from the single-stranded end of a DNA strand	dephosphorylation of vectors cleaved with a restriction enzyme, to prevent ligation of the ends of the vector with itself, to increase the recovery of vectors with inserts
deoxyribonuclease I (DNase I)	cleaves double or single DNA strands by breaking phosphodiester bonds of the DNA backbone	• nick translation • Dnase I footprinting
DNA polymerase I	copies a template DNA strand from an annealed primer	• preparation of labeled DNA probes by nick translation • site-directed mutagenesis
exonuclease III	cleavage of single nucleotides one at a time from the 5′ end of double-stranded DNA, which has a non-phosphorylated 3′ end	creation of nested deletions for sequencing
Klenow enzyme	a subfragment of DNA polymerase I lacking the 5→3′ exonuclease of the intact enzyme	creation of blunt ends for blunt-end cloning of restriction fragments
polynucleotide kinase	catalyzes the transfer of a the terminal phosphate group from ATP to the free 3′ end of a polynucleotide	end labeling of oligonucleatides to be used as probes
Restriction enzyme(s)	cleaves double-stranded DNA at sites defined by specific palindromic nucleotide sequences leaving "sticky" single-stranded ends	standard "sticky end" cloning for creation of recombinant DNAs
reverse transcriptase	catalyzes the synthesis of DNA copied from an RNA template	• cDNA cloning • generation of cDNA probes for microarray analyses • reverse transcription polymerase chain reactions (RT-PCR)

Enzymes Used in Gene Cloning, con'd

Enzyme	Activity	Applications
T4 DNA ligase	catalyzes the formation of a phosphodiester bond between 5´-phosphate and 3´-hydroxyl ends of two termini in double-stranded DNA	• joining of vector and insert in most cloning procedures • site-directed mutagenesis
Taq polymerase	a class of thermostabile DNA polymerase(s)	polymerase chain reaction
terminal deoxynucleotide transferase	catalyzes the addition of deoxyribonucleotides from deoxyribonucleotide triphophates (dNTPs) to a free 3´ hydroxyl end of double- or single-stranded DNA	cloning of cDNAs and blunt-ended DNA fragments